FROM EQUILIBRIUM TO CHAOS

Practical Bifurcation and Stability Analysis

RÜDIGER SEYDEL

Institut für Angewandte Mathematik und Statistik
University of Würzburg
Würzburg, Federal Republic of Germany

ELSEVIER

New York · Amsterdam · London

Elsevier Science Publishing Co., Inc.
52 Vanderbilt Avenue, New York, New York 10017

Sole distributors outside the United States and Canada:

Elsevier Applied Science Publishers Ltd.
Crown House, Linton Road, Barking, Essex IG11 8JU, England

Library of Congress Cataloging-in-Publication Data

Seydel, R. (Rüdiger), 1947–
From equilibrium to chaos: practical bifurcation and stability analysis / Rüdiger Seydel.
p. cm.
Bibliography: p.
Includes index.
1. Differential equations, Nonlinear. 2. Bifurcation theory.
3. Stability. I. Title.
QA372.S415 1988
515.3′52—dc19 87-28818 CIP

ISBN 0-444-01250-8

Current printing (last digit):
10 9 8 7 6 5 4 3 2 1

Manufactured in the United States of America

Contents

Preface

Nonlinear phenomena are responsible for a variety of effects. Jumping between modes, sudden onset or vanishing of periodic oscillations, loss or gain of stability, buckling of frames and shells, ignition, combustion, and chaos are but a few examples. Nonlinear phenomena arise in all fields of physics, chemistry, biology, and engineering. The classical mathematical discipline that treats nonlinear phenomena is bifurcation theory. In 1744, Euler studied the bending of a beam subjected to longitudinal compressive forces, a problem that has become a standard example in bifurcation theory. At the end of the nineteenth century, Poincaré carried out studies in bifurcation theory that have had a profound influence on the subject. More recently, singularity theory and catastrophe theory have joined bifurcation theory, providing a more detailed classification of nonlinear phenomena and more qualitative insight. Numerical analysis and the theory of differential equations (including dynamic systems) are related to bifurcation theory and are greatly needed when studying nonlinear phenomena. In the literature, there are a number of books that summarize bifurcation theory and related disciplines [16, 17, 62, 105, 113, 123, 136, 159, 228, 262, 295]. While theoretical and analytical results are well documented by these works, there is a lack of books devoted to pragmatic questions. This book gives an introduction to nonlinear phenomena on a practical level as well as an account of computational methods.

The analysis of nonlinear phenomena requires, on the one hand, tools that provide quantitative results and, on the other hand, the theoretical knowledge of nonlinear behavior that allows one to interpret these quantitative results. The tools of which we speak are analytical and numerical. Without questioning the importance of analytical methods, this book places emphasis on numerical methods. It is expected that most future investigations will be based on numerical methods.

There is a broad spectrum of such methods, ranging from elementary strategies to sophisticated algorithms. Underlying these numerous methods are but a few basic principles. This book emphasizes basic principles and shows the reader how the methods result from combining and, on occasion, modifying the underlying principles.

This book is written to address the needs of scientists and engineers and to attract mathematicians as well. Mathematical formalism is kept to a minimum; the style is motivating rather than proving. Phenomena and methods are illustrated by many examples and numerous figures; exercises and projects complete the text. The book is self-contained, assuming only basic knowledge in calculus. The extensive bibliography includes many references for analytical and numerical methods, applications in science and engineering, and software.

The first part of the book consists of two chapters that introduce stability and bifurcation. The second part, consisting of Chapters 3–7, concentrates on practical aspects and numerical methods. Chapter 3 shows what computational difficulties can arise and what kind of numerical methods are required to get around these difficulties. Chapter 4 gives an account of the principles of continuation, the procedure by which "parameter studies" for nonlinear problems will be carried out. Chapters 5 and 6 treat basic computational methods for handling bifurcations, both for systems of algebraic equations and for ordinary differential equation boundary-value problems. In Chapter 7, branching phenomena of periodic solutions are handled and related numerical methods are outlined. The third and final part, Chapters 8 and 9, focuses on qualitative aspects. Singularity theory and catastrophe theory, which help in the interpretation of numerical results, are introduced in Chapter 8. Chapter 9 is an introduction to chaotic behavior.

In summary, the object of this book is to provide an introduction to the nonlinear phenomena of bifurcation theory. Motivating examples and geometrical interpretations are essential ingredients in the style. The book attempts to provide a practical guide for performing parameter studies. While not all nonlinear phenomena are treated (delay equations are not touched upon), the spectrum covered by the book is wide. For readers who have no immediate interest in computational aspects, a path is outlined listing those sections that provide the general introduction. Readers without urgent interest in computational aspects may wish to concentrate on:

sections 1.1–1.3

all of Chapter 2

section 3.4

part of sections 5.3.3 and 5.3.5
for those who like algebra: section 5.4.1
sections 5.4.4 and 5.4.5
section 6.1
section 6.2.2
example in section 6.4
sections 7.1–7.4
section 7.7
all of Chapter 8
all of Chapter 9.

The manuscript grew out of teaching materials for courses given by the author at the State University of New York at Buffalo and at the University of Würzburg, and was enriched by some years of computational experience in bifurcation. The author thanks William N. Gill, Victor K. H. Chen, Dietrich Flockerzi, Nicholas D. Kazarinoff, and Knut Schaper, who provided encouragement and help in various stages of the manuscript. Special thanks go to John Lavery who carefully studied the manuscript and suggested countless invaluable improvements.

Notation

PROBLEM-INHERENT VARIABLES

λ: scalar parameter to be varied (branching parameter)
$\mathbf{y}$: vector of state variables, vector function, solution of an equation
n: number of components of $\mathbf{y}$ and $\mathbf{f}$
$\mathbf{f}$: vector function, defines the dynamics of the problem that is to be solved; typical equation: $\mathbf{f}(\mathbf{y}, \lambda) = \mathbf{0}$
t: independent variable, often time
$\dot{\mathbf{y}}$: derivative of $\mathbf{y}$ with respect to time, $\dot{y} = d\mathbf{y}/dt$
$\mathbf{y}'$: derivative of $\mathbf{y}$ with respect to a general independent variable
$a \leq t \leq b$: interval in which t varies
x: spatial variable, may be scalar or vector with up to three components
$\mathbf{r}$: vector function, often used to define boundary conditions as in $\mathbf{r}(\mathbf{y}(a), \mathbf{y}(b)) = \mathbf{0}$
T: period in case of a periodic oscillation
γ: additional scalar parameter

NOTATIONS FOR A GENERAL ANALYSIS

In particular examples, some of the following meanings are sometimes superseded by a local meaning.

Specific Versions of y and λ

λ_0: specific parameter value of a branch point
$\mathbf{y}_0$: specific n-vector of a branch point
y_i: i-th component of vector $\mathbf{y}$

$\mathbf{y}^i$: i-th continuation step (i is not an exponent here), specific solution
λ_i: specific parameter value, corresponds to $\mathbf{y}^i$
$\mathbf{y}^{(i)}$: Newton iterate; for $i = 1, 2, \ldots$ sequence of vectors converging to a solution $\mathbf{y}$
$\mathbf{y}^s$: stationary solution ($\mathbf{y}_s$ in sections 6.6 and 6.7)

Integers

k: frequently, the k-th component has a special meaning
N: number of nodes of a discretization
i, j, l, m, ν: other integers

Scalars

$[\mathbf{y}]$: a scalar measure of $\mathbf{y}$ (cf. section 2.2)
ρ: radius
ϑ: angle
ω: frequency
ϵ: accuracy, error tolerance
δ: distance between two solutions, or parameter
η: value of a particular boundary condition
τ: test function indicating bifurcation
Δ: increment or decrement, sometimes acting as operator on the following variable; for instance, $\Delta\lambda$ means an increment in λ
s: arclength
u, v: functions, often solutions of scalar differential equations
σ: step length
p: parameterization, or phase condition or polynomial
c_i: constants
$\mu = \alpha + i\beta$: complex-conjugate eigenvalue
ζ, ξ: further scalars with local meaning

Vectors

$\mathbf{z}$: n-vector in various roles: tangent, or initial values of a trajectory, or emanating solution
$\mathbf{d}$: difference between two n-vectors
$\mathbf{h}$: n-vector, solution of a linearization; $\mathbf{h}_0$ or $\overline{\mathbf{h}}$ are related n-vectors
$\mathbf{e}_i$: i-th unit vector (cf. Appendix 3)

$\varphi(t; \mathbf{z})$: trajectory starting at $\mathbf{z}$ (Eq. (7.7))
$\mathbf{w} = \mathbf{h} + i\mathbf{g}$: eigenvector
$\boldsymbol{\mu} = \alpha + i\beta$: vector of eigenvalues
$\boldsymbol{\Lambda}$: vector of parameters
$\mathbf{Y}$: vector with more than n components, contains $\mathbf{y}$ as subvector
$\mathbf{F}$: vector with more than n components, contains $\mathbf{f}$ as subvector
$\mathbf{R}$: vector with more than n components, contains $\mathbf{r}$ as subvector
$\mathbf{P}$: Poincaré map
$\mathbf{q}$: argument of Poincaré map

n^2 Matrices (n Rows, n Columns)

$\mathbf{I}$: identity matrix
$\mathbf{J} = \mathbf{f_y}$: Jacobian matrix of first-order partial derivatives of $\mathbf{f}$ with respect to $\mathbf{y}$
$\mathbf{M}$: monodromy matrix
$\mathbf{A}$, $\mathbf{B}$: derivatives of boundary conditions (Eq. (6.12))
$\mathbf{E}$, $\mathbf{G}_j$: special matrices of multiple shooting (Eq. (6.21), Eq. (6.22))
Φ, $\mathbf{Z}$: fundamental solution matrices (cf. section 7.2)

Further Notations and Abbreviations

Ω: hypersurface
M: manifold
tr: as superscript means "transposed"
ODE: ordinary differential equations
PDE: partial differential equations
Re: real part
Im: imaginary part
∂: partial derivative
$\bar{\mathbf{y}}$, $\bar{\lambda}$, $\bar{\mathbf{h}}$, $\bar{\mathbf{z}}$: overbar characterizes approximations
∇u: gradient of u (∇ is the "del" operator)
$\nabla^2 u$: Laplacian operator (summation of second-order derivatives)

1 Introduction and Prerequisites

1.1 A NONMATHEMATICAL INTRODUCTION

Every day of our lives we experience changes that occur either gradually or suddenly. We often characterize these changes as quantitative or qualitative, respectively. For example, consider the following simple experiment (Figure 1.1). Imagine a board supported at both ends with a load on top. If the load λ is small enough, the board will take a bended shape with a deformation depending on the magnitude of λ and the board's material properties (such as stiffness, K). This state of the board will remain "stable" in the sense that a small variation in the load λ (or in the stiffness K) leads to a state that is only slightly perturbed. Such a variation (described by Hooke's law) would be referred to as a quantitative change. The board is deformed within its elastic regime and will return to its original shape when the perturbation (λ) is removed.

The situation changes abruptly when the load λ is increased beyond a certain *critical level* λ_0, at which the board breaks (Figure 1.2b). This sudden action is an example of a qualitative change; it will also take place when the material properties are changed beyond a certain limit (see Figure 1.2a). Suppose the shape of the board is modeled by some function (solution of an equation). Loosely speaking, we may say that there is a solution for load values $\lambda < \lambda_0$ and that this solution ceases to exist for $\lambda > \lambda_0$. The load λ and stiffness K are examples of *parameters*. The outcome of any experiment and any event is controlled by parameters. Varying a parameter can result in a transition from a quantitative change to a qualitative change. The following pairs of verbs may serve as illustrations:

bend $\rightarrow$ break

incline $\rightarrow$ tilt over

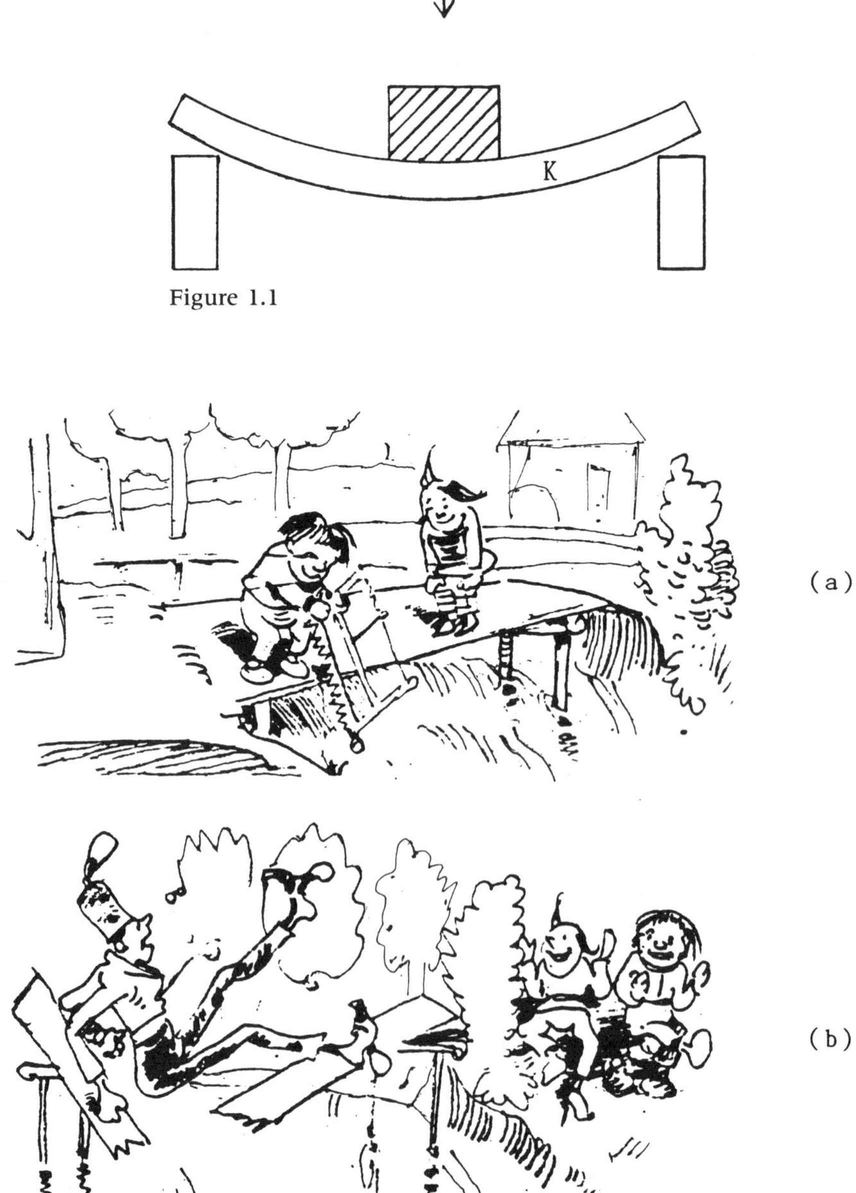

Figure 1.1

Figure 1.2 (From [50].)

stretch $\rightarrow$ tear

inflate $\rightarrow$ burst

The verbs on the left side stand for states that are stable under small *perturbations*; the response of each system is a quantitative one. This behavior ends abruptly at certain critical values of underlying parameters. The related drastic and irreversible change is reflected by the verbs on the right side.

The above-mentioned problems are much too limited to cover phenomena that we will later want to denote with the term "bifurcation." The extended range of phenomena we have in mind is indicated by the pair

stationary state $\leftrightarrow$ motion.

Let us mention a few examples. The electric membrane potential of nerves is stationary as long as the stimulating current remains below a critical threshold; if this critical value is passed, the membrane potential begins to oscillate, resulting in nerve impulses. The motion of a semitrailer is straight for moderate speeds (assuming the rig is steered straight); if the speed exceeds a certain critical value, the vehicle tends to sway. Or take the fluttering of a flag, which will occur only if the moving air passes fast enough. Similarly, the vibration of tubes depends on the speed of the internal fluid flow and on the speed of an outer flow. This type of oscillation also occurs when obstacles, such as bridges and other high structures, are exposed to strong winds. Many other examples—too complex to be listed here—occur in combustion, fluid dynamics, and geophysics. Reference will be made to these later in the text.

The transition from a stationary state to motion, and vice versa, is also a qualitative change. Here, speaking again in terms of "solutions"—of governing equations—we have a different quality of solution on either "side" of a critical parameter. Let the parameter in question again be denoted by λ, with critical value λ_0. Thinking, for instance, in terms of wind speed, the state (for example, of a flag or bridge) is stationary for $\lambda < \lambda_0$ and oscillatory for $\lambda > \lambda_0$. Qualitative changes may come in several steps, as indicated by the sequence

stationary state

regular motion

irregular motion.

The transition from regular to irregular motion is related to the onset of turbulence, or "chaos." As a first tentative definition, we will denote a qualitative change caused by the variation of some physical (or chemical or biological, etc.) parameter λ as *branching* or *bifurcation.* We will use the same symbol λ for various kinds of parameters. Some examples of parameters are listed below:

Phenomenon	Controlled by a Typical Parameter
bending of a rod	load
vibration of an engine	frequency or imbalance
combustion	temperature
nerve impulse	generating potential
superheating	strength of external magnetic field
oscillation of an airfoil	speed of plane relative to air

Some important features that may change at bifurcations have been already mentioned. The following list summarizes various kinds of qualitative changes:

stable ↔ unstable
symmetric ↔ asymmetric
stationary ↔ periodic (regular) motion
regular ↔ irregular
order ↔ chaos

Several of these changes may take place simultaneously in complicated ways. A thorough discussion of such branching phenomena requires the language of mathematics. Before proceeding to the mathematical analysis of stability and bifurcation, we shall review some important mathematical tools and concepts.

1.2 FUNDAMENTALS OF STATIONARY POINTS AND STABILITY (ODEs)

1.2.1 Trajectories and Equilibria

Many kinds of qualitative changes can be described by systems of ordinary differential equations (ODEs). In this section some elementary facts are recalled and notation is introduced.

Suppose the state of the system is described by functions $y_1(t)$ and $y_2(t)$. These *state variables* may represent, for example, the position and velocity of a particle, concentrations of two substances, or electric potentials. The independent variable t is often "time." Let the physical (or chemical, etc.) law that governs $y_1(t)$ and $y_2(t)$ be represented by two ordinary differential equations:

$$\begin{aligned}\dot{y}_1 &= f_1(y_1,y_2)\\ \dot{y}_2 &= f_2(y_1,y_2).\end{aligned} \tag{1.1}$$

In vector notation Eq. (1.1) can be written as $\dot{\mathbf{y}} = \mathbf{f}(\mathbf{y})$. This system is called *autonomous*, because the functions f_1 and f_2 do not depend explicitly on the independent variable t. As illustrated in Figure 1.3, solutions $y_1(t)$, $y_2(t)$ of this *two-dimensional* system form a set of *trajectories* or *flow lines* covering part or all of the (y_1,y_2) plane. A figure like Figure 1.3, depicting selected trajectories, is called a *phase diagram*; the (y_1,y_2) plane is the *phase plane*. A specific trajectory is selected by requesting that it pass through a prescribed point (z_1,z_2) of *initial values*

$$y_1(t_0) = z_1, \quad y_2(t_0) = z_2.$$

Because the system is autonomous, we can take $t_0 = 0$. Imagine a tiny particle floating over the (y_1,y_2) plane, the tangent of its path being

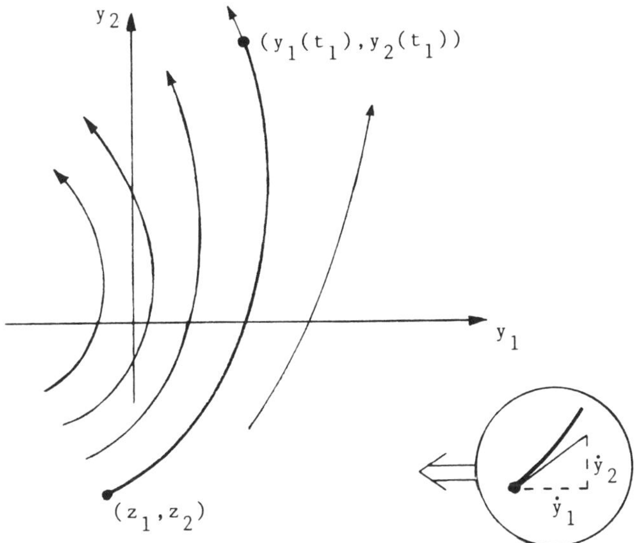

Figure 1.3

given by the differential equations. After time t_1 has elapsed, the particle will be at the point

$$(y_1(t_1), y_2(t_1))$$

(see Figure 1.3). The set of all trajectories that start from some part of the **y**-plane is called the *flow* from that part.

Of special interest are *equilibrium* points (y_1^s, y_2^s), which are defined by

$$\dot{y}_1 = 0, \quad \dot{y}_2 = 0.$$

In equilibrium points, the system is at rest. These points (y_1^s, y_2^s) are also called *stationary* solutions, and sometimes singular points, critical points, rest points, or fixed points. Stationary points are solutions of the system of equations given by the *right-hand side* of the differential equation

$$f_1(y_1^s, y_2^s) = 0$$

$$f_2(y_1^s, y_2^s) = 0.$$

In vector notation, this system of equations is written as $\mathbf{f}(\mathbf{y}^s) = 0$. For a practical evaluation of stationary points in two dimensions, we observe that each of these two equations defines a curve named *null cline.* Hence, stationary solutions are intersections of null clines. This suggests starting with drawing the null clines in a phase plane. In this way, not only the equilibria are obtained, but also some information on the global behavior of the trajectories. Recall that trajectories intersect the null cline defined by $f_1(y_1, y_2) = 0$ vertically and intersect the null cline of $f_2(y_1, y_2) = 0$ horizontally.

1.2.2 Deviations

Every differential equation and every solution thereof is affected by certain "deviations." Deviations can influence a final "solution" in several ways. First, there may be *fluctuations* of the model. Physical laws (represented by the right-hand side **f** in Eq. (1.1)) are formulated in terms of a few macrovariables, such as number of microorganisms, concentration, pressure, or temperature. Macroscopic variables simplify and average the physics of countless microscopic particles. The individual behavior of these particles is subject to molecular interaction and infinitesimal random perturbations ("noise"), causing variations in the function **f** that represents the average dynamics. More significant are the deviations caused by inadequacies in the measurements upon which **f** is based. That is, coefficients involved in **f** must be regarded as being subject to ever-present deviations. The class of deviations acting on **f** may be called "interior" perturbations.

A second type of deviation occurs when **y** is affected by an "external" force, which is not represented by the differential equation. As an example, think about touching an oscillating pendulum, thereby causing momentarily a sudden change in the dynamic behavior. Mathematically speaking, such "outer" perturbations can be described as a jump in the trajectory at some instant t_1,

$$(y_1(t_1), y_2(t_1)) \rightarrow (z_1,z_2).$$

When the perturbation ends at $t_2 > t_1$, the system (e.g., pendulum, chemical reaction) is again governed by the differential equation (1.1) starting from the initial values

$$y_1(t_2) = z_1, \quad y_2(t_2) = z_2$$

and traveling along a neighboring trajectory (Figure 1.4).

Further deviations occur when the differential equation is integrated numerically. Discretization errors of a numerical approximation scheme and rounding errors are encountered. These inevitable errors can be seen as a kind of noise affecting the calculated trajectory or interpreted as a great number of small perturbations of the kind illustrated by Figure 1.4.

Now assume that our "traveling particle" **y**, its itinerary given by differential equation (1.1), is in a stationary state, $(z_1,z_2) = (y_1^s,y_2^s)$. Theoretically, the particle would remain there forever, but in practice there is rarely such a thing as "exactness." Due to perturbations, either the position (z_1,z_2) of the particle or the location of the equilibrium

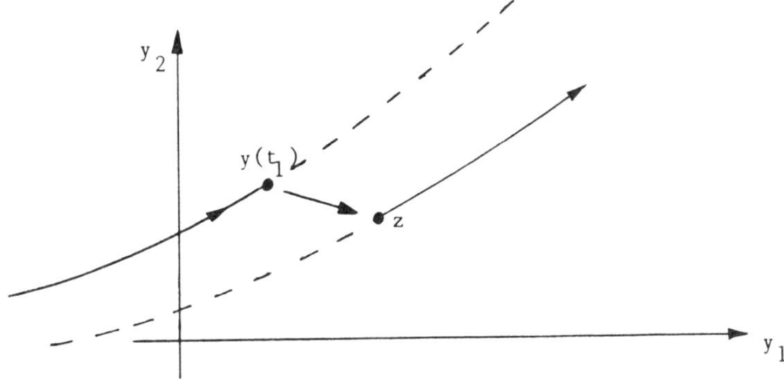

Figure 1.4

(y_1^s, y_2^s) varies slightly. This results in a deviation at, say, time $t_0 = 0$,

$$(z_1, z_2) \neq (y_1^s, y_2^s).$$

For simplicity, we assume that for $t > 0$ no further perturbations or fluctuations arise. Then it is the differential equation that decides whether trajectories starting from $\mathbf{z}$ remain near the equilibrium or depart from it. The question is how solutions $\mathbf{y}(t)$ of the *initial-value problem* (in vector notation)

$$\dot{\mathbf{y}} = \mathbf{f}(\mathbf{y}), \quad \mathbf{y}(\mathbf{0}) = \mathbf{z}$$

behave for $t \geq 0$.

1.2.3 Stability

Definition 1.1. A stationary solution $\mathbf{y}^s$ is said to be asymptotically stable if the response to a small perturbation approaches zero as the time approaches infinity. An asymptotically stable equilibrium is also called sink.

More formally, this means that

$$\mathbf{y}(t) \xrightarrow[t\to\infty]{} \mathbf{y}^s$$

for $\mathbf{y}(t)$ starting from initial values that are close to the equilibrium. Asymptotically stable equilibria are *attractors*.

Definition 1.2. The stationary solution $\mathbf{y}^s$ is said to be stable if the response to a small perturbation remains small as the time approaches infinity. Otherwise the stationary solution is called unstable (the deviation grows). An unstable equilibrium is also called source and is an example for a repellor.

As we see, asymptotic stability implies stability. The above classical definitions apply to general systems of ODEs, including the simplest "system" of one scalar differential equation. Let us illustrate stability by some scalar examples; the simplest is as follows (we omit the subscript 1):

$$\dot{y} = \lambda y; \text{ solution: } \quad y(t) = \exp(\lambda t)$$
$$\text{equilibrium: } y^s(t) \equiv 0$$

y^s is stable for $\lambda \leq 0$, asymptotically stable for $\lambda < 0$, and unstable for $\lambda > 0$.

Because the distinction between stability and asymptotic stability is artificial, we shall often use the term "stable" in the sense "asymptotically stable."

It is crucial to realize that the above definitions of stability are *local* in nature. An equilibrium may be stable for a small perturbation but unstable for a large perturbation. In the following example, the *domain of attraction* can be calculated (y is scalar):

$$\dot{y} = y(y^2 - \alpha^2), \quad y(0) = z.$$

The closed-form solution is

$$y \equiv 0 \qquad \text{for } z = 0$$

$$y(t) = (\alpha^{-2} + (z^{-2} - \alpha^{-2}) \exp(2\alpha^2 t))^{-1/2} \qquad \text{for } z \neq 0$$

(Exercise 1.1, below). For $|z| < |\alpha|$ the radicand approaches infinity. From this we infer that the equilibrium $y^s = 0$ is stable. The local character of this result is illustrated by the observation that for $|z| > |\alpha|$ solutions diverge. The domain of attraction $|z| < |\alpha|$ of the stable equilibrium $y^s = 0$ is bounded by two unstable equilibria $y^s = \pm\alpha$. Although locally stable, the equilibrium $y^s = 0$ is globally unstable when "large" perturbations occur. In a practical situation where the radius of the domain of attraction is smaller than the magnitude of possible perturbations, the local stability result is meaningless. This situation is made plausible by a model of a hill with a cross section, as depicted in Figure 1.5. Imagine a ball (rock) placed in the dip on top of the hill. Apparently the ball is stable under small vibrations but unstable under the influence of a violent quake. The undulating curve in Figure 1.5 may be seen as representing the potential energy. Figure 1.5 depicts two "weakly" stable equilibria and one "strongly" stable equilibrium.

The above scalar examples and the global stability behavior of their solutions were analyzed using a closed-form solution. In general, prob-

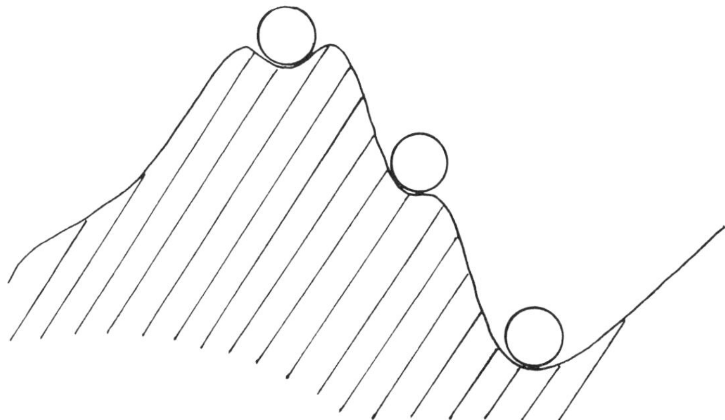

Figure 1.5

lems are so complicated that no such solution can be calculated. Accordingly, global stability results ("How large is the domain of attraction?") are usually difficult to obtain. The situation is much better with local stability results. Here the principle of *linearized stability* provides insight into what happens "close" to an equilibrium.

Exercise 1.1.

(a) Go through the procedure of calculating the closed-form solutions of the scalar differential equation $\dot{y} = y(y^2 - \alpha^2)$ given above. (Hint: Bernoulli differential equation; see Appendix 1.)
(b) Show that solutions of the initial-value problem with $y(0) = z$ possess a singularity for $|z| > |\alpha|$, that is, say, $|y(t)| \to \infty$ for $t \to t_\infty$.
(c) Calculate t_∞ and note that it is a *movable singularity*, $t_\infty = t_\infty(z)$.
(d) For $\alpha = 1$, sketch some solutions ($|z| < 2$).

A two-dimensional system that further illustrates various stability concepts is presented in Exercise 1.2.

Exercise 1.2.
Solve the system of two differential equations

$$\dot{y}_1 = y_2$$
$$\dot{y}_2 = -y_1 - \lambda y_2$$

for $\lambda = -1$, $\lambda = 0$, $\lambda = 1$. When is the equilibrium stable, asymptotically stable, unstable? (Hint: Appendix 1 summarizes how to solve a linear system.)

1.2.4 Linear Stability; Duffing Equation

A Taylor series expansion of f_1 about (y_1^s, y_2^s) gives

$$\dot{y}_1 = f_1(y_1, y_2) = f_1(y_1^s, y_2^s) + \frac{\partial f_1}{\partial y_1}(y_1^s, y_2^s)(y_1 - y_1^s)$$
$$+ \frac{\partial f_1}{\partial y_2}(y_1^s, y_2^s)(y_2 - y_2^s) + \text{terms of higher order.}$$

Expanding also the f_2 of the second differential equation, observing that $f_1(y_1^s, y_2^s) = f_2(y_1^s, y_2^s) = 0$ and dropping the high-order terms, gives two differential equations that are *linear* in $y_1 - y_1^s$ and $y_2 - y_2^s$. This system of equations is easily written down using the notation of the *Jacobian matrix* $\mathbf{f_y}$ of the first-order partial derivatives. Evaluated at the equilibrium, the Jacobian is

$$\mathbf{f}_\mathbf{y}^s = \frac{\partial \mathbf{f}}{\partial \mathbf{y}}(\mathbf{y}^s) = \begin{pmatrix} \frac{\partial f_1}{\partial y_1}(y_1^s, y_2^s) & \frac{\partial f_1}{\partial y_2}(y_1^s, y_2^s) \\ \frac{\partial f_2}{\partial y_1}(y_1^s, y_2^s) & \frac{\partial f_2}{\partial y_2}(y_1^s, y_2^s) \end{pmatrix}$$

Now the linearized system in vector notation reads

$$\dot{\mathbf{h}} = \mathbf{f}_\mathbf{y}^s \mathbf{h}. \tag{1.2}$$

The vector $\mathbf{h}$ represents first-order approximations

$$h_1(t) \approx y_1(t) - y_1^s$$
$$h_2(t) \approx y_2(t) - y_2^s$$

of small distances between the points $\mathbf{y}(t)$ on the trajectories and the stationary point $\mathbf{y}^s$. The smaller the distance $\mathbf{y} - \mathbf{y}^s$, the better the approximation $\mathbf{h}$. That is, under certain assumptions to be stated later, $\mathbf{h}(t)$ describes the *local* behavior of the solution. The approximation $\mathbf{h}(t)$ indicates how the process evolves when the initial state deviates slightly from its equilibrium values. Thus the question of local stability is reduced to a discussion of the linear system Eq. (1.2). The standard procedure is to insert the *ansatz* (hypothesis)

$$\begin{aligned} h_1(t) &= e^{\mu t} w_1 \\ h_2(t) &= e^{\mu t} w_2 \end{aligned} \tag{1.3}$$

into Eq. (1.2). A straightforward calculation shows that the *eigenvalue problem*

$$(\mathbf{f}_\mathbf{y}^s - \mu \mathbf{I})\mathbf{w} = \mathbf{0} \tag{1.4}$$

results, with eigenvalue μ, constant eigenvector $\mathbf{w}$, and identity matrix

$$\mathbf{I} = \begin{pmatrix} 1 & 0 \\ 0 & 1 \end{pmatrix}.$$

First, the eigenvalues μ_1 and μ_2 are calculated. The existence of a nontrivial solution $\mathbf{w} \neq 0$ of Eq. (1.4) requires μ_1 and μ_2 to be the roots of the *characteristic equation*

$$0 = \det(\mathbf{f}_\mathbf{y}^s - \mu \mathbf{I}).$$

In the two-dimensional situation, this is a quadratic equation; its roots μ_1 and μ_2 are easily obtained. Then eigenvectors $\mathbf{w}$ can be calculated as solutions of the specific system (1.4) of linear equations that results from inserting μ_1 and μ_2. From this point on, the procedure is best illustrated by studying an example.

Example 1.1. Duffing Equation (Without External Forcing). Consider the second-order scalar differential equation

$$\ddot{u} + \dot{u} - u + u^3 = 0, \tag{1.5}$$

which is a special case of the Duffing equation

$$\ddot{u} + a\dot{u} + bu + cu^3 = d \cos \omega t.$$

Duffing equations are frequently used to describe oscillations in series-resonance circuits, where a nonlinear inductor or capacitor is modeled by a cubic polynomial [83, 245]. For our purpose of illustrating some calculus, we choose in Eq. (1.5) simple coefficients and omit the excitation ($d = 0$). Later, more general Duffing equations will be treated.

We introduce a vector $\mathbf{y}$ via $y_1 = u$, $y_2 = \dot{u}$. The equation (1.5) is written as a first-order system of ODEs,

$$\dot{y}_1 = y_2 = f_1(y_1,y_2)$$

$$\dot{y}_2 = y_1 - y_1^3 - y_2 = f_2(y_1,y_2).$$

The phase plane associated with the scalar second-order equation (1.5) displays u and $\dot{u}$. Calculating stationary points, one finds that $y_2^s = 0$ and that y_1^s is a solution of

$$0 = y_1(1 - y_1^2).$$

This equation has three distinct roots, and the above system thus has three distinct stationary points (y_1^s,y_2^s),

$$(0,0), \quad (1,0), \quad (-1,0).$$

The Jacobian matrix of the first-order partial derivatives is

$$\mathbf{f_y} = \begin{pmatrix} 0 & 1 \\ 1 - 3y_1^2 & -1 \end{pmatrix}.$$

The three equilibrium points (y_1^s,y_2^s) produce three eigenvalue problems (see Eq. (1.4)). In this particular example, the analysis is simple because the Jacobian $\mathbf{f_y}$ and therefore Eq. (1.4) are identical for the two equilibria $(\pm 1,0)$. Pursuing the analysis in more detail, we obtain the following results:

$(y_1^s, y_2^s) = (0,0)$: $\qquad \mathbf{f}_\mathbf{y}^s = \begin{pmatrix} 0 & 1 \\ 1 & -1 \end{pmatrix}.$

characteristic equation: $0 = \mu^2 + \mu - 1$
roots: $\mu_{1,2} = \frac{1}{2}(-1 \pm \sqrt{5})$
Two distinct roots imply the existence of two eigenvectors, which are obtained by Eq. (1.4),

$$\begin{pmatrix} -\mu & 1 \\ 1 & -1-\mu \end{pmatrix} \begin{pmatrix} w_1 \\ w_2 \end{pmatrix} = \begin{pmatrix} 0 \\ 0 \end{pmatrix}.$$

The first component of this system of linear equations is

$$-\mu w_1 + w_2 = 0,$$

which suggests choosing $w_1 = 1$. This implies that $w_2 = \mu$ and generates the (two) eigenvectors:

$$\mathbf{w} = \begin{pmatrix} 1 \\ \mu \end{pmatrix}, \quad \mu = \mu_1, \mu_2.$$

$(y_1^s, y_2^s) = (\pm 1, 0)$: $\qquad \mathbf{f}_\mathbf{y}^s = \begin{pmatrix} 0 & 1 \\ -2 & -1 \end{pmatrix}$

characteristic equation: $0 = \mu^2 + \mu + 2$
roots: $\mu_{1,2} = \frac{1}{2}(-1 \pm \sqrt{-7})$ (complex conjugates)
We omit calculation of the complex eigenvectors because we do not need them.

This example will be analyzed further at the end of this section. In the meantime, we discuss important types of qualitative behavior of trajectories close to an equilibrium. What follows is based on the ansatz Eq. (1.3) with the associated eigenvalue problem Eq. (1.4).

Case 1.1. μ_1, μ_2 **real,** $\mu_1 \cdot \mu_2 > 0$, $\mu_1 \neq \mu_2$.

In this case, where both eigenvalues are real and have the same sign, the stationary point $\mathbf{y}^s$ is called a *node*. This case splits into two subcases:

(a) $\mu < 0$ implies $\lim_{t \to \infty} e^{\mu t} = 0$.
Therefore, $\mathbf{h}(t)$ tends to zero and $\mathbf{y}(t)$ converges to $\mathbf{y}^s$ in a sufficiently small neighborhood of the node. This type of

node is called a *stable node*. Small perturbations die out at stable nodes.

(b) $\mu > 0$ implies $\lim_{t\to\infty} e^{\mu t} = \infty$.
As a consequence, $\mathbf{h}(t)$ "explodes" locally, which means that the trajectories $\mathbf{y}(t)$ leave the neighborhood of the node. This type of node is an *unstable node*.

The two real eigenvectors that exist in both these subcases have a geometrical meaning (see Figure 1.6, which depicts a stable node). They define two straight lines passing through the stationary point. Each half-ray is a trajectory of the linear system (1.2) (see Exercise 1.3 below). If $|\mu|$ is large (small), we call the corresponding motion fast (slow). A negative eigenvalue μ means that the trajectory is directed toward the equilibrium point $\mathbf{y}^s$ (stable node), while a positive eigenvalue is associated with motion away from $\mathbf{y}^s$ (unstable node). For the nonlinear problem, the situation is as in Figure 1.7, where a stable node is illustrated. At the stationary point, the four heavy solid trajectories are tangent to the lines (*eigenspaces*) of Figure 1.6. For an unstable node, the directions in Figure 1.6 and Figure 1.7 are reversed.

Case 1.2. μ_1, μ_2 real, $\mu_1 \cdot \mu_2 < 0$.

When the real eigenvalues have different signs, the stationary point is called a *saddle point*. Because one of the eigenvalues is

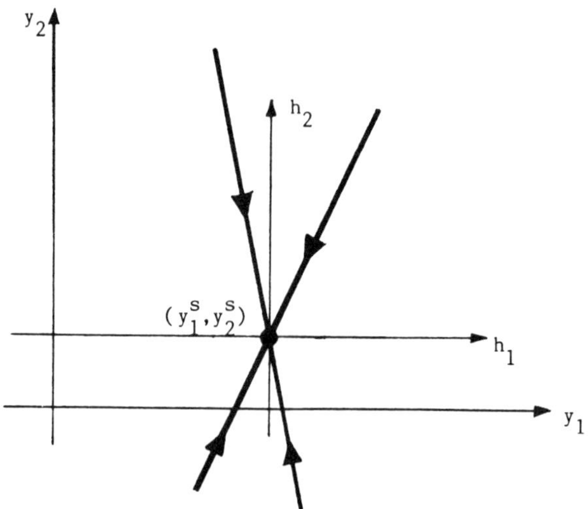

Figure 1.6

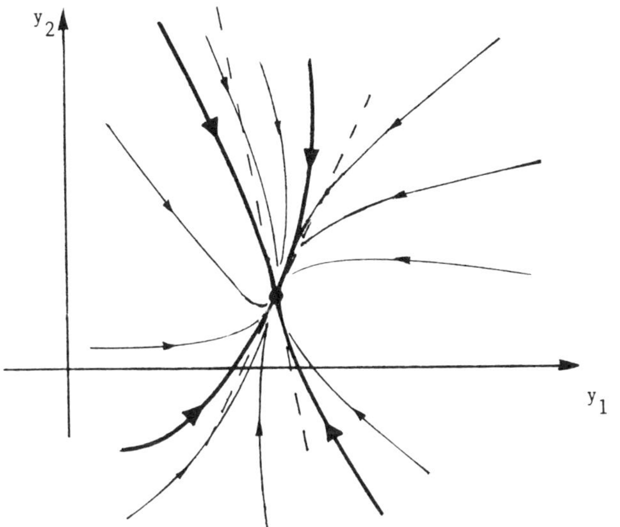

Figure 1.7

positive, two of the four trajectories associated with the eigenspaces leave the equilibrium. A saddle point, therefore, is always unstable (see Figure 1.8). The two trajectories that enter the saddle point are the *stable manifolds* of the saddle.

Case 1.3. μ_1, μ_2 complex conjugates with nonzero real part.

Set

$$\mu_1 = \alpha + i\beta, \quad \mu_2 = \alpha - i\beta$$

(i: imaginary unit) where α and $\pm\beta$ denote the real and imaginary parts, respectively. Going back to ansatz Eq. (1.3), the time-dependent part of $\mathbf{h}(t)$ is

$$e^{(\alpha + i\beta)t} = e^{\alpha t}e^{i\beta t}.$$

The factor $e^{i\beta t} = \cos \beta t + i \sin \beta t$ represents a rotation counterclockwise through β radians if $\beta > 0$, and clockwise through $-\beta$ radians if $\beta < 0$. The factor $\exp(\alpha t)$ is a radius, which is increasing if $\alpha > 0$ and decreasing if $\alpha < 0$. Any trajectory close to the equilibrium thus resembles a *spiral.* The corresponding equilibrium is called a *unstable focus* ($\alpha > 0$) or a *stable focus* ($\alpha < 0$). Figure 1.9 depicts an unstable focus. In order to determine whether the rotation is clockwise or counterclockwise, one picks a test point

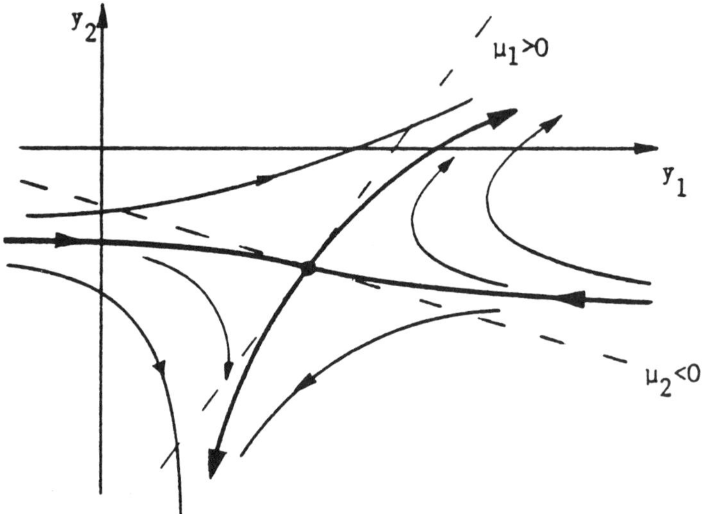

Figure 1.8

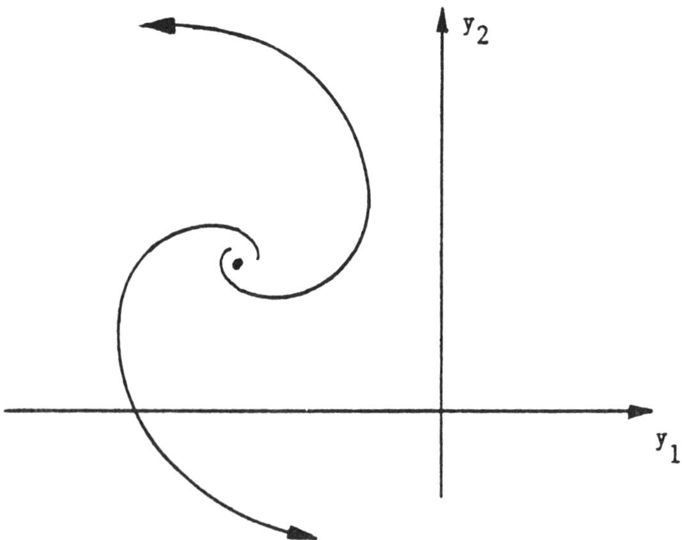

Figure 1.9

$\mathbf{z}$ close to the equilibrium and evaluates

$$\dot{z}_1 = f_1(z_1,z_2)$$
$$\dot{z}_2 = f_2(z_1,z_2).$$

For instance, study the signs of $f_j(z + \epsilon e_k)$ for arbitrary small ϵ and the k-th unit vector e_k.

Example 1.2. Duffing Equation (1.5), Continued.
Having recalled the above three classes of stationary points, we continue with the discussion of Eq. (1.5). The above analysis reveals that

(0,0) is a saddle point ($\mu_1 \approx 0.62$, $\mu_2 \approx -1.62$)
(± 1,0) are stable foci.

The test point (1.1,0) close to the focus (1,0) will drift "downward," since $\dot{y}_1 = 0$, $\dot{y}_2 < 0$. This indicates a clockwise rotation around the focus, as illustrated in Figure 1.10. The two stable manifolds that approach the saddle point (0,0) for $t \rightarrow \infty$ are examples of *separatrices*. Separatrices divide (parts of) the phase space into attracting basins. Here the phase plane is separated into two basins, the shaded and unshaded regions in Figure 1.10. All particles that start in the hatched region ultimately approach the focus (1,0); all the other trajectories end up at the focus (-1,0). What is depicted in Figure 1.10 extends the local results of the analysis of the stationary points to a global picture. Thereby, global results

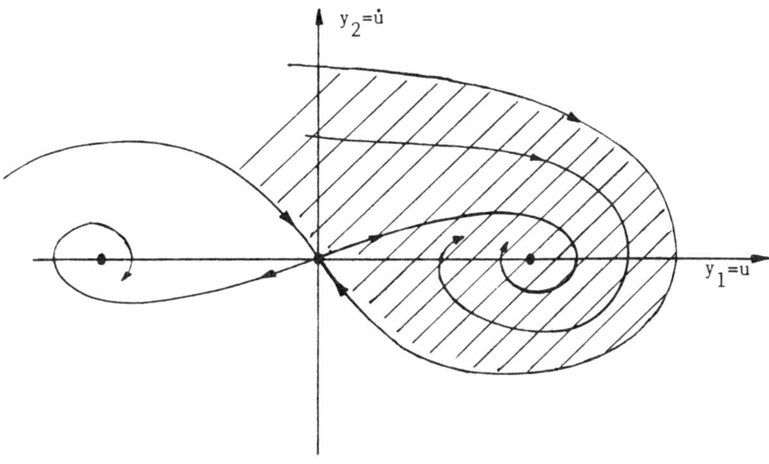

Figure 1.10

that can be obtained by numerical integration of the differential equation (1.1) are anticipated. In particular, the calculation of a separatrix provides global insight. Because only the final limit point (the saddle) of a separatrix is known, the integration must be *backward*, $t \to -\infty$, starting from suitable initial values $\mathbf{y}(0) = \mathbf{z}$. An obvious choice for $\mathbf{z}$ is to take the coordinates of the saddle (0,0), trusting that rounding and discretization errors cause $\mathbf{f}(\mathbf{y}) \neq 0$, thereby enabling the calculations to leave the saddle point. This initial phase tends to be slow. Also, the resulting trajectory may be quite far away from the separatrix. If rounding and discretization errors are zero, this method will not work at all. This situation suggests starting from $\mathbf{z}$ not identical but close to the saddle point and close also to the separatrix. Such a starting point is provided by a point on the separatrix of the linearized problem—that is, by the eigenvector associated with the negative eigenvalue. This eigenvector

$$\mathbf{w} = \begin{pmatrix} 1 \\ -\frac{1}{2}(1 + \sqrt{5}) \end{pmatrix}$$

produces the starting point

$$z_1 = \epsilon, \quad z_2 = -\tfrac{1}{2}\epsilon(1 + \sqrt{5})$$

for, say, $\epsilon = 0.0001$. The trajectory obtained this way can be expected to be close to the exact separatrix in the neighborhood of the saddle point. In this example, the separatrices spiral around the "core" part of the plane (see Figure 1.10 again), and the two attracting basins spiral around, encircling each other.

1.2.5 Degenerate Cases, Parameter Dependence

As seen above, the eigenvalues of the Jacobian matrix evaluated at an equilibrium point determine the dynamic behavior in the neighborhood of the equilibrium. This holds for all three types of equilibrium points (nodal points, saddle points, and focal points). These type of points are *generic* in that the assumptions ($\mu_1 \neq \mu_2$, $\mu_1 \cdot \mu_2 \neq 0$, or $\alpha = \text{Re}(\mu) \neq 0$) almost always apply. For an analysis of the exceptional cases

$\mu_1 = \mu_2$ (a special node)

$\mu_1 \cdot \mu_2 = 0$ (require nonlinear terms)

$\mu_{1,2} = \pm\beta i$ (a *center* of concentric circles), we refer to some hints in Section 8.5 and to ODE textbooks, for instance [15, 132, 147]. The equilibrium is called *hyperbolic* or *nondegenerate* when the Jacobian has no eigenvalue with zero real part. The exceptional cases $\mu_1\mu_2 = 0$ and $\mu_{1,2} = \pm i\beta$ are *nonhyperbolic* or *degenerate.*

In any given case it is unlikely that there will be a degenerate equilibrium point. Therefore one might forget about the exceptional degenerate cases. But that would oversimplify matters. Usually a differential equation describing a real-life problem involves one or more parameters. Denoting one such parameter by λ, the differential equation reads

$$\dot{y}_1 = f_1(y_1, y_2, \lambda)$$

$$\dot{y}_2 = f_2(y_1, y_2, \lambda).$$

Because the systems $\dot{\mathbf{y}} = \mathbf{f}(\mathbf{y}, \lambda)$ depend on λ, we speak of a *family* of differential equations. Solutions now depend both on the independent variable t and on the parameter λ,

$$\mathbf{y}(t;\lambda).$$

Consequently, stationary points, Jacobian matrices, and the eigenvalues μ depend on λ,

$$\mu(\lambda) = \alpha(\lambda) + i\beta(\lambda).$$

Upon varying the parameter λ, the position and the qualitative features of a stationary point can vary. For example, imagine a stable focus ($\alpha(\lambda) < 0$) for some values of λ. When λ passes some critical value λ_0, the real part $\alpha(\lambda)$ may change sign and the stationary point may turn into an unstable focus. During this transition, a degenerate focus (center) is encountered for an instant (see Figure 1.11). Other metamorphoses of

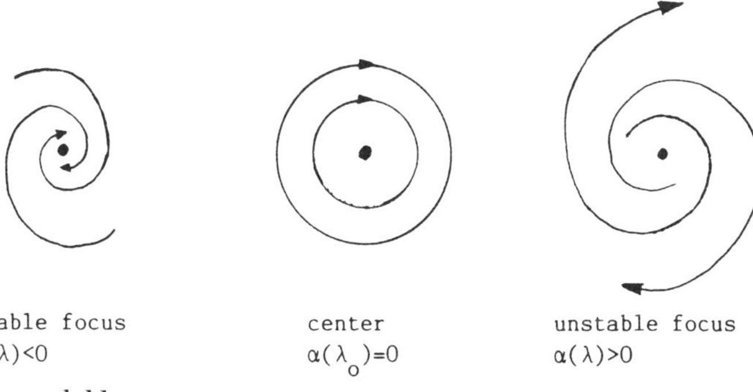

Figure 1.11

a stable spiral to an unstable spiral are possible (see Exercise 1.10). Summarizing, we point out that degenerate cases can occur "momentarily." Often, qualitative changes such as a loss of stability are encountered when a degenerate case is passed. This makes the *occurrence* of "degenerate" cases especially interesting. There will be many examples and exercises in the following section (Exercises 1.8 and 1.9). A main concern of this book is to treat problems related to degenerate solutions.

1.2.6 Generalizations

So far our analysis has concentrated on a system of two first-order differential equations—that is, the vector functions $\mathbf{y}(t)$ and $\mathbf{h}(t)$, the initial vector $\mathbf{z}$, and the eigenvectors $\mathbf{w}$ have consisted of two components. Most of the above results generalize to n-dimensional vectors. In the general case, the Jacobian matrix $\mathbf{f_y}$ consists of n^2 first-order partial derivatives $(\partial f_i/\partial y_j)$, $i, j = 1, \ldots, n$. The equilibria can be characterized based on the eigenvalues $\mu_1, \ldots, \mu_n$ of the Jacobian $\mathbf{f}^s_\mathbf{y}$. The number of cases generated by the various combinations of the eigenvalues increases dramatically. For $n = 3$, there are already 10 different types of nondegenerate stationary points. Classification is not as simple as in the planar situation [15]. The following general stability result is attributed to Liapunov (1892) [213]:

Theorem 1.1. Suppose $\mathbf{f}(\mathbf{y})$ is two times continuously differentiable and $\mathbf{f}(\mathbf{y}^s) = 0$. The real parts of the eigenvalues μ_j $(j = 1, \ldots, n)$ of the Jacobian evaluated at the stationary solution $\mathbf{y}^s$ determine stability in the following way:

(a) *$Re(\mu_j) < 0$ for all j implies asymptotic stability*
(b) *$Re(\mu_k) > 0$ for one (or more) k implies instability.*

Theorem 1.1 establishes the principle of linearized stability. In the case of nondegenerate (hyperbolic) equilibria, *local stability* is determined by the eigenvalues of the Jacobian. In order to stress the local character of this stability criterion, this type of stability is also called "conditional stability" [159] or "linear stability."

Exercise 1.3.
Consider a node or a saddle point in the plane with two intersecting straight lines determined by the eigenvectors (see Figure 1.6). Show that a trajectory of Eq. (1.2) starting on such a line remains there.

Exercise 1.4. (cf. Exercise 1.2).
Discuss the stability of the equilibria of

$$\dot{y}_1 = y_2$$
$$\dot{y}_2 = -y_1 - \lambda y_2$$

for all values of the parameter λ.

Exercise 1.5.
Perform the stability analysis of the three equilibria of Eq. (1.5) in detail.

Exercise 1.6.
A simple mathematical model of heartbeat (Zeeman [371]) is given by

$$\dot{y}_1 = -a(y_1^3 - by_1 + y_2)$$
$$\dot{y}_2 = y_1 - c.$$

($y_1(t)$: length of a muscle fiber (plus a constant); $y_2(t)$: electrochemical control.)

Discuss the equilibrium (diastole) for $a = 100$, $b = 1$, $c = 1.1$. Sketch some trajectories.

Exercise 1.7.
Consider the differential equations

$$\dot{y}_1 = y_2 + \tfrac{1}{2}\exp(y_1^2 - 1)$$
$$\dot{y}_2 = y_1^2 + y_1.$$

Discuss equilibria and stability and sketch trajectories for $-1.5 \leq y_1 \leq 0.5$, $-1 \leq y_2 \leq 0$.

1.3 FUNDAMENTALS OF LIMIT CYCLES AND WAVES

Again we consider the system of ODEs, Eq. (1.1), $\dot{\mathbf{y}} = \mathbf{f}(\mathbf{y})$, concentrating on the situation of two scalar equations. As emphasized before, stability and instability results are of local nature. Even if one is able to find all the equilibria of a particular problem, it is not guaranteed that putting all the local pieces together will give a complete global picture. There are other attractors that are not as easy to obtain as equilibria. One such attractor is the *limit cycle.* Limit cycles represent regular motions. Examples are nerve impulses, currents in electrical circuits, vibrations of violin strings, the flutter of panels, and laser light. Additional examples will be mentioned later.

1.3.1 Van der Pol Equation

Many electrical and biological oscillations can be modeled by the van der Pol equation [355]. This equation is a differential equation of second order where the nonlinearity results from damping,

$$\ddot{u} - \lambda(1 - u^2)\dot{u} + u = 0. \tag{1.6}$$

The van der Pol equation, Eq. (1.6), can be transformed into a system by means of, for instance, $y_1 = u$, $y_2 = \dot{u}$, leading to

$$\begin{aligned} \dot{y}_1 &= y_2 \\ \dot{y}_2 &= \lambda(1 - y_1^2)y_2 - y_1. \end{aligned}$$

This system has one stationary solution at $(y_1^s, y_2^s) = (0,0)$. For $0 < \lambda < 2$, this point is an unstable focus (Exercise 1.8). To discuss the dynamics of the system for small positive values of λ, one must find a method of determining where the trajectories tend as $t \to \infty$.

One method of studying this global behavior is a *numerical simulation.* Such a simulation may proceed as follows. Starting from various initial points, a numerical integration method is applied to calculate trajectories for finite time intervals $0 \le t \le t_f$.

Algorithm 1.1. Simulation.

Choose $\mathbf{z}$ and t_f
Integrate Eq. (1.1), starting from $\mathbf{y}(0) = \mathbf{z}$ and terminating at t_f.

The choice of $\mathbf{z}$ and t_f strongly affects the amount of information provided by the numerical simulation. It makes sense to analyze possible stationary points first. The unstable focus at (0,0) of the van der Pol equation suggests starting simulations near (0,0). The trajectory for the data,

$$z_1 = 0.1, \quad z_2 = 0, \quad t_f = 50, \quad \lambda = 0.5,$$

is plotted in Figure 1.12. As is seen in Figure 1.12, the trajectory approaches a closed curve and remains there. Such a closed curve toward which the trajectory winds is termed a *limit cycle.* A limit cycle is a periodic solution (orbit). That is, after some *period T* is elapsed, the solution values are the same,

$$\mathbf{y}(t + T) = \mathbf{y}(t).$$

Trajectories starting outside the limit cycle also wind toward it. Limit cycles that are approached by nearby trajectories are called stable; orbits are unstable when trajectories leave their neighborhoods (see Exercise

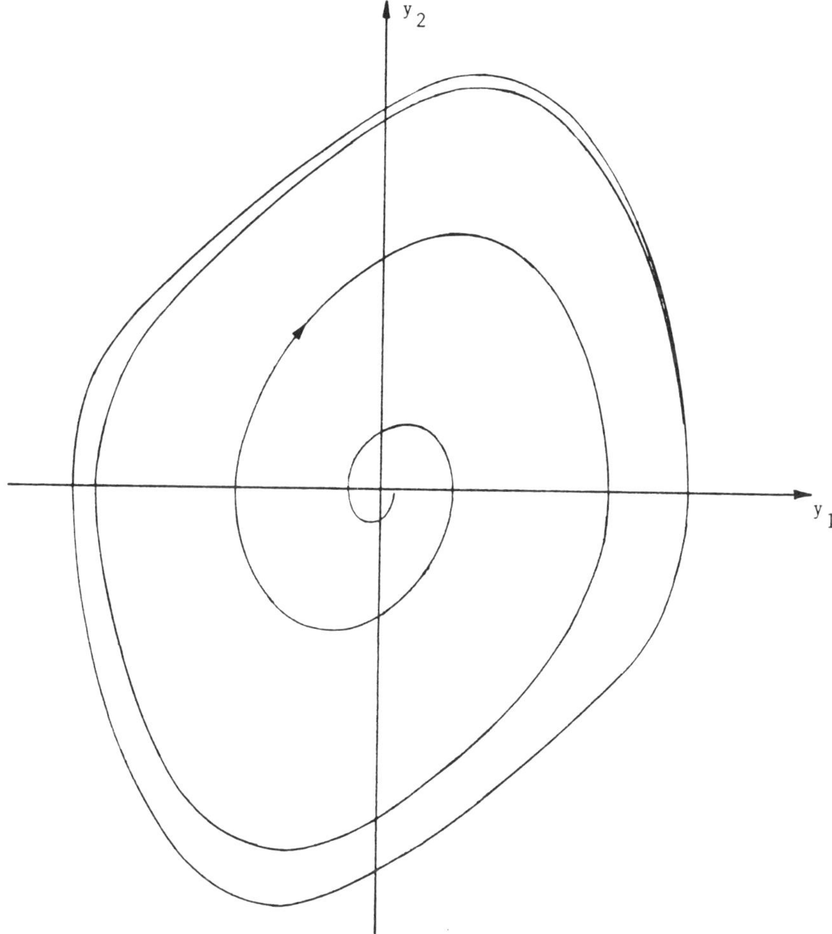

Figure 1.12

1.12). We have for the van der Pol equation a stable limit cycle. Generating this particular limit cycle by means of a simulation is fairly easy, but in general simulation is not necessarily the best way to create limit cycles; other methods will be discussed later.

Analytical methods for approximating limit cycles dominated for several decades. Since one object of this book is to make use of modern computers, we confine ourselves to illustrating one of the simplest analytical methods. We briefly illustrate van der Pol's *method of slowly varying amplitude* for Algorithm 1.1. In this method, one seeks a solution in the form

$$u(t) = A(t) \cos \omega t$$

with frequency ω and amplitude $A(t)$. The amplitude $A(t)$ is assumed to be slowly varying,

$$|\dot{A}| \ll |A| \omega.$$

This assumption justifies approximating $\dot{u}$ and $\ddot{u}$ by the expressions

$$-\omega A \sin \omega t \approx \dot{u}$$

$$-2\omega\dot{A} \sin \omega t - \omega^2 A \cos \omega t \approx \ddot{u}.$$

Inserting the ansatz for u and these approximations to its derivatives into the van der Pol equation yields

$$\begin{aligned} 0 = \sin \omega t[-2\omega\dot{A} + \lambda\omega A - \tfrac{1}{4}\lambda\omega A^3] \\ + \cos \omega t[A - A\omega^2] - \tfrac{1}{4}\lambda\omega A^3 \sin 3\omega t. \end{aligned}$$

The method is based on expanding the nonlinear term $u^2\dot{u}$ into a Fourier series and neglecting subharmonics—that is, dropping the term $\sin 3\omega t$. Since the equality holds for all t, the two bracketed terms must vanish. Two equations result for amplitude A and frequency ω,

$$\dot{A} = \tfrac{1}{2}\lambda A - \lambda A^3/8$$

$$\omega = 1.$$

The equation for A is a Bernoulli differential equation. Its solution describes the initial transient phase, thereby revealing the stability characteristics of the limit cycle (Exercise 1.8). The limit cycle itself is obtained by studying the asymptotic behavior of A. After transient effects have died out, we have $\dot{A} = 0$, which implies $A^2 = 4$. In summary, the method of slowly varying amplitude yields

$$u(t) \approx 2 \cos(t)$$

as an approximation of the limit cycle.

For an extensive discussion of how to apply analytical methods in the context of limit cycles, refer to [188, 335].

In the planar situation there are criteria for the existence of limit cycles. Two famous criteria attributed to Poincaré and Bendixson are:

(a) Let D be a finite domain that contains no stationary point and from which no trajectory departs. Then D contains a limit cycle.

(b) If the expression

$$\frac{\partial f_1}{\partial y_1} + \frac{\partial f_2}{\partial y_2}$$

does not change sign within a domain D, then no limit cycle can exist in D.

These criteria refer to domains D that are part of the phase plane. The above results can be formulated in this two-dimensional situation because trajectories do not intersect; no trajectory can escape the domain inside a limit cycle. In more than two dimensions, this "escaping" is possible and the dynamics can be much richer; phenomena such as chaos or trajectories on tori are possible. Accordingly, the above criteria do not hold for $n \geq 3$.

1.3.2 Waves

So far we have discussed two types of attractors—namely, stable equilibria and limit cycles. These particular solutions form *limit sets* for neighboring trajectories. In addition to stationary trajectories and limit cycles, there are limit sets that consist of trajectories connecting saddles and nodes. When distinct saddles are connected, one encounters a *heteroclinic orbit* (illustrated by the phase diagram of Figure 1.13a); a heteroclinic

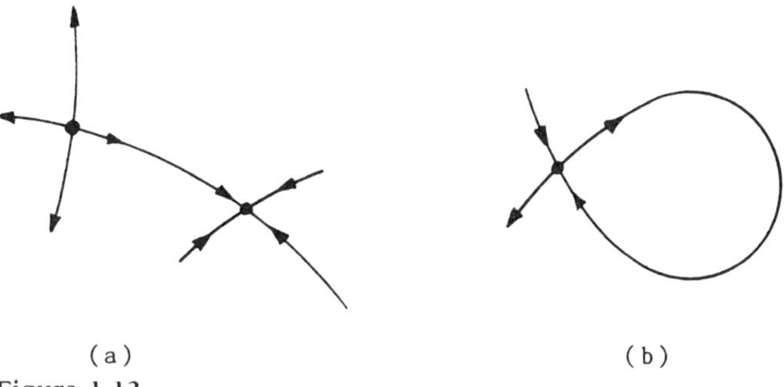

Figure 1.13

orbit may also join a saddle to a node, or vice versa. A *homoclinic orbit* connects a saddle point to itself (see Figure 1.13b); such orbits have an infinite period. Several heteroclinic orbits may form a closed path, called a homoclinic cycle. In a nonplanar two-dimensional case (say, on a torus or on a cylinder, in the case of a pendulum) even further limit sets are possible [123].

Heteroclinic and homoclinic orbits are of great interest in applications because they form the profiles of *traveling wave* solutions of many reaction-diffusion problems. As an example, consider the scalar partial differential equations

$$u_t = u_{xx} + f(u).$$

Solutions $u(x,t)$ depend on the space variable x and on the time t. Every zero of f constitutes a stationary solution of the partial differential equation (PDE). For instance, Fisher's model for the migration of advantageous genes is described by

$$f(u) = \zeta u(1 - u)$$

[19, 94, 127]. The PDE then possesses two stationary solutions—namely, $u = 0$ and $u = 1$. A traveling wave solution is a profile $U(x)$ that travels along the x-axis with *propagation speed* λ. Neither the shape of the wave nor the speed of propagation changes. To find traveling waves, we seek solutions

$$u(x,t) = U(x - \lambda t).$$

Substituting this expression into the PDE yields the ODE of second order

$$0 = U'' + \lambda U' + f(u).$$

By solving this ODE, one can calculate asymptotic states for the PDE. Let u_1 and u_2 be zeros of f and hence stationary solutions for both the PDE and the ODE. The asymptotic behavior of solutions U for $x \to \pm\infty$ determines the type of traveling wave. Every solution with

$$U(\infty) = u_1, \quad U(-\infty) = u_2, \quad u_1 \neq u_2$$

is a *front wave* of the ODE. This corresponds to a heteroclinic orbit of the ODE, connecting the two stationary points u_1 and u_2. In case $u_1 = u_2$, for nonconstant U one encounters a *pulse wave*; the reacting medium returns to its original state after a pulse traverses it. Pulses are characterized by homoclinic orbits in the (U,U') phase plane. A *wave train* occurs if the medium "fires" pulse-type spikes at regular intervals; related solutions U of the ODE are periodic.

Waves are nonequilibrium modes that persist in time. Permanent patterns of solutions are also observed in systems of PDEs and in PDEs with two or three space variables. Cases of interest include

plane waves: $u(\mathbf{x},t) = U(\mathbf{x}^{tr}\mathbf{e}_V - |\mathbf{V}|t)$,
$|\mathbf{V}|$ is the length of the velocity vector $\mathbf{V}$, $\mathbf{e}_V$ the unit vector in direction of $\mathbf{V}$.
target patterns: $u(\mathbf{x},t) = U(|\mathbf{x}|, t)$
rotating spiral: $u(\mathbf{x},t) = U(\rho, \vartheta - \lambda t)$,
$\mathbf{x} = (\rho \cos\vartheta, \rho \sin\vartheta)$, U periodic in the second argument.

For a review and a classification of wave solutions, see [92, 93]; further references will be given in Chapter 7.

Exercise 1.8.
Consider the van der Pol equation, Eq. (1.6). Discuss the stability properties of the equilibria as they vary with the parameter λ. Solve the Bernoulli differential equation for $A(t)$ that results from the method of slowly varying amplitude.

Exercise 1.9.
The system

$$\dot{y}_1 = 3(y_1 + y_2 - y_1^3/3 + \lambda)$$
$$\dot{y}_2 = -(y_1 - 0.7 + 0.8y_2)/3,$$

describing impulses of nerve potentials, has been discussed by FitzHugh [95] (see also [125]). The variable y_1 represents the membrane potential, while y_2 is a measure of accommodation and refractoriness. The parameter λ is the stimulating current.

(a) For $y_1 = 0.4$, calculate the values of λ and y_2 of the equilibrium.
(b) Prove that there is a stable limit cycle for the λ of part (a). (Hint: Construct a domain D from which no trajectory departs. Choose a rectangle.)

Exercise 1.10.
Figure 1.11 shows a metamorphosis of a stable spiral into an unstable spiral. Find two more ways in which a spiral can lose its stability. Draw "snapshots" of these phase planes illustrating the metamorphoses. (Hint: Encircle the spiral with a limit cycle of varying radius.)

Exercise 1.11.
Consider the system of ODEs

$$\dot{y}_1 = -y_2 + y_1(y_1^2 + y_2^2)^{-1/2}(1 - y_1^2 - y_2^2)$$
$$\dot{y}_2 = y_1 + y_2(y_1^2 + y_2^2)^{-1/2}(1 - y_1^2 - y_2^2).$$

Use polar coordinates ρ, ϑ,

$$y_1 = \rho \cos \vartheta, \quad y_2 = \rho \sin \vartheta,$$

to transform the above system into

$$\dot{\rho} = 1 - \rho^2, \quad \dot{\vartheta} = 1.$$

Solve the transformed system to show that the unit circle in the (y_1,y_2)-plane is a stable limit cycle.

Exercise 1.12.
Consider the system of ODEs

$$\dot{y}_1 = -y_2 + y_1(y_1^2 + y_2^2 - 1)$$
$$\dot{y}_2 = y_1 + y_2(y_1^2 + y_2^2 - 1).$$

Show that the periodic orbit is unstable.

Exercise 1.13.
In the (y_1,y_2)-plane ($y_1 = \rho \cos \vartheta$, $y_2 = \rho \sin \vartheta$), two differential equations are given by

$$\dot{\rho} = \rho - \rho^2$$
$$\dot{\vartheta} = \lambda - \cos 2\vartheta.$$

The values of $\lambda = 1$ and $\lambda = -1$ separate regions of different dynamic behavior. Calculate all stationary states and sketch five (y_1,y_2) phase portraits illustrating the dynamics.

1.4 SOME FUNDAMENTAL NUMERICAL METHODS

In the previous sections we defined stationary solutions and mentioned numerical integration as a method of investigating dynamic behavior near such solutions. In the present section, we briefly review a minimum number of numerical methods that one must know to analyze nonlinear phenomena. Most of the problems of stability analysis can be reduced or transformed in such a way that their solution requires only a small number of standard tools of numerical analysis. Every student and researcher should put together his own toolbox and not rely solely on "packages." The reader familiar with the Newton method and related iterative procedures for nonlinear equations may want to skip this section.

Stationary solutions $\mathbf{y}^s$ of systems of ODEs are defined as solutions of the (nonlinear) vector equation $\mathbf{f}(\mathbf{y}) = \mathbf{0}$. By means of a linearization of $\mathbf{f}$, one obtains

$$\mathbf{0} = \mathbf{f}(\mathbf{y}^s) \approx \mathbf{f}(\mathbf{y}^{(0)}) + \mathbf{f}_{\mathbf{y}}^0 \cdot (\mathbf{y}^s - \mathbf{y}^{(0)}).$$

Here, $\mathbf{y}^{(0)}$ is an approximation of $\mathbf{y}^s$, and $\mathbf{f}_{\mathbf{y}}^0$ is the Jacobian matrix evaluated at $\mathbf{y}^{(0)}$. The above relation establishes a way of calculating a better approximation $\mathbf{y}^{(1)}$ by solving

$$\mathbf{0} = \mathbf{f}(\mathbf{y}^{(0)}) + \mathbf{f}_{\mathbf{y}}^0 \cdot (\mathbf{y}^{(1)} - \mathbf{y}^{(0)}). \tag{1.7}$$

Equation (1.7) is a system of linear equations. Solving it for the quantity $\mathbf{y}^{(1)} - \mathbf{y}^{(0)}$ and adding this quantity to $\mathbf{y}^{(0)}$ gives a better approximation $\mathbf{y}^{(1)}$. Applying Eq. (1.7) repeatedly produces a series of approximations

$$\mathbf{y}^{(1)}, \mathbf{y}^{(2)}, \mathbf{y}^{(3)}, \ldots .$$

This method is called the Newton method or the Newton-Raphson method. Introducing vectors $\mathbf{d}^{(i)}$ for the difference between two iterates, namely, $\mathbf{y}^{(i)} - \mathbf{y}^{(i+1)}$, we have

$$\frac{\partial \mathbf{f}(\mathbf{y}^{(i)})}{\partial \mathbf{y}} \mathbf{d}^{(i)} = \mathbf{f}(\mathbf{y}^{(i)}) \tag{1.8a}$$

$$i = 0, 1, 2, \ldots$$

$$\mathbf{y}^{(i+1)} = \mathbf{y}^{(i)} - \mathbf{d}^{(i)}. \tag{1.8b}$$

Under certain assumptions (the most important of which is nonsingularity of the Jacobian at the solution), one can guarantee *quadratic* convergence in a neighborhood of the solution. That is, both $\mathbf{f}(\mathbf{y}^{(i)})$ and

$\mathbf{d}^{(i)}$ quickly approach zero. Roughly speaking, the number of correct decimal digits doubles on each step. In practice, however, the usual assumptions ensuring this rate of convergence are not always satisfied. Therefore we often encounter limitations on the rate of convergence.

First, locally, we cannot expect quadratic convergence when the Jacobian is replaced by an approximation (see below). The second limitation is by far more severe. For many difficult problems an arbitrary initial guess $\mathbf{y}^{(0)}$ does not lead to convergence of the sequence produced by Eq. (1.8), because $\mathbf{y}^{(0)}$ may be too far away from a solution. In order to enlarge the domain of convergence, one modifies Newton's method as follows: A minimization technique replaces Eq. (1.8b) by

$$\mathbf{y}^{(i+1)} = \mathbf{y}^{(i)} - \delta \mathbf{d}^{(i)} \tag{1.8b'}$$

with δ such that $|\mathbf{f}(\mathbf{y}^{(i+1)})|$ is minimal or at least closer to zero than $|\mathbf{f}(\mathbf{y}^{(i)})|$. The resulting *damped* Newton method, Eqs. (1.8a) and (1.8b′), has reasonable and often global convergence properties. One must be aware, however, that a fast quadratic convergence is still restricted to a (small) neighborhood of the solution; the global initial phase may be slow (linear convergence).

The evaluation of the Jacobian in Eq. (1.8) mostly is expensive. Because analytic expressions for the partial derivatives of $\mathbf{f}$ are often unavailable, implementation of the Newton method must make use of *numerical differentiation*. Let $\mathbf{z}^k$ be the k-th column of the Jacobian $\mathbf{f_y}$. The following algorithm can be used to calculate an approximation of $\mathbf{f_y}(\mathbf{y})$ ($\mathbf{U}$ and $\mathbf{V}$ are two auxiliary n-vectors).

Algorithm 1.2. Evaluate $\mathbf{U} = \mathbf{f}(\mathbf{y})$

Choose a small $\epsilon > 0$

Loop for $k = 1, 2, \ldots, n$:

$$\epsilon_1 = \epsilon \max\{1, y_k\} y_k / |y_k|, \quad y_k = y_k + \epsilon_1$$

Evaluate the vector $\mathbf{V} = \mathbf{f}(\mathbf{y})$

$$y_k = y_k - \epsilon_1$$

$$\mathbf{z}^k = (\mathbf{V} - \mathbf{U})/\epsilon_1.$$

The choice of ϵ depends on the number of mantissa digits ν available on a particular computer. Usually,

$$\epsilon = 10^{-\nu/2}$$

leads to a reliable approximation of the Jacobian. The cost of approximating the n^2 matrix $\mathbf{f_y}(\mathbf{y})$ by means of Algorithm 1.2 is that of evaluating the vector function $\mathbf{f}$ n times (the evaluation of $\mathbf{U}$ usually was carried out earlier).

Evaluating the Jacobian or approximating it by Algorithm 1.2 is so expensive that remedies are needed. The simplest strategy is to replace Eq. (1.8a) by

$$\mathbf{f}_{\mathbf{y}}^{0}\cdot\mathbf{d}^{(i)} = \mathbf{f}(\mathbf{y}^{(i)}).$$

This *chord method* involves the same approximate Jacobian for all iterations, or at least for a few iterations. The chord method is a slow procedure. There are other variants of the Newton method with convergence rates ranging between the linear convergence of the chord method and the quadratic convergence of the original Newton method. The Jacobian can be approximated by algorithms that are cheaper (and coarser) than the procedure in Algorithm 1.2. Procedures, called update methods (for example, Broyden's method [46]), are in wide use.

In such methods, a matrix of rank one, which requires only one function evaluation, is added to the approximate Jacobian of the previous step. For details we refer to textbooks on numerical analysis, for instance, [72, 334]. The use of rank-one updates reduces the convergence rate per iteration step to *superlinear* convergence. The saving in function evaluations is, however, significant, and the overall computing time is usually reduced. This strategy of combining numerical differentiation with rank-one updates is excellent. One takes advantage of rank-one updates especially when the convergence is satisfactory. This kind of *quasi-Newton method* is a compromise between the locally fast but expensive Newton method and the cheap but more slowly converging chord method [78], which is highly recommended.

A general treatment of methods for solving nonlinear equations is given by Ortega and Rheinboldt [249]. The linear equation (1.8a) can be solved by Gaussian elimination or by **LU** decomposition. Recall that the latter means the calculation of a lower triangular matrix $\mathbf{L}$ and an upper triangular matrix $\mathbf{U}$ such that

$$\mathbf{f_y} = \mathbf{LU}.$$

The diagonal elements of $\mathbf{L}$ are normalized to unity. Because this **LU** decomposition does not always exist, a permutation of the rows of $\mathbf{f_y}$, which reflects the pivoting of the Gaussian elimination, may be necessary. A sophisticated algorithm for solving linear equations is given by Businger and Golub in [367].

Another basic tool of numerical analysis is numerical methods for integration of the initial-value problem

$$\dot{\mathbf{y}} = \mathbf{f}(t,\mathbf{y}), \quad \mathbf{y}(t_0) = \mathbf{z}.$$

The solution $\mathbf{y}(t)$ is approximated at discrete values t_k of the independent variable t,

$$\mathbf{y}(t_1), \mathbf{y}(t_2), \mathbf{y}(t_3), \ldots .$$

There are numerous algorithms for numerical integration. Among the most reliable integrators are the Runge-Kutta-Fehlberg methods (RKF). Runge-Kutta-Fehlberg methods are one-step methods in that calculation of the approximation to $\mathbf{y}(t_{k+1})$ involves only data of the previous step $\mathbf{y}(t_k)$. Here, for notational convenience, we use the letter $\mathbf{y}$ also for the numerical approximation. The difference between two consecutive t_k's is the *step size*, $\Delta_k = t_{k+1} - t_k$. One RKF step for approximating $\mathbf{y}(t + \Delta)$, starting from $\mathbf{y}(t)$, has the form

$$\mathbf{y}(t + \Delta) = \mathbf{y}(t) + \Delta \sum_{i=1}^{m} c_i \mathbf{f}(\tilde{t}_i, \tilde{\mathbf{y}}^i). \tag{1.9}$$

In this formula, $\mathbf{f}$ is evaluated at certain values of $\tilde{t}_i$ and $\tilde{\mathbf{y}}^i$ (see Appendix 4). The step size Δ is varied from step to step so that certain accuracy requirements are met [88]. Besides RKF integrators, there are many other integrators, such as multistep methods and extrapolation methods. For details, see textbooks on numerical analysis.

Above we mentioned some of the standard devices of numerical analysis, but the list is far from complete. A routine that calculates eigenvalues of a matrix is also required. Here we mention the QR algorithm as the standard procedure [109]; for references to programs, see Appendix 6. ODE boundary-value problems can be solved by using *shooting methods*. Shooting methods integrate initial-value problems in a systematic way in order to find the initial values of the particular trajectory that satisfies the boundary conditions.

The initial and final values of a trajectory in the shooting method do not in general satisfy the boundary conditions. To see this, suppose that the boundary conditions are given by $\mathbf{r}(\mathbf{y}(a), \mathbf{y}(b)) = \mathbf{0}$. We denote the trajectory that starts from an initial vector $\mathbf{z}$ by $\varphi(t;\mathbf{z})$, $a \le t \le b$. One wants to find $\mathbf{z}$ such that the *residual* $\mathbf{r}(\mathbf{z},\varphi(b;\mathbf{z}))$ vanishes. This establishes an equation of the type $\mathbf{f}(\mathbf{z}) = \mathbf{0}$. A combination of Newton's method with an integrator serves as the simplest shooting device. Note that each evaluation of $\mathbf{f}$ means the integration of a trajectory (Exercises 1.16, 1.17); $\mathbf{f}$ is not given explicitly. Professional shooting codes use the *multiple shooting* approach, in which the integration interval is suitably subdivided. The resulting shortening of the integration intervals enlarges the domain of attraction of the Newton method and dampens the influence of "bad" initial guesses [219, 334].

We conclude this section with a list of the main tools required in the numerical analysis of nonlinear phenomena. These tools are:

linear equation solver
Newton method (with modifications)
ODE integrator
QR method for calculating eigenvalues
solver for ODE boundary-value problems (such as shooting)
plotting devices

Exercise 1.14.
Consider the example of the Duffing equation, Eq. (1.5). Integrate the separatrix numerically as described in Section 1.2.

Exercise 1.15.
Consider the van der Pol equation, Eq. (1.6). For $\lambda = 1$, carry out a numerical simulation (Algorithm 1.1), starting from several points z inside and outside the limit cycle (see Figure 1.12).

Exercise 1.16.
Suppose some second-order differential equation is given together with boundary conditions $u(0) = 0$, $u(1) = 0$. Give a geometrical interpretation of the shooting method. To this end, consider the initial vector $\mathbf{z} = (u(0),u'(0)) = (0,\delta)$ for various δ. Draw a sketch and indicate $\mathbf{f}$.

Exercise 1.17 (project).
Formulate a driving algorithm for a simple shooting method. To this end, use the Newton method and an ODE integrator as black boxes.

2 Basic Nonlinear Phenomena

Beginning with this chapter, nonlinearity and parameter dependence will play a crucial role. We shall assume throughout that λ is a real parameter, and we shall study solutions of a system of ODEs,

$$\dot{\mathbf{y}} = \mathbf{f}(\mathbf{y},\lambda), \tag{2.1}$$

or solutions of a system of "algebraic" equations,

$$\mathbf{0} = \mathbf{f}(\mathbf{y},\lambda). \tag{2.2}$$

Sometimes boundary conditions must be attached to Eq. (2.1). As in Chapter 1, the vectors $\mathbf{y}$ and $\mathbf{f}$ have n components. If a particular example involves more than one parameter, we assume for the time being that all except λ are kept fixed. Clearly, solutions $\mathbf{y}$ of Eq. (2.1) or Eq. (2.2) in general vary with λ. We shall assume throughout that $\mathbf{f}$ depends smoothly on $\mathbf{y}$ and λ—that is, $\mathbf{f}$ is to be sufficiently often continuously differentiable. This hypothesis is usually met by practical examples.

2.1 A PREPARATORY EXAMPLE

We first study a simple scalar example (illustrated by Figure 2.1). We have drawn a parabola and a straight line. Both curves are solutions of the scalar equation

$$0 = f(y,\lambda) = [1 - \lambda + (y - 2)^2](y - \lambda + 2.75). \tag{2.3}$$

Assume now that we want to investigate the solutions of Eq. (2.3) without knowing the result depicted in Figure 2.1. That is, we start from

$$0 = y^3 - y^2(\lambda + 1.25) + y(3\lambda - 6) + \lambda^2 - 7.75\lambda + 13.75. \tag{2.4}$$

We shall not try to calculate a solution $y(\lambda)$ of Eq. (2.4) analytically. The structure of Eq. (2.4) enables us to calculate easily λ-values for

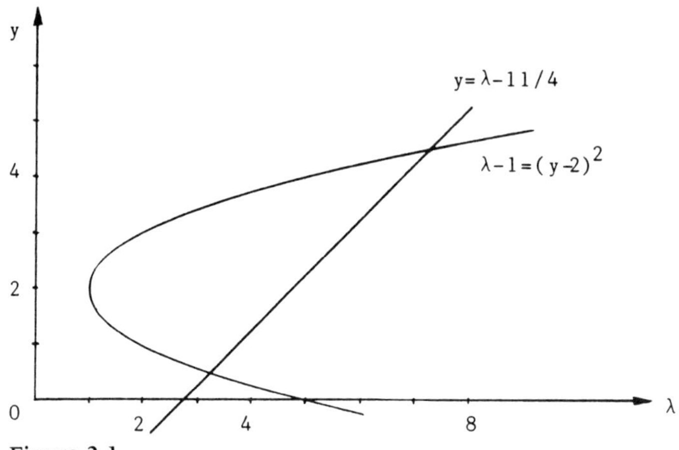

Figure 2.1

given values of y. For example, $y = 1$ implies $0 = (\lambda - 3.75)(\lambda - 2)$. This produces two points on the solution curve(s); we choose the point $\lambda = 2, y = 1$. Starting from this point, we can generate a solution curve by calculating the derivative $dy/d\lambda$. Differentiating the identity

$$0 \equiv f(y(\lambda),\lambda)$$

with respect to λ by using the chain rule yields

$$0 = \frac{\partial f(y,\lambda)}{\partial y}\frac{dy}{d\lambda} + \frac{\partial f(y,\lambda)}{\partial \lambda} . \tag{2.5}$$

In our Eq. (2.4), differentiation gives

$$0 = \{3y^2 - y(2\lambda + 2.5) + (3\lambda - 6)\}\frac{dy}{d\lambda} - (y^2 - 3y - 2\lambda + 7.75),$$

which leads to the explicit formula

$$\frac{dy}{d\lambda} = \frac{y^2 - 3y - 2\lambda + 7.75}{3y^2 - y(2\lambda + 2.5) + 3\lambda - 6} = -\frac{\partial f/\partial\lambda}{\partial f/\partial y} . \tag{2.6}$$

This expression establishes a differential equation for $y(\lambda)$. Starting from the chosen point ($\lambda = 2$, $y = 1$), we can integrate this initial-value problem in order to generate nearby solutions. The slope of the "first" tangent is $dy/d\lambda = -0.5$. A small step in the direction of this tangent produces a new point, and we can evaluate a new tangent there. In this way we can approximate the curve until a point with $f_y = 0$ is reached (Exercise 2.1). A vanishing denominator in Eq. (2.6) produces a singularity of the differential equation hindering a successful *tracing* of the entire solution curve.

The situation is as follows (see Figure 2.2): Let (y^*,λ^*) be a solution of Eq. (2.2). We encounter three different cases along a solution curve:

(a) $f_y(y^*,\lambda^*) \neq 0$: the curve can be uniquely continued in a λ-neighborhood—that is, there are nearby values $\lambda \neq \lambda^*$ such that $f(y,\lambda) = 0$ has a solution y close to y^*.

(b) $f_y(y^*,\lambda^*) = 0$ and $f_\lambda(y^*,\lambda^*) \neq 0$: the tangent is not defined.

(c) $f_y(y^*,\lambda^*) = 0$ and $f_\lambda(y^*,\lambda^*) = 0$: the tangent is not unique. With additional effort, tangents can be calculated.

Apparently something "happens" in cases (b) and (c), where f_y becomes zero. A significant part of our attention will be focused on cases (b) and (c). The standard situation (a) is characterized by the implicit function theorem (Appendix 2).

Exercise 2.1.
Integrate the differential equation Eq. (2.6) numerically, starting from $\lambda = 2$, $y = 1$. Integrate "to the left" ($\lambda < 2$) and "to the right" ($\lambda > 2$) and see what happens for $\lambda \approx 1$ and $\lambda \approx 3.25$.

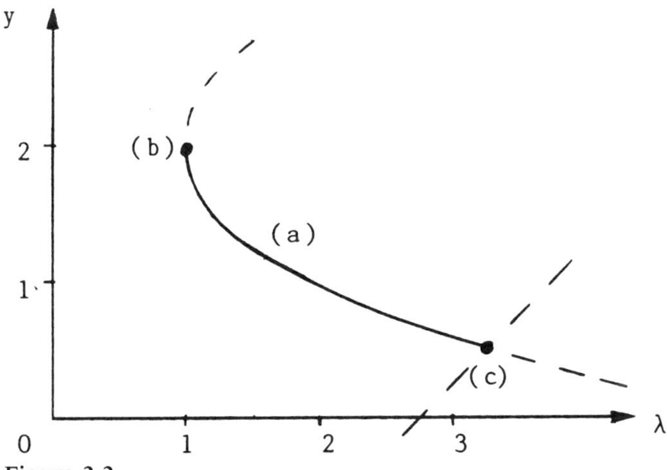

Figure 2.2

Exercise 2.2.
Calculate all solutions $y_1(\lambda)$, $y_2(\lambda)$ of the two equations

$$y_1 + \lambda y_1(y_1^2 - 1 + y_2^2) = 0$$

$$10y_2 - \lambda y_2(1 + 2y_1^2 + y_2^2) = 0.$$

Sketch $N(\lambda) = y_1^2 + 2y_2^2$ versus λ for all solutions.

2.2 ELEMENTARY DEFINITIONS

We now return to the general vector equations (2.1) and (2.2). In order to illustrate graphically the dependence of $\mathbf{y}$ on λ in a two-dimensional diagram, we require a scalar measure of a the n-vector $\mathbf{y}$. We shall use the notation $[\mathbf{y}]$ for such a scalar measure of $\mathbf{y}$. Examples are

$$[\mathbf{y}] = y_k \text{ for any of the } n \text{ indices, } 1 \leq k \leq n$$

$$[\mathbf{y}] = (y_1^2 + y_2^2 + \cdots + y_n^2)^{1/2} \quad (\textit{euclidian vector norm}) \tag{2.7}$$

$$[\mathbf{y}] = \max\{|y_1|, |y_2|, \ldots, |y_n|\}.$$

In Exercise 2.2, $[\mathbf{y}]$ was taken to be yet another scalar measure. The three examples in Eq. (2.7) are appropriate for stationary solutions $\mathbf{y}$. The formula are easily adapted to solutions $\mathbf{y}(t)$ of ODEs for $\mathbf{y}(t)$ evaluated at some specific value $t = t_0$. The first of the scalar measures in Eq. (2.7) then reads $[\mathbf{y}] = y_k(t_0)$. Note that the values of $[\mathbf{y}]$ vary with λ. The various choices of $[\mathbf{y}]$ are not equal in the amount of information they provide. As we shall see later, a bad choice might conceal rather than clarify the branching behavior in a particular example. In practical problems involving ODEs, the choice $[\mathbf{y}] = y_k(t_0)$ usually turns out to be good. Some equations involve a variable that reflects essential physical properties in a natural way. Choosing that variable for the scalar measure $[\mathbf{y}]$ is to be preferred to choosing the second or third of the measures in Eq. (2.7), which are somewhat artificial. For differential equations with oscillating solutions, amplitude is often a good candidate. Other naturally occurring scalar measures are energy functionals and physical moments.

Equipped with a scalar measure $[\mathbf{y}]$ of the vector $\mathbf{y}$, we are able to depict the solutions of Eq. (2.1) or Eq. (2.2) in a diagram like Figure 2.1. For a particular example, such a diagram may look like Figure 2.3.

Definition 2.1. A diagram depicting $[\mathbf{y}]$ versus λ, where $(\mathbf{y},\lambda)$ solves Eq. (2.1) or Eq. (2.2), will be called a branching diagram (or bifurcation diagram or response diagram).

Branching diagrams will turn out to be extremely useful. As Figure 2.1 and Figure 2.3 illustrate, solutions may form continua, reflected by smooth curves in branching diagrams. These curves are called *branches*. Upon varying λ, the number of solutions may change. New branches may emerge, branches may end, or branches may intersect. The following list reports on the *multiplicity* of solutions as shown in the fictitious branching diagram of Figure 2.3:

λ-Interval	Number of Solutions y
$\lambda < \lambda_1$	1
$\lambda_1 \leq \lambda < \lambda_2$	2
λ_2	3
$\lambda_2 < \lambda < \lambda_3$	4
λ_3	3
$\lambda_3 < \lambda < \lambda_4$	2
λ_4	1
etc.	

The significance of the values λ_1, λ_2, . . . gives rise to the following informal definition of a branch point:

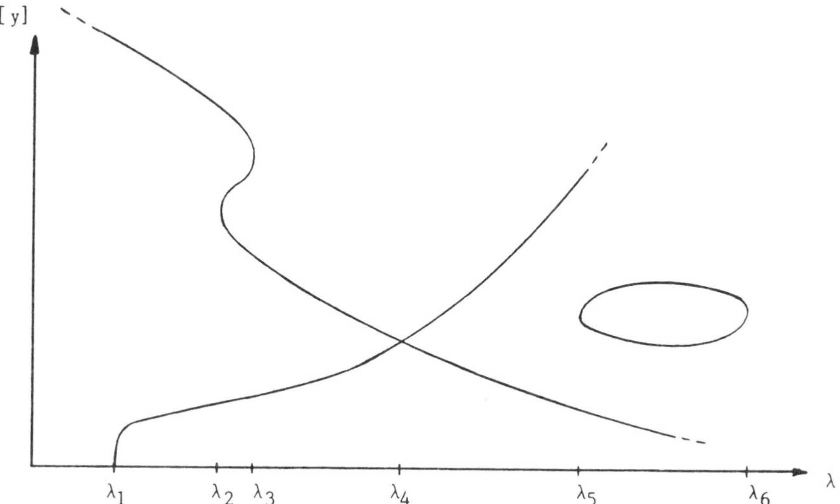

Figure 2.3

Definition 2.2. A branch point (with respect to λ) is a solution $(\mathbf{y}_0,\lambda_0)$ of Eq. (2.1) or Eq. (2.2) where the number of solutions changes when λ passes λ_0.

Exercise 2.3.
Consider Eq. (2.3) with Figure 2.1. Establish a table of the multiplicity of solutions for various ranges of λ.

2.3 BUCKLING AND OSCILLATION OF A BEAM

Consider the following situation (see Figure 2.4). A beam is subjected at its ends to a compressive force Γ along its axis. The beam may be excited by a harmonic excitation $P(x,t)$ that depends on the spatial variable x (with $0 \le x \le 1$) and on the time t. This experiment may represent a beam supporting a machine with a rotating imbalance (see Figure 2.4). We denote the viscous damping by δ and the membrane stiffness by K. Small deflections $v(x,t)$ of the beam are described by solutions of the partial differential equation

$$v_{xxxx} + (\Gamma - K \int_0^1 (v_x(\xi,t))^2 \, d\xi) v_{xx} + \delta v_t + v_{tt} = P$$

[153, 158]. We simplify the analysis by assuming perfect symmetry of $P(x,t)$ around $x = 0.5$, in which case the response of the beam is likely to be in the "first mode." That is, we take both P and v to be sinusoidal in x,

$$P(x,t) = \gamma \cos \omega t \sin \pi x$$

$$v(x,t) = u(t) \sin \pi x.$$

The ansatz for P reflects the additional assumption of a harmonic excitation with frequency ω. Inserting these expressions into the PDE leads to a Duffing equation describing the temporal behavior of the displacement of the beam,

$$\ddot{u} + \delta \dot{u} - \pi^2(\Gamma - \pi^2)u + \tfrac{1}{2}K\pi^4 u^3 = \gamma \cos \omega t. \tag{2.8}$$

We first discuss the stability of deflections when the external force is zero ($\gamma = 0$). It turns out (Exercise 2.4) that for $\Gamma > \pi^2$ there are two stable equilibria with $u \neq 0$, whereas for $\Gamma < \pi^2$ there is only one equilibrium ($u = 0$, stable). We interpret the force Γ as our branching

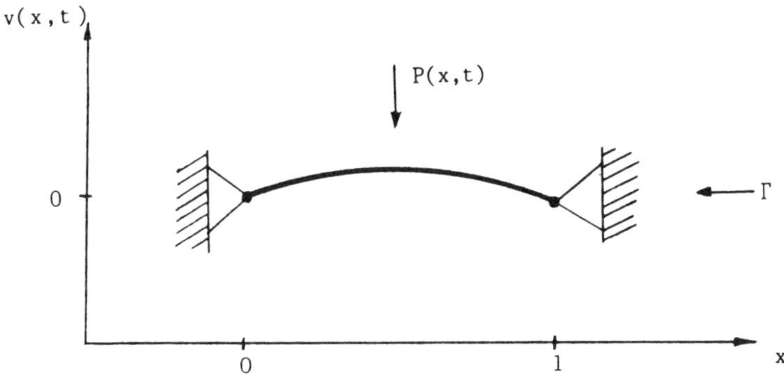

Figure 2.4

parameter λ and depict the results in a branching diagram (Figure 2.5). The stationary solution $(u,\Gamma) = (0,\pi^2)$ of Eq. (2.8) with $\gamma = 0$ is an example of a branch point. As Figure 2.5 indicates, this branch point separates domains of different qualitative behavior. In particular, the "trivial" solution $u = 0$ loses its stability at $\Gamma = \pi^2$. The value $\Gamma = \pi^2$ is called *Euler's first buckling load,* because Euler calculated the critical load where a beam buckles (this will be discussed in some detail in Section 6.5). The symmetry of the branching diagram with respect to the Γ-axis reflects the basic assumption of perfect symmetry.

So far no external energy has entered the system ($\gamma = 0$). Now we study Eq. (2.8) with excitation ($\gamma \neq 0$). The possible responses of

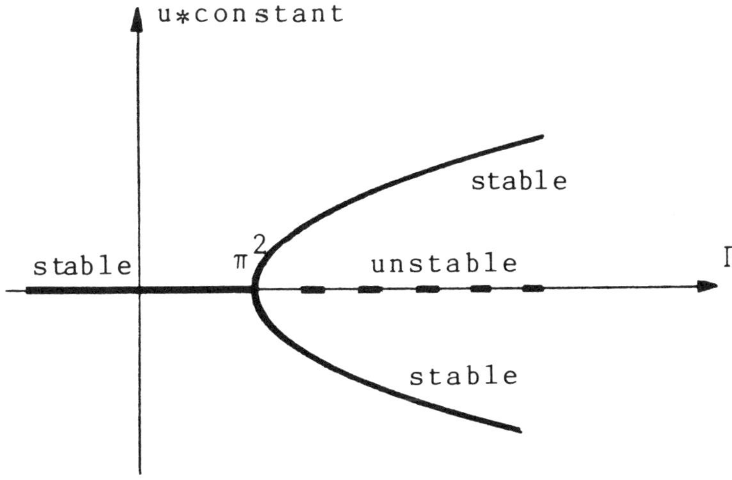

Figure 2.5

the beam can be explained most easily by discussing the experiment illustrated in Figure 2.6. Imagine a vehicle with a ball rolling inside on a cross section with one minimum ($\Gamma < \pi^2$) or two minima ($\Gamma > \pi^2$). Moving the vehicles back and forth in a harmonic fashion corresponds to the term $\gamma \cos \omega t$. We expect interesting effects in the case $\Gamma > \pi^2$. Choose (artificially)

$$\Gamma = \pi^2 + 0.2\pi^{-2}, \quad K = 16\pi^{-4}/15, \quad \gamma = 0.4.$$

This choice leads to the specific Duffing equation

$$\ddot{u} + \tfrac{1}{25}\dot{u} - \tfrac{1}{5}u + \tfrac{8}{15}u^3 = \tfrac{2}{5}\cos \omega t. \tag{2.9}$$

The harmonic responses can be calculated numerically; we defer a discussion of the specific methods to Section 6.1 and focus our attention on the results. The parameter Γ is now kept fixed, and we choose the frequency ω as branching parameter λ. This parameter is considered to be constant; it is varied in what is called a *quasistatic* way. The main response of the oscillator to the sinusoidal forcing is shown in Figure 2.7. As a scalar measure of $u(t)$, amplitude A is chosen. The branching diagram Figure 2.7 shows a typical response of an oscillator with a *hardening spring*. We find two branch points with parameter values

$$\omega_1 = 0.748, \quad \omega_2 = 2.502.$$

Hysteresis effects, such as that depicted in Figure 2.7, are ubiquitous in science. Characteristic for hysteresis effects are jump phenomena, which here take place at ω_1 and ω_2. This can be explained as follows. Starting with an oscillation represented by a point on the upper branch for $\omega < \omega_2$, $\omega \approx \omega_2$ (that is, close to B), the amplitude is "large." Slowly

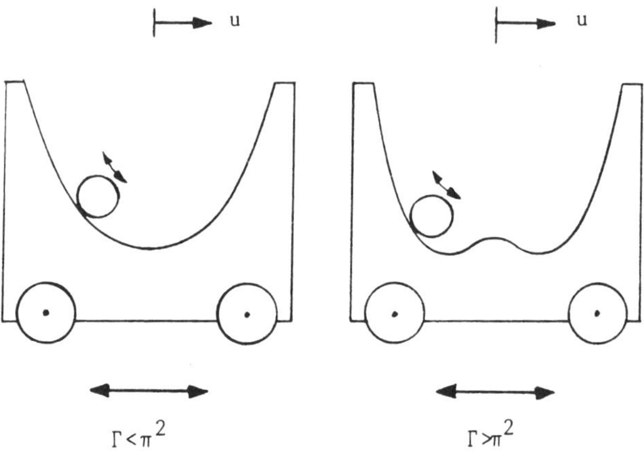

Figure 2.6

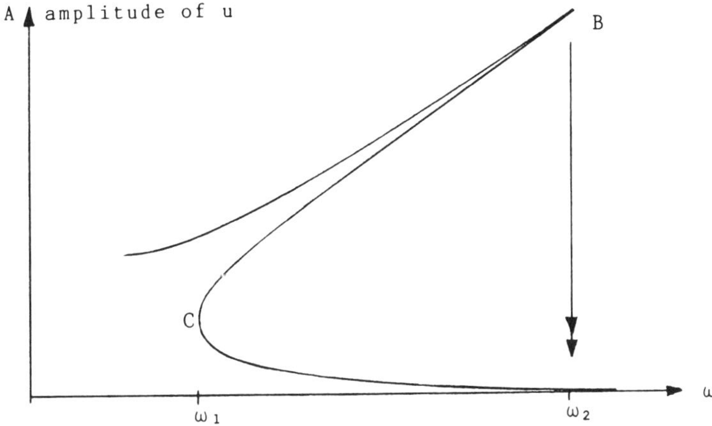

Figure 2.7

increasing ω beyond the value ω_2 causes a jump to the lower branch—that is, to small amplitude. A jump from small amplitude to large amplitude will occur for $\omega \approx \omega_1$ (Figure 2.7C, Figure 2.8).

There is another phenomenon in our experiment that is not shown in Figure 2.7. Consider very high values of the frequency ω (significantly larger than the *natural frequency* of the system). Then, obviously, the system (the ball in Figure 2.6) does not have enough time for a full oscillation to pass both stable equilibria. The system oscillates with small amplitude around one of the stable equilibria. The overall amplitude is close to the x-value of the equilibrium,

$$A \approx \sqrt{(3/8)}.$$

Such a solution corresponds to a "new" branch, which merges with the lower branch of Figure 2.7 at $\omega = \omega_3 = 0.8427$. Based on our results, we expect the amplitude A to behave as indicated in Figure 2.8. Note that we assume that ω is reduced in an almost stationary way ($\dot{\omega} \approx 0$), starting from high values. Transient phenomena are not shown in Figure 2.8. The response of the middle part of the branching diagrams Figure 2.7 and Figure 2.8 (between B and C) is unstable and thus cannot be realized in any experiment. Jump phenomena of oscillations caused by varying frequencies occur in centrifuges and in many other commonly used instruments.

The most important results of this section are summarized in the two branching diagrams, Figure 2.5 and Figure 2.7. Comparing these two figures, we notice the different qualities of the branching phenom-

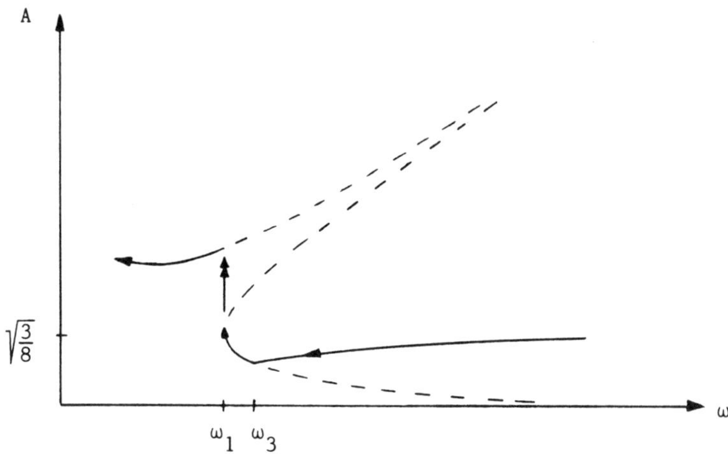

Figure 2.8

ena. At this point, a classification of various branching phenomena is needed.

Exercise 2.4.
Carry out the analysis of Eq. (2.8) in detail. First derive Eq. (2.8) using the ansatz for P and v. Then discuss the type of equilibria using Γ as the parameter ($\gamma = 0$).

2.4 TURNING POINTS AND BIFURCATION POINTS: THE GEOMETRIC VIEW

Two commonly encountered nonlinear phenomena are *turning points* and *bifurcation points*. We encountered them already in the previous section, without defining them precisely. It is easiest to define these branch points geometrically. We introduce these phenomena by simple examples, point out differences between them, and continue with an algebraic definition in the next section. The terms "turning point" and "bifurcation point" are used in mathematics. In applications, other names—which place more emphasis on practical features—are often used. In structural engineering, such phenomena are called *snapping* and *buckling* [276]. We shall first concentrate on turning points and bifurcation points, deferring an introduction to Hopf bifurcation to Section 2.6.

The simplest examples of turning points can be introduced using the scalar equations

$$0 = y^2 - \lambda \qquad \text{or} \qquad 0 = y^2 + \lambda.$$

The solutions $y(\lambda)$ of $0 = y^2 - \lambda$ form a parabola that is defined only for $\lambda \geq 0$. At $\lambda = 0$ there is only one solution ($y = 0$), whereas for $\lambda > 0$ there are two solutions, $y = +\sqrt{\lambda}$ and $y = -\sqrt{\lambda}$. The point where solutions begin to exist ($\lambda = 0$, $y = 0$ in this example) is a turning point. The branching diagram in Figure 2.9 depicts a typical situation and prompts the name "turning point." The branch comes from one side ("side" refers to the λ-direction) and turns back at the turning point (emphasized by the dot in Figure 2.9). In (a), the branch opens to the right. The case of a turning point on a branch opening to the left is depicted in Figure 2.9(b), the simplest example of which is $0 = y^2 + \lambda$. Terms "right" or "left" in connection with branching diagrams refer to larger or smaller values of the branching parameter λ. Turning points frequently arise in pairs, resulting in hysteresis effects, as in Eq. (2.9); see Figure 2.7. Other names for turning points are *limit points* or *saddle nodes* and, less respectfully, *noses* or *knees*.

The name "saddle node" is motivated by the stability behavior of the solutions when they are interpreted as equilibria of differential equations. We study this via the scalar differential equation

$$\dot{y} = \lambda - y^2,$$

which is related to the above scalar examples. Clearly, the equilibrium $+\sqrt{\lambda}$ is stable, whereas $-\sqrt{\lambda}$ is unstable. The standard stability analysis recalled in Chapter 1 reduces in the scalar case to the investigation of the sign of f_y. The stability results of this example can be presented graphically, as in Figure 2.10. This figure follows the well-established

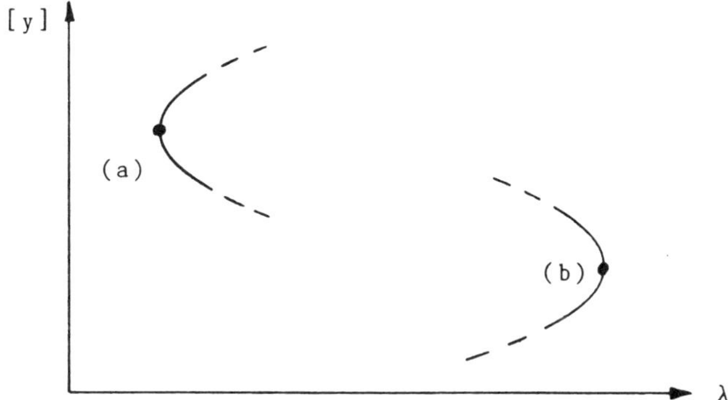

Figure 2.9

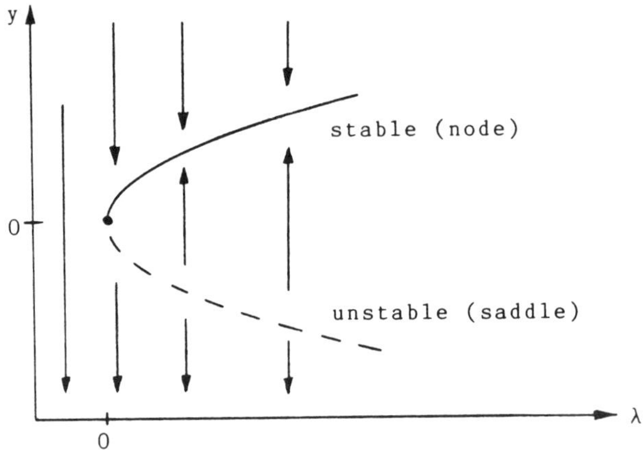

Figure 2.10

convention of distinguishing between stable equilibria and unstable equilibria by using a heavy continuous curve for the former and a dashed curve for the latter.

The question arises as to whether a turning point always separates stable equilibria from unstable equilibria. The answer is no. It is easy to construct a counterexample in higher dimensions:

$$\dot{y}_1 = \lambda - y_1^2, \quad \dot{y}_2 = cy_2. \tag{2.10}$$

A stability analysis of the equilibria of Eq. (2.10) reveals that one equilibrium is a saddle and the other is a node. The stability of the node is determined by the sign of c, the result being shown in Figure 2.11. We

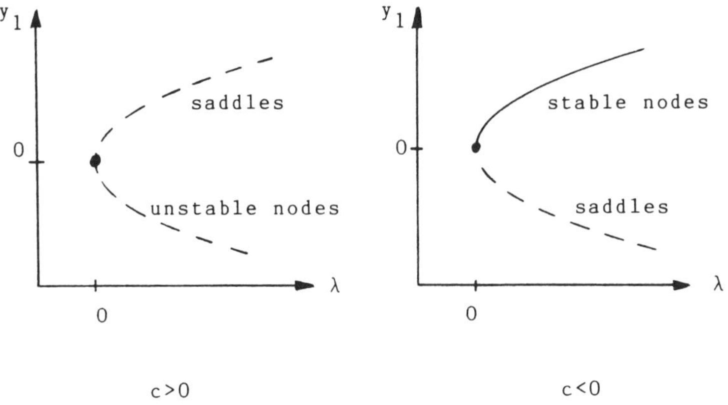

Figure 2.11

learn from this example that both the half-branches meeting at a turning point can be unstable. Later we shall see examples of this type that are important in practical applications (for example, the problem of a continuous stirred tank reactor in Section 2.8 and Section 8.4). The following terminology will turn out to be useful.

We call a branch or a part of a branch stable (unstable) if all its solutions are stable (unstable). A branch will be called periodic or symmetric if its solutions are periodic or symmetric.

The stability analysis discussed in this book is valid under small perturbations that conform with the underlying model or assumptions. It can easily happen that a branch that was claimed to be stable becomes unstable when the model is enhanced. For example, suppose that, based on some similarity assumption, a three-dimensional flow has been modeled by a two-dimensional PDE. In order to obtain valid stability results for two-dimensional solutions, three-dimensional disturbances should be analyzed. Unfortunately this stability analysis may be more expensive than the calculation of the solutions themselves. Various types of instabilities can be defined for various three-dimensional patterns (see [53, 63]). Other examples of how changes in models or assumptions affect bifurcation and stability results will come up later in this book. The reader should bear in mind that subsequent stability claims will generally be of a limited nature because of restrictions of particular models or assumptions.

In summary, locally there are no solutions on one side of a turning point and two solutions on the other side. At a turning point two solutions are born or two solutions annihilate each other.

A bifurcation point is easy to characterize geometrically: at a bifurcation point, two branches with distinct tangents intersect. Two types of bifurcation are shown in Figure 2.12. These are

(a) transcritical bifurcation

(b) pitchfork bifurcation.

The points of bifurcation are emphasized by dots. As in Figure 2.12, bifurcation points have a two-sided character. *Pitchfork bifurcations* are points at which one of the two branches is one-sided.

Model examples of bifurcation are furnished by the following scalar differential equations.

Example 2.1. Transcritical Bifurcation.

$$\dot{y} = \lambda y - y^2$$

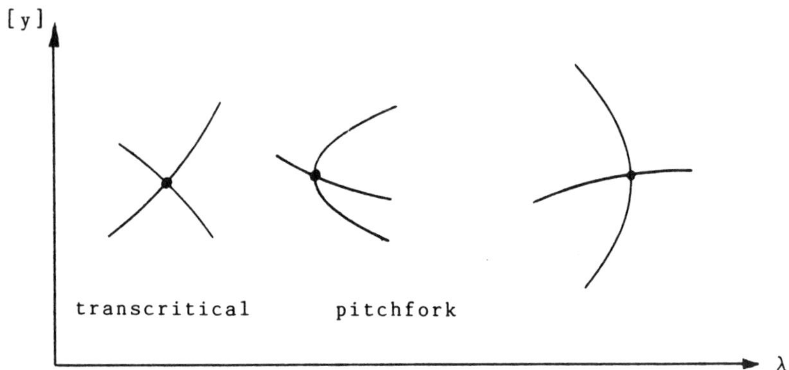

Figure 2.12

The equilibria are $y = 0$ and $y = \lambda$. The stable branches are indicated in Figure 2.13 by heavy lines. The *trivial branch* $y = 0$ loses stability at the bifurcation point $(y,\lambda) = (0,0)$. Simultaneously, there is an *exchange of stability* to the other branch.

Example 2.2. Supercritical Pitchfork.

$$\dot{y} = \lambda y - y^3$$

For $\lambda > 0$ there are two nontrivial equilibria, $y = \pm\sqrt{\lambda}$. The transition of stability is illustrated by Figure 2.14. We recall the example of the buckling of a beam, which exhibits this kind of loss of stability, from Section 2.3 (Figure 2.5).

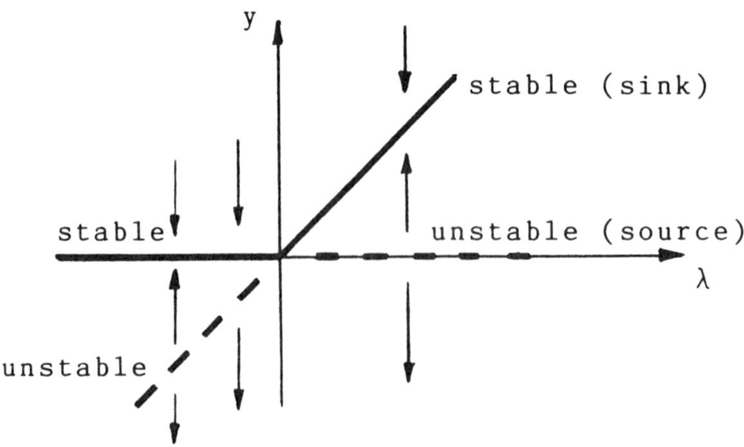

Figure 2.13

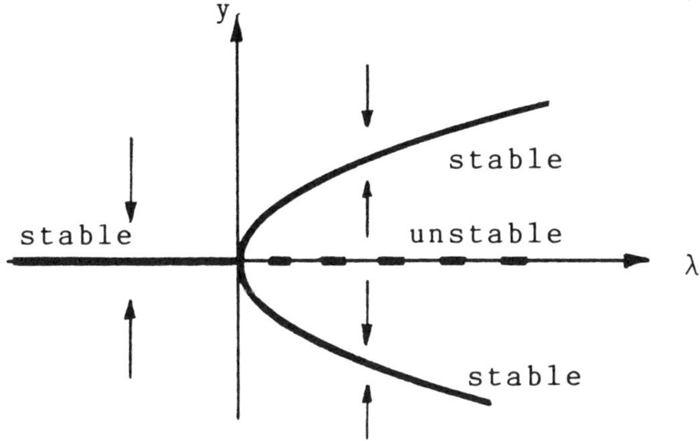

Figure 2.14

Example 2.3. Subcritical Pitchfork.

$$\dot{y} = \lambda y + y^3$$

Equilibria and stability behavior for this equation are shown in Figure 2.15. There is again a loss of stability at the bifurcation point $(y,\lambda) = (0,0)$. In contrast to the example of Figure 2.14, there is no exchange of stability. Instead, the stability is lost locally at the bifurcation point.

Each of Figures 2.13, 2.14, and 2.15 shows only one possible situation. In another example, λ may be replaced by $-\lambda$, resulting in

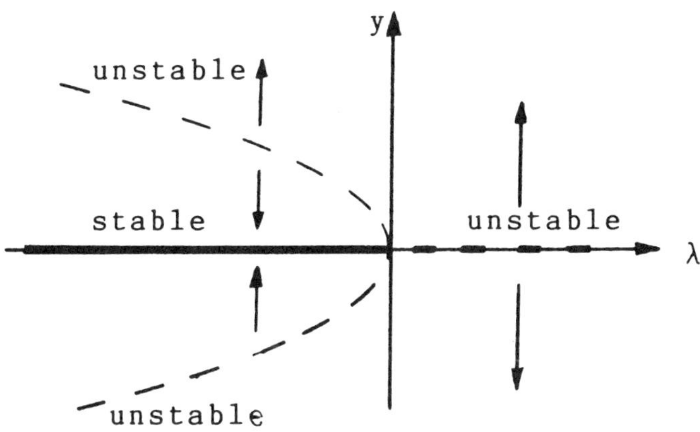

Figure 2.15

a stability distribution, as in Figure 2.16a. For ease of comparison, the situations of Figures 2.13, 2.14, and 2.15 are repeated in Figure 2.16b. Finally, in Figure 2.16c we indicate that all the *half-branches* meeting at a bifurcation point can be unstable. This can be achieved by attaching additional differential equations to the above scalar equations, as in Eq. (2.10). The terms "supercritical" and "subcritical" are not defined when all the half-branches are unstable. Supercritical bifurcations have locally stable solutions on both sides of the bifurcation [159].

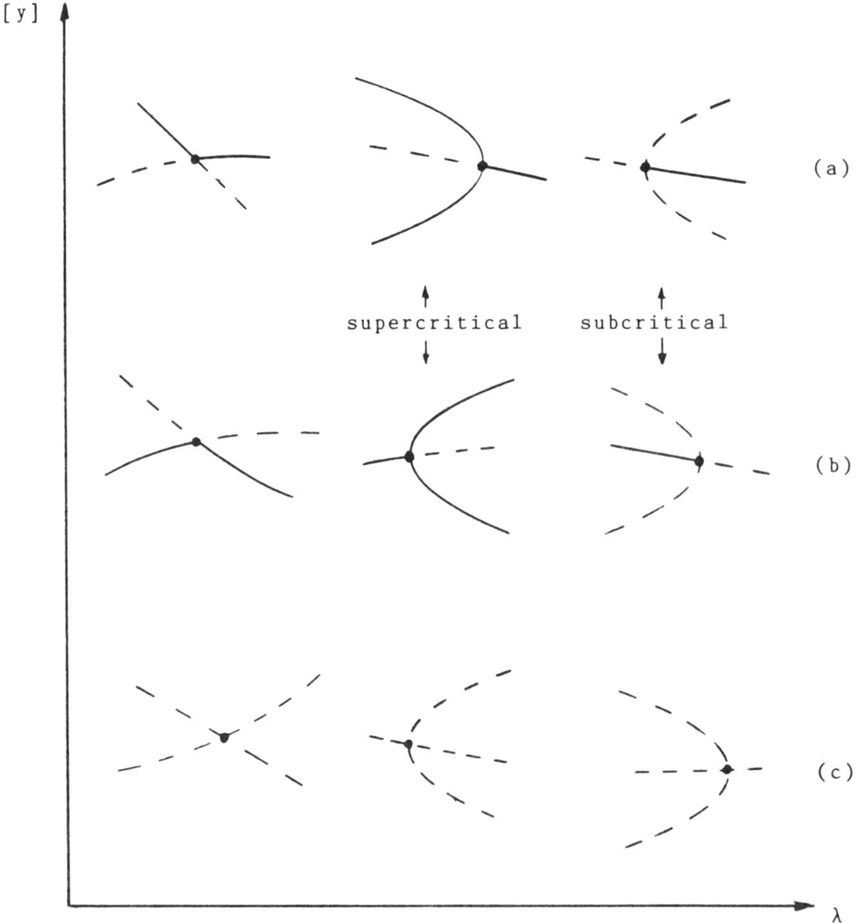

Figure 2.16

In the literature there is some variance in the use of the words "supercritical" and "subcritical." These terms are occasionally used in a fashion solely oriented toward decreasing or increasing values of the parameter λ [112] and not related to the concept of stability. In this vein a pitchfork bifurcation at $\lambda = \lambda_0$ is said to be supercritical if there is locally only one solution for $\lambda < \lambda_0$—that is, "supercritical" bifurcates to the right and "subcritical" bifurcates to the left (the λ-axis points to the right). This book will follow the more physical convention indicated in Figure 2.16.

So far we have introduced the concepts of turning point and *simple* bifurcation point, the latter denoting a point where precisely two branches intersect. (Note that this interpretation of the term "simple bifurcation point" does not rule out the possibility that globally the two branches might be connected.) Turning points and bifurcation points are examples of branch points. We mention in passing that there are additional branching phenomena. For example, a hysteresis phenomenon may collapse into a situation like that shown in Figure 2.17a. This kind of branch point is called a *hysteresis point* or *cusp point.* A bifurcation point where more than two branches intersect (Figure 2.17b) is called a *multiple bifurcation point.* Finally, in Figure 2.17c an *isola center* is depicted (also called *point of isola formation* or *hermit* [36]).

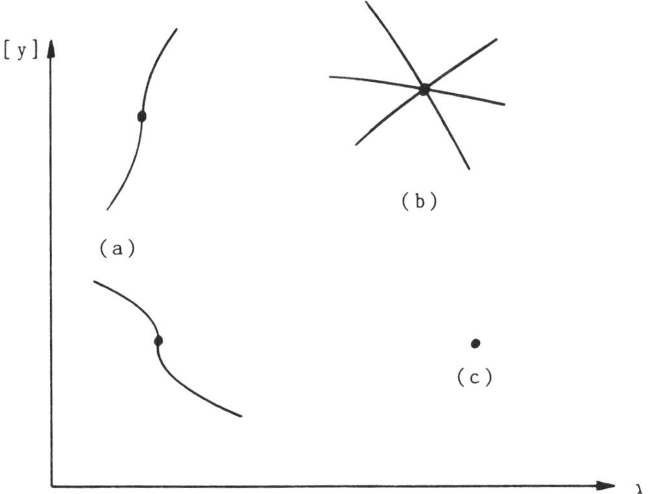

Figure 2.17

A neighboring solution cannot be found either to the right or to the left of an isola center. The occurrence of hysteresis points or isola centers will be discussed in Section 2.9.

Exercise 2.5.
Consider the Hammerstein equation

$$\lambda y(t) = \frac{2}{\pi}\int_0^{\pi} (3 \sin \tau \sin t + 2 \sin 2\tau \sin 2t)(y(\tau) + y^3(\tau))\, d\tau.$$

Solutions $y(t;\lambda)$ of this integral equation depend on the independent variable t and on the parameter λ.

(a) Show that every solution is of the form

$$y = a(\lambda) \sin t + b(\lambda) \sin 2t.$$

(b) Are there nontrivial solutions for $\lambda = 0$?
(c) Calculate all solutions for $\lambda > 0$ (for $\lambda = 10$ there are nine different solutions).
(d) For each solution, calculate

$$[y] = \int_0^{\pi} y^2(t;\lambda)\, dt$$

and sketch a branching diagram.

Exercise 2.6 (from [159]).
Check the stability of all equilibrium solutions of the scalar differential equation

$$\dot{y} = y(9 - \lambda y)(\lambda + 2y - y^2)([\lambda - 10]^2 + [y - 3]^2 - 1)$$

and draw a branching diagram using dashed lines for unstable equilibria.

2.5 TURNING POINTS AND BIFURCATION POINTS: THE ALGEBRAIC VIEW

In order to understand some of the numerical methods for calculating branch points, we must characterize turning points and bifurcation points algebraically. In particular, we must answer the question of how an observer traveling along a one-sided branch can distinguish between a turning point and a pitchfork bifurcation (Figure 2.18). We first study a simple example that exhibits both phenomena.

Example 2.4.
We consider the two scalar equations ($n = 2$)

$$\begin{aligned} 0 &= 1 + \lambda(y_1^2 + y_2^2 - 1) = f_1(y_1,y_2,\lambda) \\ 0 &= 10y_2 - \lambda y_2(1 + 2y_1^2 + y_2^2) = f_2(y_1,y_2,\lambda). \end{aligned} \tag{2.11}$$

The second equation is satisfied for $y_2 = 0$, which implies by means of the first equation

$$y_1^2 = \frac{\lambda - 1}{\lambda}.$$

This establishes two branches

$$(y_1,y_2) = \left(\pm \left(\frac{\lambda - 1}{\lambda} \right)^{1/2}, 0 \right),$$

which are defined for $\lambda \geq 1$ and $\lambda < 0$. It remains to discuss the case $y_2 \neq 0$. Substituting the expression

$$y_1^2 + y_2^2 = 1 - \frac{1}{\lambda}$$

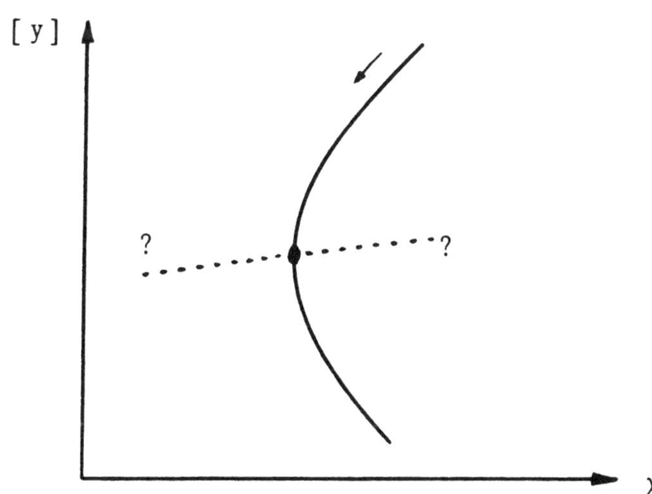

Figure 2.18

from the first equation into the second equation yields

$$0 = 10 - \lambda\left(1 + y_1^2 + \left(1 - \frac{1}{\lambda}\right)\right) = 11 - 2\lambda - \lambda y_1^2.$$

This gives

$$(y_1,y_2) = \left(\pm\left(\frac{11 - 2\lambda}{\lambda}\right)^{1/2}, \pm\left(\frac{3\lambda - 12}{\lambda}\right)^{1/2}\right)$$

as additional branches, defined for $4 \leq \lambda \leq 5.5$. These branches form a smooth loop in the three-dimensional (y_1,y_2,λ) space. Figure 2.19 shows the branches for $\lambda > 0$ and $y_2 \geq 0$ as solid lines and those for $y_2 < 0$ as dotted lines. The branches are symmetric with respect to the planes defined by $y_1 = 0$ and $y_2 = 0$, the latter plane indicated by parallel lines. There are three turning points and two bifurcation points—namely, those with (y_1,y_2,λ) values

$(0, 0, 1)$	turning point
$(\pm\frac{1}{2}\sqrt{3}, 0, 4)$	bifurcation points
$(0, \pm 3(11)^{-1/2}, 5.5)$	turning points.

We postpone further analysis of branch points to illustrate problems that may arise when the scalar measure [ ] is not chosen properly.

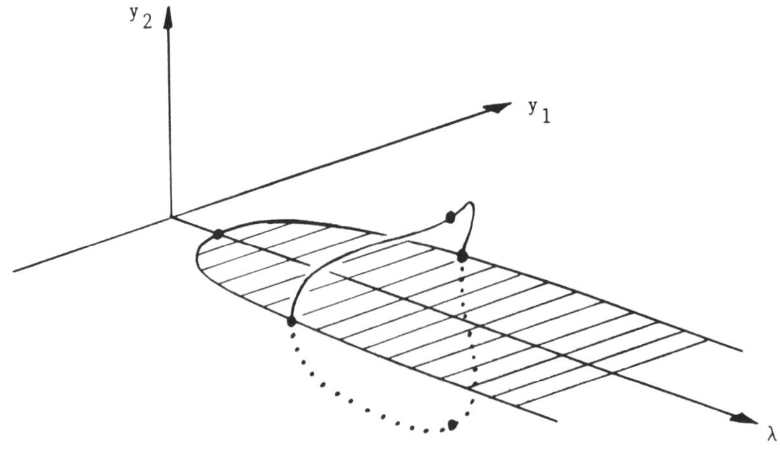

Figure 2.19

As the scalar measure for the branching diagram, we choose

$$[\mathbf{y}] = y_1^2 + 2y_2^2$$

and use Figure 2.20 for the example under consideration. In Figure 2.20 we see an effect that is typical for branching diagrams of equations where symmetries are involved. The two half-branches with $1 \leq \lambda \leq 4$ have the same values of $[\mathbf{y}]$, and Figure 2.20 does not distinguish between them. Part (a) is *covered twice.* This phenomenon is even more significant with the four different parts of the loop branch with $4 \leq \lambda \leq 5.5$. All four parts have the same values of $[y]$, and part (b) in Figure 2.20 is covered four times. Thus the branching diagram Figure 2.20 conceals the fact that the bifurcations at $\lambda = 4$ are of the pitchfork type and that the dots at $\lambda = 1$ and $\lambda = 5.5$ represent turning points. For a better understanding of this effect, compare Figure 2.20 with Figure 2.19. Certain choices of $[\mathbf{y}]$ lead to better insight into the branching behavior of a particular equation, other choices provide much less information (Exercise 2.7). Figure 2.20 illustrates that in branching diagrams there need not be a curve that passes pitchfork bifurcations or turning points perpendicularly to the λ-axis.

We now continue with Eq. (2.11), concentrating on characteristic algebraic properties of turning points and bifurcation points. Motivated by the example in Section 2.1 and the implicit function theorem, we study the derivatives of $\mathbf{f}(\mathbf{y},\lambda)$. The Jacobian matrix is

$$\mathbf{f_y}(\mathbf{y},\lambda) = \begin{pmatrix} 2\lambda y_1 & 2\lambda y_2 \\ -4\lambda y_1 y_2 & 10 - \lambda - 2\lambda y_1^2 - 3\lambda y_2^2 \end{pmatrix}$$

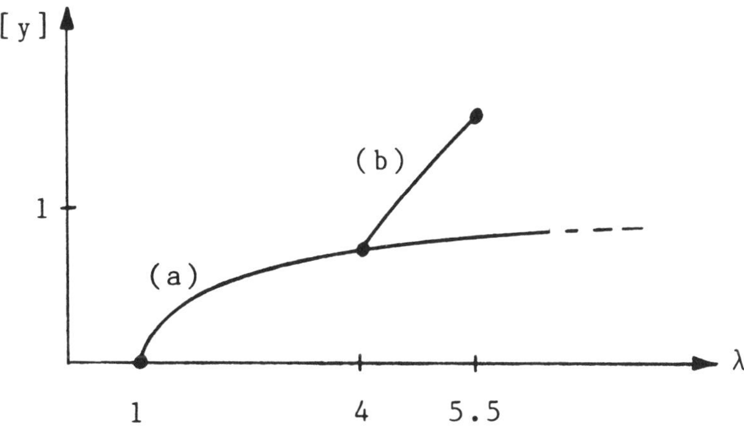

Figure 2.20

and the vector of the partial derivatives with respect to the parameter λ is

$$\mathbf{f}_\lambda(\mathbf{y},\lambda) = \begin{pmatrix} y_1^2 + y_2^2 - 1 \\ -y_2(1 + 2y_1^2 + y_2^2) \end{pmatrix}.$$

Substituting the $(\mathbf{y},\lambda)$ values of one of the bifurcation points and one of the turning points into these derivatives, we obtain:

for the bifurcation point $(\frac{1}{2}\sqrt{3}, 0, 4)$:

$$\mathbf{f}_\mathbf{y} = \begin{pmatrix} 4\sqrt{3} & 0 \\ 0 & 0 \end{pmatrix}, \quad \mathbf{f}_\lambda = \begin{pmatrix} -\frac{1}{4} \\ 0 \end{pmatrix};$$

for the turning point $(0, 3(11)^{-1/2}, 5.5)$:

$$\mathbf{f}_\mathbf{y} = \begin{pmatrix} 0 & 3(11)^{1/2} \\ 0 & -9 \end{pmatrix}, \quad \mathbf{f}_\lambda = \begin{pmatrix} -2/11 \\ -60\cdot(11)^{-3/2} \end{pmatrix}.$$

The Jacobian is *singular* in both cases—that is, the *rank* is less than $n = 2$, or equivalently (Appendix 3),

$$\det \mathbf{f}_\mathbf{y}(\mathbf{y}_0,\lambda_0) = 0. \tag{2.12}$$

Here and in the remainder of the book, branch points will be denoted by $(\mathbf{y}_0,\lambda_0)$; $\mathbf{y}_0$ being an n-vector. To distinguish between a bifurcation point and a turning point, we need more information. Upon attaching the vector $\partial\mathbf{f}/\partial\lambda$ to the Jacobian matrix, one obtains the augmented matrix

$$\begin{pmatrix} 4\sqrt{3} & 0 & -\frac{1}{4} \\ 0 & 0 & 0 \end{pmatrix}$$

for the bifurcation point and

$$\begin{pmatrix} 0 & 3\cdot(11)^{1/2} & -2/11 \\ 0 & -9 & -60\cdot(11)^{-3/2} \end{pmatrix}$$

for the turning point. The matrix of the bifurcation point is still singular (rank $< n$), whereas the matrix evaluated at the turning point has full rank (rank $= 2 = n$).

This difference between a turning point and a bifurcation point is worth deeper analysis. To this end we return to Eq. (2.2),

$$\mathbf{0} = \mathbf{f}(\mathbf{y},\lambda).$$

Previously, $\mathbf{y}$ was a vector with n components. In what follows, it will be convenient to make the parameter λ into the $(n + 1)$-st component of $\mathbf{y}$, that is,

$$y_{n+1} = \lambda,$$

and to consider Eq. (2.2) to be a set of n equations in $n + 1$ unknowns,

$$0 = f_i(y_1, y_2, \ldots, y_n, y_{n+1}) \qquad (i = 1, 2, \ldots, n).$$

The rectangular matrix of the partial derivatives consists of $n + 1$ columns $\mathbf{z}^i$,

$$(\mathbf{f_y} \mid \mathbf{f}_\lambda) = (\mathbf{z}^1 \mid \mathbf{z}^2 \mid \cdots \mid \mathbf{z}^k \mid \cdots \mid \mathbf{z}^n \mid \mathbf{z}^{n+1}) \qquad (2.13)$$

We are free to interpret any of the $n + 1$ components (say, the k-th component) as a parameter. Call this parameter γ:

$$\gamma = y_k.$$

The dependence of the remaining n components

$$y_1, y_2, \ldots, y_{k-1}, y_{k+1}, \ldots, y_n, y_{n+1}$$

on γ is characterized by the "new" Jacobian that results from Eq. (2.13) by removing the k-th column. For a turning point, it is possible to find an index k such that the new Jacobian is nonsingular (full rank $= n$), whereas for a bifurcation point no such k exists (rank $< n$ for all choices of γ).

This exchange of columns of the matrix Eq. (2.13) has a striking geometrical interpretation. The transition from parameter λ to parameter γ means that the branching diagram [$\mathbf{y}$]-versus-γ is obtained from the diagram [$\mathbf{y}$]-versus-λ by a rotation of 90°. As illustrated in Figure 2.21, the turning point disappears when the branch is parametrized by $\gamma = y_k$. This removal of the singularity by change of parameter is reflected algebraically in the nonsingularity of the "new" Jacobian. After

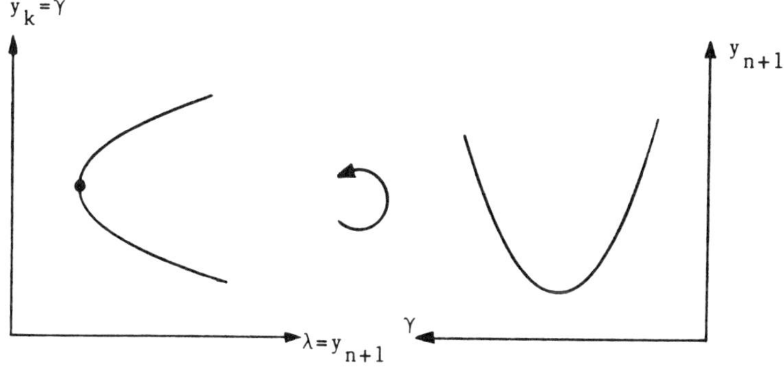

Figure 2.21

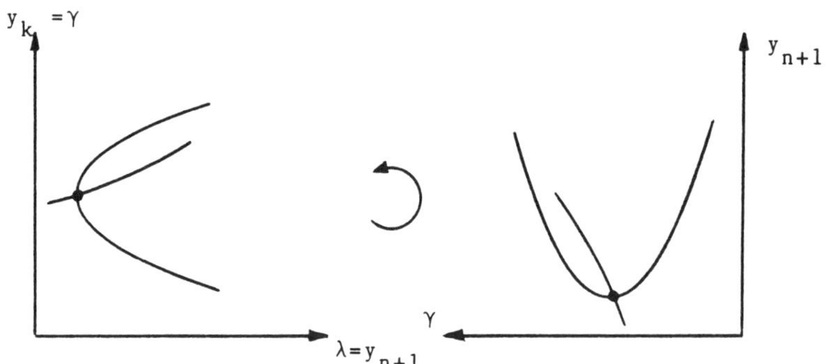

Figure 2.22

this new choice of parameter and rearrangement of components, the implicit function theorem works again.

Trying the same exchange with the matrix associated with a bifurcation point, we obtain Figure 2.22. Here the singularity is not removed by changing the parameter; the bifurcation point remains.

The preserving or increasing the rank by attaching the column $\mathbf{f}_\lambda$ to the Jacobian $\mathbf{f}_\mathbf{y}$ can be described in terms of the *range* of a matrix. Assume that the rank of the Jacobian is $n - 1$. We first consider the situation of a bifurcation point. Here there are constants c_i such that

$$\mathbf{f}_\lambda = \mathbf{z}^{n+1} = \sum_{i=1}^{n} c_i \mathbf{z}^i, \tag{2.14}$$

that is,

$$\mathbf{f}_\lambda \in \text{range } \mathbf{f}_\mathbf{y}, \tag{2.15}$$

because the vectors $\mathbf{z}^i$ span the range of $\mathbf{f}_\mathbf{y}$. In contrast, for a turning point there are no constants c_i such that $\mathbf{f}_\lambda$ can be expressed as a linear combination of the n column vectors $\mathbf{z}^i$ of the singular matrix $\mathbf{f}_\mathbf{y}$, that is,

$$\mathbf{f}_\lambda \notin \text{range } (\mathbf{f}_\mathbf{y}). \tag{2.16}$$

(Loosely speaking, $\mathbf{f}_\lambda$ carries full-rank information that is lost by the $\mathbf{z}^i$.) This is a characteristic feature of a turning point.

Now we are prepared to give the following formal definitions of bifurcation point and turning point.

Definition 2.3. (y_0,λ_0) is a simple bifurcation point if the following four conditions hold:

(1) $f(\mathbf{y}_0,\lambda_0) = 0$
(2) $f_y(\mathbf{y}_0,\lambda_0)$ *has a simple eigenvalue* 0 *or, equivalently, rank* $(f_y(\mathbf{y}_0,\lambda_0)) = n - 1$
(3) $f_\lambda(\mathbf{y}_0,\lambda_0) \in$ *range* $f_y(\mathbf{y}_0,\lambda_0)$
(4) *exactly two branches intersect with two distinct tangents.*

Hypothesis (4) will be formalized based on second-order derivatives in Section 5.4.1.

Definition 2.4. $(\mathbf{y}_0,\lambda_0)$ *is a turning point if the following four conditions hold:*

(1) $f(\mathbf{y}_0,\lambda_0) = 0$
(2) $f_y(\mathbf{y}_0,\lambda_0)$ *has a simple eigenvalue* 0 *or, equivalently, rank* $f_y(\mathbf{y}_0,\lambda_0) = n - 1$
(3) $f_\lambda(\mathbf{y}_0,\lambda_0) \notin$ *range* $f_y(\mathbf{y}_0,\lambda_0)$, *that is, rank* $(f_y(\mathbf{y}_0,\lambda_0) \mid f_\lambda(\mathbf{y}_0,\lambda_0)) = n$
(4) *there is a parameterization* $\mathbf{y}(\sigma)$, $\lambda(\sigma)$ *with* $\mathbf{y}(\sigma_0) = \mathbf{y}_0$, $\lambda(\sigma_0) = \lambda_0$, *and* $d^2\lambda(\sigma_0)/d\sigma^2 \neq 0$.

Hypotheses (1), (2), and (3) guarantee that the tangent to the branch at $(\mathbf{y}_0,\lambda_0)$ is perpendicular to the λ-axis in the $(n + 1)$-dimensional $(\mathbf{y},\lambda)$ space. Hypothesis (4) implies the "turning property" and prevents $(\mathbf{y}_0,\lambda_0)$ from being a hysteresis point (see Figure 2.17a). In the literature, condition (4) is sometimes replaced by other equivalent conditions.

Exercise 2.7.
Draw branching diagrams depicting the solutions of Eq. (2.11) using all the scalar measures in Eq. (2.7).

In classical bifurcation theory, the terms "branching" and "bifurcation" were used as synonyms. They designated the bifurcation points where two or more branches intersect [192, 333, 354]. The terms *primary bifurcation* (for a bifurcation from a trivial solution) or *secondary bifurcation* (for a bifurcation from a nontrivial branch) reflected the order in which these phenomena were analyzed. Limitations of the classical terminology have become apparent only recently, influenced by such subjects as *catastrophe theory* and by the discovery of new nonlinear effects. The omnipresence of turning points has suggested the use of classical terms in broader senses. However, the great variety of nonlinear phenomena makes it difficult to find convincing names. The terms "regular singular point" [120] or "isolated nonisolated solution" [180] may serve as examples of the difficulties in finding evident names. The standard turning point (Definition 2.4) was sometimes called *simple quadratic turning point*, and our hysteresis point was also called *simple cubic turning point* [331]. There is no consensus about which name is to be used for a particular phenomenon. The lack of striking names may explain why many accomplishments of bifurcation theory and singularity theory have become popular only through catastrophe theory, chaos theory, or synergetics.

In this book we distinguish between branch point and bifurcation point, a distinction that several authors in the recent literature use [176, 216, 312, 353]. The term "branching" will be used to include both bifurcation and turning-point phenomena. Our tendency will be to use simple labels for widespread phenomena, restricting complex names to less important effects.

2.6 HOPF BIFURCATION

The bifurcation points we have handled so far have had one common feature. All the branches have represented equilibria—that is, both branches intersecting in a bifurcation point have consisted of (stationary) solutions of the equation

$$\dot{\mathbf{y}} = \mathbf{0} = \mathbf{f}(\mathbf{y},\lambda).$$

The term "equilibrium" characterizes a physical situation. Mathematically speaking, we say that the solutions on the emanating branch remain in the same "space"—namely, in the space of n-dimensional vectors. We shall call bifurcation that is characterized by intersecting branches of equilibrium points *stationary bifurcation* or *steady-state bifurcation* (Figure 2.23).

Physically, an equilibrium represents a situation without "life." It may mean no motion of a pendulum, no reaction in a reactor, no nerve

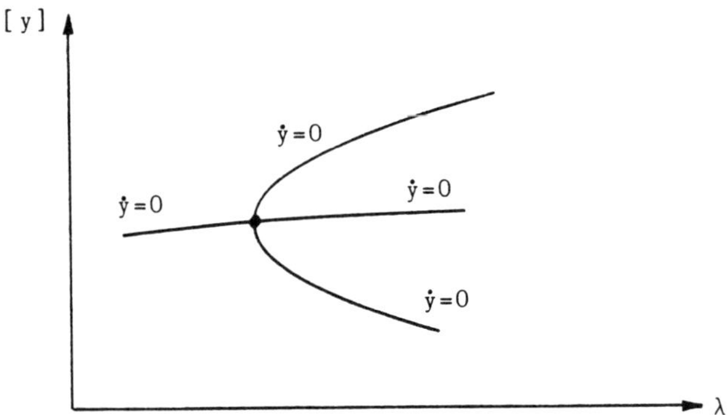

Figure 2.23

activity, no flutter of an airfoil, no laser operation, or no circadian rhythms of biological clocks [368]. A stationary bifurcation does not give rise to such exciting phenomena; the branching solutions show no more life than the previously known solutions. The full richness of the nonlinear world is not found at equilibria.

The analysis for $\dot{\mathbf{y}} = \mathbf{0}$ is far simpler than the analysis of the situation with $\dot{\mathbf{y}} \neq \mathbf{0}$. The assumption that $\dot{\mathbf{y}} = \mathbf{0}$ (when in reality $\dot{\mathbf{y}} = \epsilon$ for small $|\epsilon|$) may be justified because equilibria do occur in nature. But nature does not care about simplifying assumptions, and the equilibrium may suddenly be superseded by motion. When the motion can be represented by limit cycles, the oscillations are regular and hence manifestations of an order. In case $\mathbf{y}$ is a macroscopic variable describing, for example, the dynamic behavior of a flow, its oscillation is seen as self-organization of microscopic parts toward a coherent structure; fluctuations around an equilibrium are no longer damped [106, 129].

The mathematical vehicles that describe this kind of "life" are the functions $\mathbf{y}(t)$. The corresponding function spaces include equilibria as the special-case constant solutions (Figure 2.24). The type of bifurcation that connects equilibria with periodic motion is *Hopf bifurcation*. Hopf bifurcation is the door that opens from the small room of equilibria to the large hall of periodic solutions.

We met periodic solutions in the discussion of limit cycles in Section 1.3; recall especially Exercises 1.9, 1.10, 1.11, and 1.12. We now

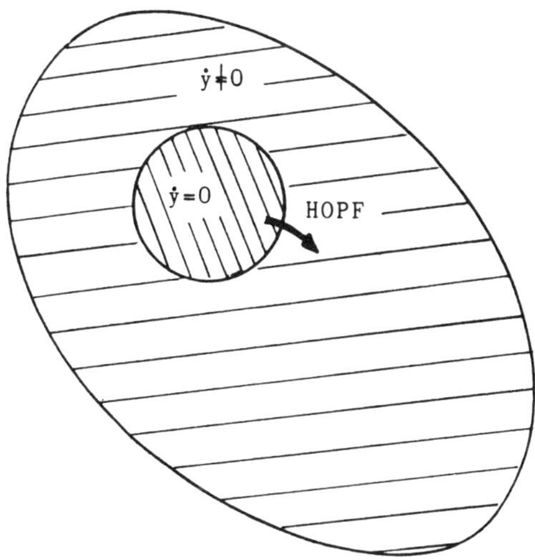

Figure 2.24

discuss a simple example of Hopf bifurcation (similar to Exercise 1.12):

$$\begin{aligned} \dot{y}_1 &= -y_2 + y_1(\lambda - y_1^2 - y_2^2) \\ \dot{y}_2 &= y_1 + y_2(\lambda - y_1^2 - y_2^2). \end{aligned} \tag{2.17}$$

A straightforward investigation shows that $y_1 = y_2 = 0$ is the only equilibrium for all λ; there is no stationary bifurcation. The Jacobian matrix

$$\begin{pmatrix} \lambda & -1 \\ 1 & \lambda \end{pmatrix}$$

has eigenvalues $\lambda \pm i$. We conclude that the equilibrium is stable for $\lambda < 0$ and unstable for $\lambda > 0$—that is, there is a loss of stability at $\lambda = 0$ but no exchange of stability inside the space of equilibria.

As in Exercises 1.11 and 1.12, a limit cycle can be constructed for Eq. (2.17). Using polar coordinates ρ, ϑ with

$$y_1 = \rho \cos \vartheta, \quad y_2 = \rho \sin \vartheta,$$

we obtain from Eq. (2.17) the two equations

$$\begin{aligned} \dot{\rho} \cos \vartheta - \rho\dot{\vartheta} \sin \vartheta &= -\rho \sin \vartheta + \rho \cos \vartheta(\lambda - \rho^2) \\ \dot{\rho} \sin \vartheta + \rho\dot{\vartheta} \cos \vartheta &= \rho \cos \vartheta + \rho \sin \vartheta(\lambda - \rho^2). \end{aligned}$$

Multiplying the first equation by $\cos \vartheta$ and the second equation by $\sin \vartheta$ and adding yields the differential equation

$$\dot{\rho} = \rho(\lambda - \rho^2).$$

Similarly, one obtains

$$\dot{\vartheta} = 1.$$

For $\rho = \sqrt{\lambda}$ we have $\dot{\rho} = 0$. Hence, there is a periodic orbit for $\lambda > 0$ and the amplitude of the orbit grows with the square root $\sqrt{\lambda}$. Because

$$\begin{aligned} \dot{\rho} < 0 &\quad \text{for } \rho > \sqrt{\lambda} \\ \dot{\rho} > 0 &\quad \text{for } \rho < \sqrt{\lambda} \end{aligned}$$

the orbit is stable.

Figure 2.25 summarizes the results. At $\lambda_0 = 0$ there is an exchange of stability from stable equilibrium to stable limit cycle. The limit cycle encircles the unstable equilibrium, with an amplitude growing with the square root $(\lambda - \lambda_0)^{1/2}$. Such a bifurcation from an equilibrium to a periodic oscillation is called Hopf bifurcation. We observe that the Ja-

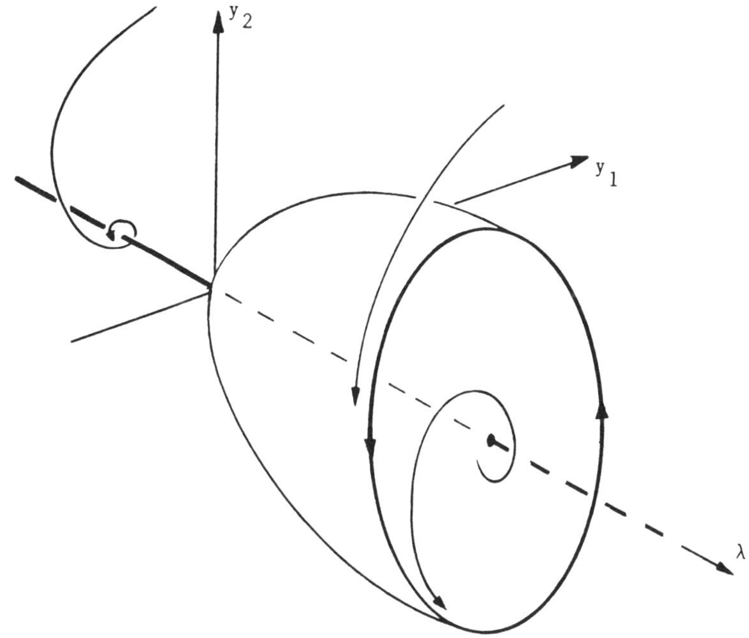

Figure 2.25

cobian, evaluated at the Hopf point, has a pair of purely imaginary eigenvalues.

We are now prepared to give a general characterization of Hopf bifurcation. The basic results were known to Poincaré; the planar case was handled by Andronov in 1929. In spite of these early results, bifurcation from equilibria to limit cycles is commonly referred to as Hopf bifurcation [136, 228], because it was E. Hopf who proved the following theorem for the n-dimensional case in 1942 [157]:

Theorem 2.1. Assume

(1) $f(\mathbf{y}_0,\lambda_0) = 0.$
(2) $f_{\mathbf{y}}(\mathbf{y}_0,\lambda_0)$ *has a simple pair of purely imaginary eigenvalues* $\mu(\lambda_0) = \pm i\beta$ *and no other eigenvalue with zero real part.*
(3) $d(Re\ \mu(\lambda_0))/d\lambda \neq 0.$

Then there is a birth of limit cycles at $(\mathbf{y}_0,\lambda_0)$. *The initial period (of the zero-amplitude oscillation) is*

$$T_0 = \frac{2\pi}{\beta}.$$

Hypotheses (1), (2), and (3) can be viewed as a definition of Hopf bifurcation. Condition (3) is the *transversality hypothesis*; it is "usually" satisfied. In order to understand Hopf bifurcation better, we visualize what happens in the complex plane of the eigenvalues $\mu(\lambda)$. (Recall that the $\mu(\lambda)$ are the eigenvalues of the Jacobian $\mathbf{f}_y$; we consider them for λ near λ_0.) Figure 2.26 shows a possible path of the particular pair of eigenvalues that satisfy $\mu(\lambda_0) = \pm i\beta$. Condition (3) is responsible for the transversal crossing of the imaginary axis. If all the other eigenvalues have strictly negative real parts, Figure 2.26 illustrates a loss of stability; for a gain of stability, just reverse the arrows. The situation of Hopf bifurcation depicted in Figure 2.26 contrasts with stationary bifurcation, where there is an eigenvalue 0 (see Figure 2.27). For stationary bifurcation, the crossing of the imaginary axis is at the real axis.

Near Hopf bifurcations there is locally only one periodic solution for each λ (apart from phase shifts). That is, only one half-branch of periodic solutions comes out from the stationary branch. Recall that in the case of stationary bifurcation two half-branches come out. This difference between stationary bifurcation and Hopf bifurcation may or may not be reflected by branching diagrams. Branching diagrams can be constructed where half-branches of a pitchfork are identified (see, for example, Figure 2.20). On the other hand, as we shall soon see, there exist branching diagrams where two curves intersect in a Hopf bifurcation point.

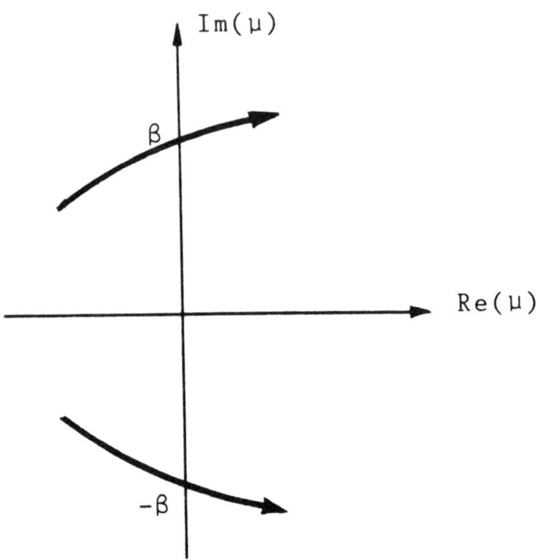

Figure 2.26

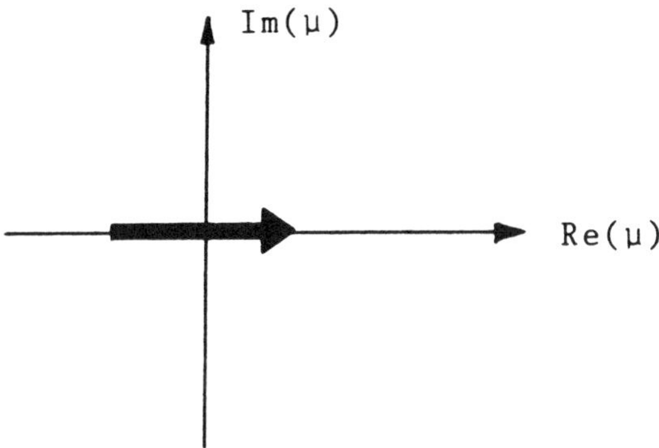

Figure 2.27

Exercise 2.8.
Consider again the FitzHugh model of the behavior of nerve membrane potential (Exercise 1.9).

(a) Calculate the equilibria. (Hint: Express λ in terms of y_1.)
(b) Calculate the Hopf bifurcation points.
(c) Draw a bifurcation diagram of y_1 versus λ. Distinguish between stable and unstable equilibria. Use the results of Exercise 1.9 and try to formulate a conjecture on how far the periodic branches extend.

Exercise 2.9 (see Eq. (2.17)).
Consider the system of differential equations

$$\dot{y}_1 = -y_2 + y_1(\lambda + y_1^2 + y_2^2)$$
$$\dot{y}_2 = y_1 + y_2(\lambda + y_1^2 + y_2^2).$$

Show that there is a Hopf bifurcation to unstable periodic orbits. Draw a picture similar to Figure 2.25.

Exercise 2.10.
By a theorem of Vieta, the roots μ_1, μ_2, μ_3 of the cubic equation $0 = -\mu^3 + a\mu^2 - b\mu + c$ satisfy the relations

$$a = \mu_1 + \mu_2 + \mu_3$$
$$c = \mu_1\mu_2\mu_3.$$

(a) Show that the eigenvalues of the Jacobian satisfy the relation

$$\mu_1\mu_2\mu_3 = \det(\mathbf{f_y}(\mathbf{y},\lambda)).$$

(b) What do you obtain for the sum $\mu_1 + \mu_2 + \mu_3$?
(c) Show that $\det(\mathbf{f_y}(\mathbf{y},\lambda)) > 0$ implies local instability for the equilibrium $(\mathbf{y},\lambda)$.
(d) Investigate which aspects of (a) and (c) might be generalized.

2.7 CONVECTION DESCRIBED BY LORENZ'S EQUATION

In this section we carry out an evaluation of Hopf bifurcation points for the famous Lorenz equations. In 1963, Lorenz published a system of ODEs [217] that has been given much attention in the literature. The physical background is a flow in a layer of fluid of uniform depth heated from below, with the temperature difference between the upper and lower surfaces maintained at a constant value. This problem is related to the Rayleigh-Bénard problem [31, 267, 52]. (We know today that Bénard's original experiments and Rayleigh's analysis are at variance; see Section 3.4.) The outcome of Rayleigh-Bénard-type experiments is governed by the values of the Rayleigh number *Ra*. For values of *Ra* below a critical value R_c, the system has a stable steady-state solution in which the fluid is motionless and the temperature varies linearly with depth. When the Rayleigh number is increased past R_c, the purely conductive state becomes unstable and convection develops. Driven by buoyancy, the fluid motion organizes in cells or rolls. The onset of certain regular cloud patterns in the atmosphere, the drift of continents, and the granulation observed on the surface of the sun have recently all been attributed to this phenomenon.

If the convective currents are organized in cylindrical rolls, the governing equations can be written as a set of PDEs in two space variables [217]. Expanding the dependent variables in a Fourier series, truncating these series to one term, and substituting into the PDE yields the three ordinary differential equations

$$\begin{aligned} \dot{y}_1 &= P(y_2 - y_1) \\ \dot{y}_2 &= -y_1y_3 + Ry_1 - y_2 \qquad (2.18) \\ \dot{y}_3 &= y_1y_2 - \mathrm{b}y_3 \end{aligned}$$

for the variables y_1, y_2, y_3. The quantities in Eq. (2.18) have physical relevance:

y_1 is proportional to the intensity of convection

y_2 is proportional to temperature difference between ascending and descending currents

y_3 is proportional to the distortion of the vertical temperature profile from linearity

$R = Ra/R_c$ is the relative Rayleigh number

$P = Pr$ is the Prandtl number

b is a constant

(see [217]). The relative Rayleigh number R is our branching parameter. We first calculate the steady state. From the first equation we obtain $y_1 = y_2$. Substituting the expression for y_3 from the third equation into the second yields the equation

$$0 = -y_1^3/\mathrm{b} + Ry_1 - y_1,$$

which has the three roots

$$y_1 = 0, \quad y_1 = \pm(\mathrm{b}R - \mathrm{b})^{1/2}.$$

For convenience, we shall denote the square root

$$(\mathrm{b}(\lambda - 1))^{1/2} = (\mathrm{b}R - \mathrm{b})^{1/2}$$

by S. Obtaining the corresponding y_2 and y_3, one finds that the equilibria are

conduction:

$$(y_1,y_2,y_3) = (0,0,0) \qquad \text{for all } R$$

convection:

$$(y_1,y_2,y_3) = (\pm S, \pm S, R - 1) \qquad \text{for } R \geq 1.$$

The convection branch bifurcates off the trivial branch at $R = 1$. In order to check for stability, we need the Jacobian matrix

$$\begin{pmatrix} -P & P & 0 \\ -y_3 + R & -1 & -y_1 \\ y_2 & y_1 & -\mathrm{b} \end{pmatrix}.$$

We leave the discussion on stability of the trivial branch to the reader (Exercise 2.11) and substitute the (positive) nontrivial branch into the

Jacobian. This gives

$$\begin{pmatrix} -P & P & 0 \\ 1 & -1 & -S \\ S & S & -\mathrm{b} \end{pmatrix}.$$

The branching parameter R is hidden in S. The eigenvalues μ are calculated via the determinant (compare Section 1.2.4)

$$\begin{aligned} 0 &= \begin{vmatrix} -P-\mu & P & 0 \\ 1 & -1-\mu & -S \\ S & S & -\mathrm{b}-\mu \end{vmatrix} \\ &= -(P+\mu)(1+\mu)(\mathrm{b}+\mu) - S^2P - (P+\mu)S^2 + P(\mathrm{b}+\mu) \\ &= -\mu^3 - \mu^2(P+1+\mathrm{b}) - \mu(\mathrm{b}+\mathrm{b}P+S^2) - 2S^2P. \end{aligned}$$

At the Hopf bifurcation point, the characteristic polynomial has a pair of purely imaginary roots $\pm i\beta$. Hence the polynomial can be written as

$$\pm(\mu^2+\beta^2)(\mu+\alpha),$$

$-\alpha$ being the third eigenvalue. The values of β and α are still unknown. They can be calculated conveniently by equating the coefficients of both versions of the characteristic polynomial because they are identical at the Hopf bifurcation. To this end we multiply out, taking the negative sign for convenience,

$$0 = -\mu^3 - \mu^2\alpha - \mu\beta^2 - \beta^2\alpha.$$

Equating the coefficients gives three equations for the three unknowns β, α, and R (via S):

$$\begin{aligned} \alpha &= P+1+\mathrm{b} \\ \beta^2 &= \mathrm{b}+\mathrm{b}P+S^2 \\ \beta^2\alpha &= 2S^2P. \end{aligned}$$

Substituting the first and second of these expressions into the third gives

$$(\mathrm{b}+\mathrm{b}P+S^2)(P+1+\mathrm{b}) = 2S^2P.$$

It follows readily that

$$S^2 = \mathrm{b}(R-1) = \frac{(1+\mathrm{b}+P)\mathrm{b}(1+P)}{P-\mathrm{b}-1},$$

which gives the parameter value

$$R_0 = 1 + \frac{(1 + b + P)(1 + P)}{P - b - 1} = \frac{P(P + 3 + b)}{P - b - 1}$$

for Hopf bifurcation. Let us take a numerical example: b = 4, P = 16. This choice of constants leads to

$$R_0 = \frac{368}{11} = 33.4545\ldots.$$

The initial period of the "first" periodic oscillation is given by

$$T_0 = \frac{2\pi}{\beta} = 2\pi(b + bP + S^2)^{-1/2} = 0.4467\ldots.$$

The third eigenvalue is $-\alpha = -21$. Provided the transversality condition holds (it does), the negative sign of this eigenvalue shows that there is a loss of stability of the equilibrium when R is increased past R_0.

In the branching diagram Figure 2.28, we anticipate results obtained by calculating periodic oscillations. The equilibria are drawn with a solid line indicating stable equilibria and a dashed line indicating unstable equilibria. Because $y_3(t)$ varies between its maxima and min-

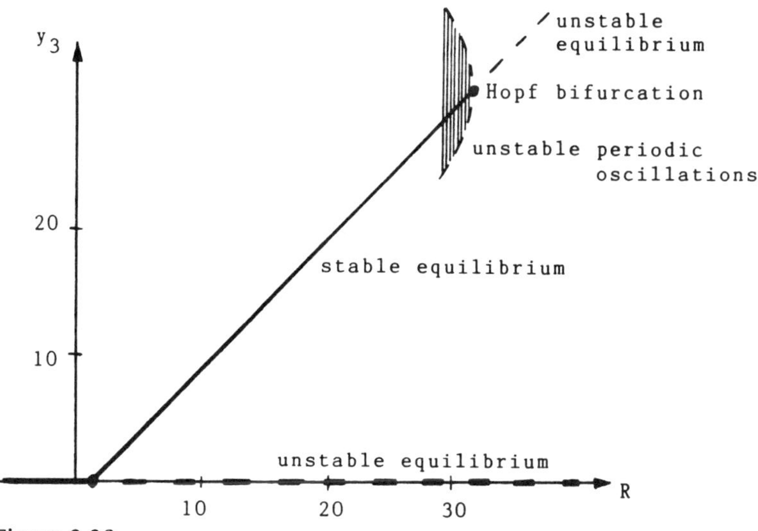

Figure 2.28

ima, oscillations do not easily fit into this diagram. A dashed curve represents the range of the $y_3(t)$ values. The oscillations are unstable; this is a result of applying methods that will be introduced in Chapter 7. Note that only one branch of periodic solutions branches off; the two half-curves represent two different scalar measures for the same periodic solutions.

The graph of the nontrivial branch of equilibria in Figure 2.28 is covered twice. That is, we have two Hopf bifurcations with the same value R_0. The "other" branch bears no new information because there is a symmetry involved in Eq. (2.18). Inspecting the Lorenz equations, one realizes that $(-y_1(t), -y_2(t), y_3(t))$ is a solution of Eq. (2.18) whenever $(y_1(t), y_2(t), y_3(t))$ is a solution. Geometrically, this is symmetry with respect to the y_3-axis.

Note that the *motion* of the convective state is represented here as an equilibrium of an ODE. This seeming contrast between physical motion and mathematical equilibrium occurs because the PDE was established for a *stream function.* The nontrivial equilibrium of Eq. (2.18) represents a regular convection with steady velocity and temperature profiles. The periodic oscillation of Eq. (2.18) represents a less-regular convection.

Exercise 2.11.
Show that the trivial solution of Eq. (2.18) is unstable for $R > 1$. (Hint: Use Exercise 2.10.)

For a rigorous treatment of Hopf bifurcation, see [136, 228]. In [228] an English translation of Hopf's original paper [157] is included. For historical remarks, see also [16]. In our discussion, we have excluded bizarre bifurcations where the transversality condition does not hold or where multiple eigenvalues cross the imaginary axis. References to treatments of such bifurcations are given in [159]. Calculation of Hopf points, as above, is limited to simple problems with not more than $n = 3$ equations. Note that this analysis did not reveal any information about stability and the "dircction" of the emerging branch. As [136] shows, it is possible to obtain such local information by hand calculation, but the effort required is immense. In general, one must resort to numerical methods (Chapter 7). For an extensive analysis of the Lorenz equation, see [329]. Because this ODE system stems from a severe truncation of a PDE system, the question of how to obtain more realistic models has been considered. More modes are included in the analysis of [70]. As shown in [370], the Lorenz equations also model a geometry where the flow is constrained in a torus-like tube.

2.8 HOPF BIFURCATION AND STABILITY

We have learned that a Hopf bifurcation point connects stationary solutions with periodic solutions. In the first example of Hopf bifurcation (see Figure 2.25), a stable orbit encircled an unstable equilibrium. In Exercise 2.9 and in the Lorenz system Eq. (2.18), unstable cycles coexisted with stable equilibria. In this section we shall discuss loss (or gain) of stability in some detail.

To illustrate bifurcation behavior graphically, we use symbols that distinguish between stable and unstable periodic oscillations. These symbols (dots and circles) are introduced in Figure 2.29. The branches in Figure 2.29 are continuous; the gaps between the small circles or dots do not indicate gaps between solutions. For every λ in the corresponding range there is a periodic orbit. Stability and instability of periodic oscillations will be formally defined in Chapter 7. For the time being it is sufficient to have the intuitive understanding that a stable periodic orbit is approached by nearby trajectories, whereas trajectories leave a neighborhood of an unstable periodic orbit.

Figure 2.30 illustrates various cases. As usual, a heavy continuous line represents stable stationary solutions, and unstable stationary solutions are indicated by dashed lines. A thick dot marks a branch point—here a Hopf bifurcation point. The rule is as follows: Locally, unstable periodic orbits encircle stable equilibria ((a) and (d)), and stable periodic orbits encircle unstable equilibria ((b) and (c)). This rule relates the direction of the emanating branch of periodic solutions to the stability properties of these solutions. Following our earlier defi-

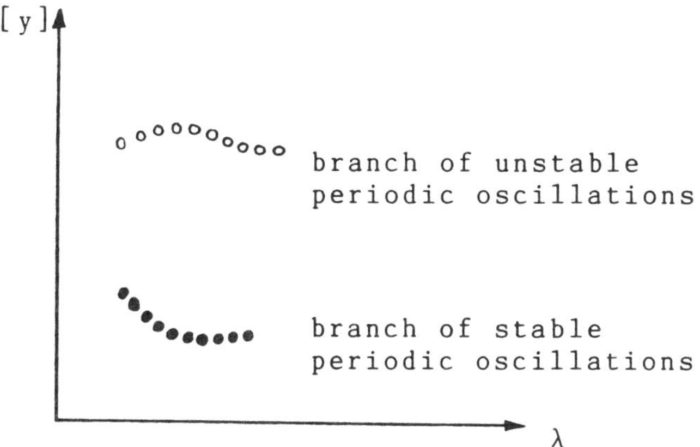

Figure 2.29

nition, cases (a) and (d) are subcritical bifurcations, whereas cases (b) and (c) are supercritical. In the supercritical cases an exchange of stability takes place between the "periodic branch" and the "stationary branch." Note again that these rules are valid only in a neighborhood of the Hopf bifurcation point. The stability (instability) of emanating periodic orbits may be lost by mechanisms that will be discussed in Chapter 7. Classical Hopf bifurcation assumes that the stationary branch loses (gains) its stability (see Figure 2.30a–d). Also, in cases (e) and (f), one eigenvalue crosses the imaginary axis transversally. Here at least one of the eigenvalues of the Jacobian has a strictly positive real part. Hopf bifurcation is then a branching to unstable periodic orbits, no matter what the direction. An example for this latter type of Hopf bifurcation will be given in Section 7.4.1.

In cases (a) and (d) in Figure 2.30, an exchange of stability may still occur globally. In order to make this clear, we digress for a moment and discuss a practical way of choosing the scalar measure $[\mathbf{y}]$. As mentioned in Section 2.2, two reasonable candidates are

$$[\mathbf{y}] = y_k(t_0) \qquad \text{for some } t_0 \text{ and an index } k,\ 1 \le k \le n$$

$$[\mathbf{y}] = \text{amplitude of } y_k(t).$$

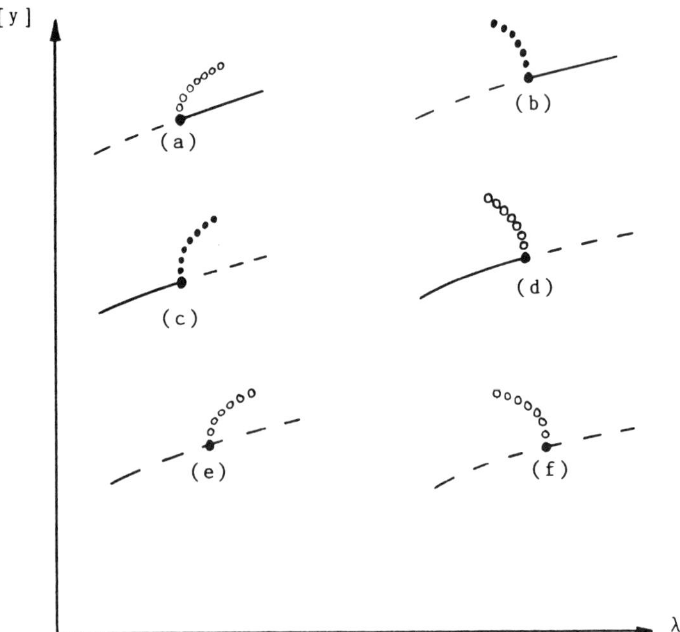

Figure 2.30

The choice of k does not cause trouble; just take $k = 1$, for example. The remaining questions of how to pick t_0 and how to calculate the amplitude can be solved by the compromise

$$[\mathbf{y}] = y_k(t_0) \text{ with } y_k(t_0) = \text{maximum or minimum of } y_k(t). \quad (2.19)$$

(This scalar measure was anticipated in Figure 2.28.) As we shall see in Section 7.1, there is an easy way to obtain Eq. (2.19) as a by-product of the numerical calculation of an orbit. The choice of Eq. (2.19) is not unique because there is at least one maximum and one minimum of $y_k(t)$. Depicting both values in a branching diagram, one includes also the amplitude information. In this way every periodic branch can be represented by two lines. The graphical illustration of a Hopf bifurcation thus resembles that of a pitchfork bifurcation. These two lines may be somewhat misleading because there is only one periodic branch, but this kind of graphical presentation illustrates the dynamics well. Even when only the maximum or only the minimum is depicted, one can still get a feeling for the evolution of amplitude.

We first illustrate a supercritical Hopf bifurcation with a branching diagram depicting the $[\mathbf{y}]$ of Eq. (2.19); see Figure 2.31. As we can see, amplitude grows continuously for increasing values of λ. This kind of Hopf bifurcation is called *soft loss of stability* or *soft generation* of limit cycles. An interesting effect occurs when, in the case of a subcritical bifurcation, the periodic branch turns back and gains stability (Figure 2.32). Then, when λ is increased past λ_0, the amplitude suddenly undergoes a jump and takes "large" values. This kind of jump from a stable equilibrium to large-amplitude periodic orbits is called *hard loss of stability* or *hard generation* of limit cycles. For every λ in the interval be-

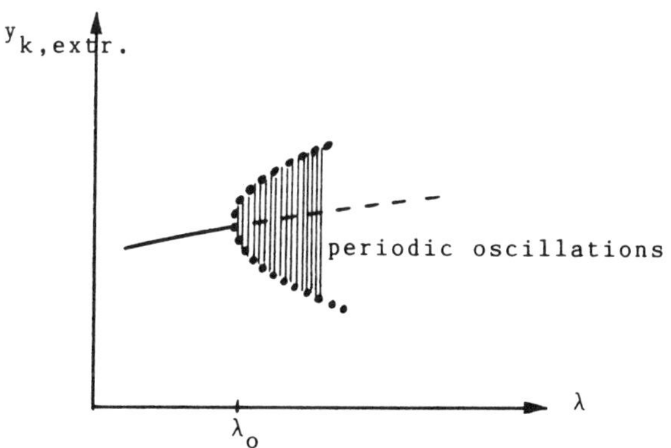

Figure 2.31

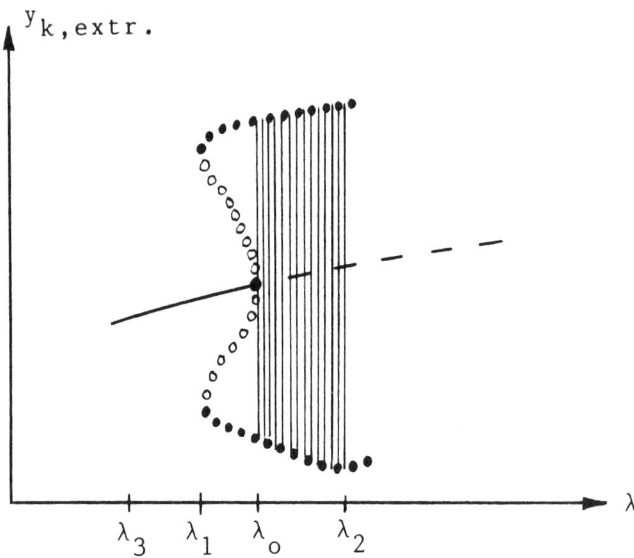

Figure 2.32

tween λ_1 (turning point) and λ_0 (Hopf bifurcation), a stable equilibrium, an unstable periodic orbit, and a stable periodic orbit coexist. Schematically, in two dimensions, this looks like the phase diagram depicted in Figure 2.33. We see two periodic orbits encircling each other as well as the equilibrium. The smaller orbit is a separatrix because it separates two attracting basins. (In n-dimensional phase spaces, separatrices are composed of $(n - 1)$-dimensional "hypersurfaces"; in general, they are not visible when the phase portrait is projected to a plane.) For parameter values in the interval $\lambda_1 \leq \lambda \leq \lambda_0$ we have a *bistable* situation. The jump to higher amplitude can occur before λ_0 is reached if the system is subjected to perturbations.

When the parameter is decreased from λ_2 (Figure 2.32), a discontinuity occurs again. This time the jump from large values of amplitude to zero amplitude (= equilibrium) takes place at the turning point ($\lambda = \lambda_1$). The bistable situation of hard generation of limit cycles is a hysteresis phenomenon similar to the situation that often occurs with pairs of turning points (see, for instance, Figure 2.7 or Figure 2.8). In a real experiment, unstable solutions are not observable. Knowledge of the existence of unstable branches comes solely from solving the equations. In order to get a feeling for what can be represented physically, disregard all dashed lines and open dots in the above figures—that is, omit all unstable solutions.

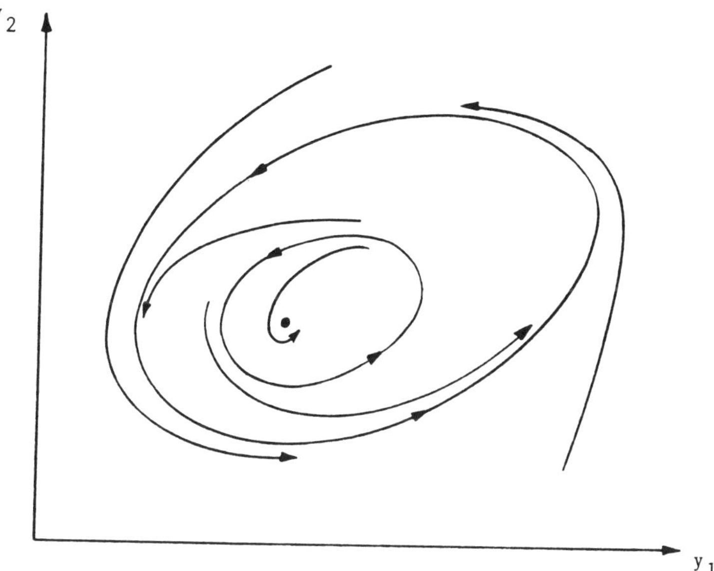

Figure 2.33

A bistable situation combined with a subcritical Hopf bifurcation enables the onset of *bursts*. A burst consists of a short oscillation of several *spikes* followed by a period of less activity (here stationary solutions). This interval of quiescence may be followed by a burst of several pulses that, after extinction, again gives way to a nearly constant value. Such repetitive burst behavior can be explained using Figure 2.32. Imagine that the control parameter λ depends on another variable or on an "outer" influence in such a way that λ sweeps slowly back and forth through the interval $\lambda_3 \leq \lambda \leq \lambda_2$. This interval contains the narrower interval of bistability where the stable stationary branch (resting mode) and the stable periodic branch (active mode) overlap. Each sweep of the larger interval causes a jump—from burst to quiescence or from quiescence to burst. Such alternation between active and inactive modes has been investigated in examples from chemistry [281] and from nerve physiology [156]. In the latter paper, the slow sweeping of the parameter interval was modeled by attaching to the equations a "slow oscillator" that controls λ. In a specific system of equations, a bursting phenomenon may be hidden because the role of the slowly and periodically varying parameter λ may be played by some variable that is controlled by a subsystem.

Two *examples* conclude this section. First we give a solution to the simple model of nerve impulses due to FitzHugh,

$$\dot{y}_1 = 3(y_1 + y_2 - y_1^3/3 + \lambda)$$
$$\dot{y}_2 = -(y_1 - 0.7 + 0.8y_2)/3.$$

The analysis of this example has already been done in Exercises 1.9 and 2.8. The branching diagram including the periodic orbits is shown in Figure 2.34. This figure depicts the branch of stationary solutions including the unstable part and two Hopf bifurcation points. At both Hopf points the bifurcation is subcritical with hard loss of stability. The bistable regions are very narrow:

$$-1.41314 \leq \lambda \leq -1.40352$$
$$-0.34647 \leq \lambda \leq -0.33685.$$

The underlying hard loss of stability is typical for the onset of nerve impulses. When the stimulating current (λ) passes a critical level (Hopf point), the amplitude immediately attains its full level. The situation depicted in Figure 2.34, where the two Hopf bifurcation points are connected by a branch of periodic orbits, is not uncommon. In Figure 2.34, only the curve (chain of dots) that represents the maxima is drawn. This gives enough information on the amplitude. We will continue to use the scalar measure Eq. (2.19) in the sequel.

The second example is a system of ODEs that models a *continuous stirred tank reactor,* commonly abbreviated CSTR. "Continuous" refers to a continuous flow entering (and leaving) the reactor—that is, a CSTR is an *open system.* Human beings and other living organism that have input of reactants (nutrients) and output of products (wastes) are complex examples of CSTRs.

The specific CSTR model we will discuss is

$$\begin{aligned}\dot{y}_1 &= -y_1 + \text{Da}(1 - y_1)\exp(y_2) \\ \dot{y}_2 &= -y_2 + B\cdot\text{Da}(1 - y_1)\exp(y_2) - \beta y_2\end{aligned} \tag{2.20}$$

[351, 352]. The exponential term reflects an infinite activation energy. The variables occurring in Eq. (2.20) are

the Damköhler number Da $= \lambda$

y_1, y_2, which describe the material and energy balances

β, the heat transfer coefficient ($\beta = 3$)

B, the rise in adiabatic temperature ($B = 16.2$).

A branching diagram is shown in Figure 2.35. We see a branch of

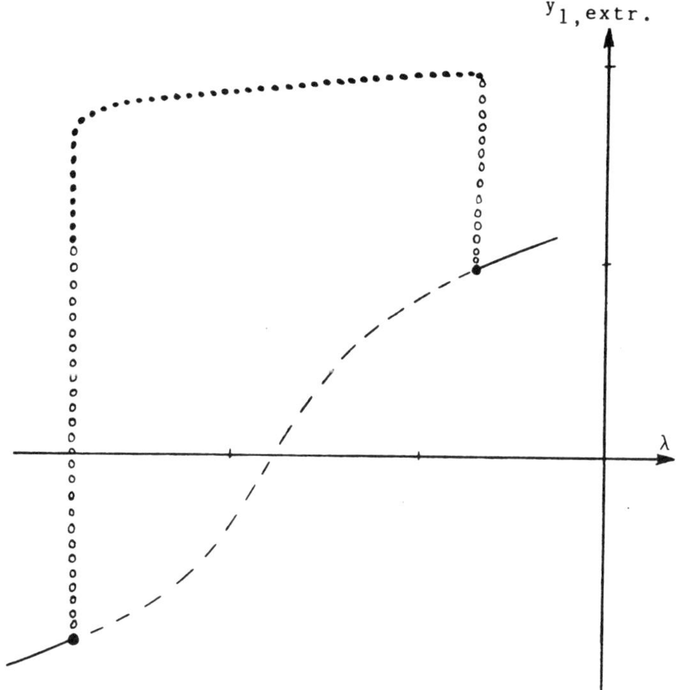

Figure 2.34

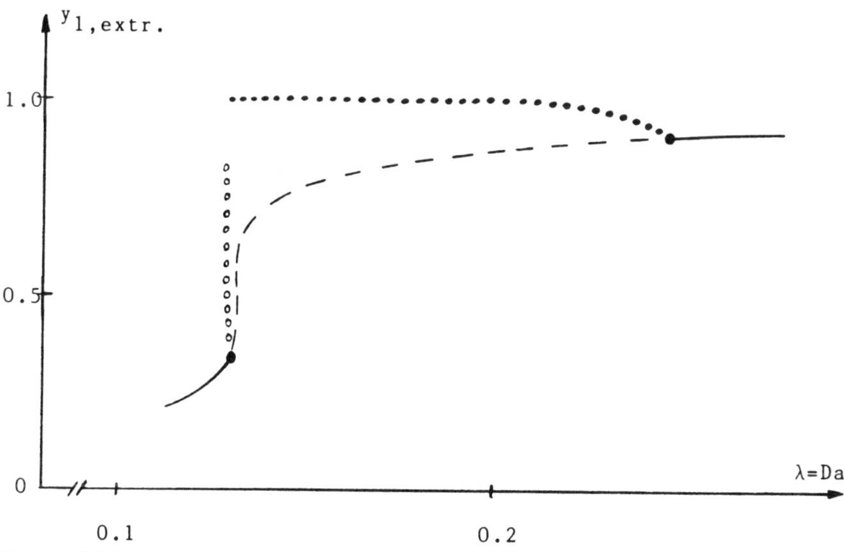

Figure 2.35

stationary solutions with two Hopf bifurcation points, for which the parameter values and initial periods are

$$\lambda_0 = 0.13001, \quad T_0 = 8.4802$$

and

$$\lambda_0 = 0.24648, \quad T_0 = 1.1863.$$

Again the two Hopf points are connected by a branch of periodic orbits. This example exhibits both subcritical and supercritical bifurcation. At the right Hopf point there is soft generation of limit cycles, whereas at the left Hopf point the loss of stability is hard. The branch of stationary solutions has two turning points close to each other with a difference in λ of $\Delta\lambda = 0.0005$. It is not clear in Figure 2.35 that the lower turning point is above the left Hopf point. The hysteresis part of the branching diagram thus lies fully in the unstable range between the Hopf points. The papers [351, 352] present results also for other values of β and B.

Exercise 2.12.
Illustrate Figure 2.32 by drawing five snapshots of phase diagrams (for $\lambda = \lambda_3$, $\lambda = \lambda_1$, $\lambda_1 < \lambda < \lambda_0$, $\lambda = \lambda_0$, $\lambda = \lambda_2$) for

(a) increasing λ and
(b) decreasing λ.
(c) Adapt Figure 2.32 to case (b).
(d) Imagine λ slowly sweeping back and forth over the interval $\lambda_3 \le \lambda \le \lambda_2$ five times. Sketch qualitatively the bursting phenomenon.

2.9 GENERIC BRANCHING

When speaking of branching, we will take the word *generic* as a synonym for "most typical." In this section we discuss what kind of branching is generic and in what senses other kinds of branching are not generic. Let us first ask which type of branching arises in nature.

We begin with a motivating example. In Section 2.4, we discussed

$$0 = \lambda y + y^3$$

as a simple scalar example exhibiting pitchfork bifurcation. The resulting branching diagram (Figure 2.15) shows a bifurcation point at $(y,\lambda) = (0,0)$ with a pitchfork, symmetric with repect to the λ-axis, opening to the left. We now perturb this model equation, considering the equation

$$0 = \lambda y + y^3 + \gamma. \tag{2.21}$$

We discussed in Section 1.2.2 that any model is subject to small perturbations $\gamma \neq 0$ caused, for instance, by inaccuracies in measurements. A straightforward analysis reveals that in the perturbed case the bifurcation is destroyed. Differentiating Eq. (2.21) with respect to λ gives

$$0 = y + \lambda \frac{dy}{d\lambda} + 3y^2 \frac{dy}{d\lambda}.$$

Because $y = 0$ is not a solution of Eq. (2.21), there is no horizontal tangent $dy/d\lambda = 0$ to the solution curve $y(\lambda)$. Checking for a vertical tangent means considering the dependence of λ on y and differentiating with respect to y. This yields

$$\frac{d\lambda}{dy} = 0 \qquad \text{for } \lambda = -3y^2,$$

which, substituted into Eq. (2.21), in turn yields the loci

$$(y,\lambda) = \left(\left(\frac{\gamma}{2}\right)^{1/3}, -3\left(\frac{\gamma}{2}\right)^{2/3}\right)$$

of the vertical tangents. There are three solutions of Eq. (2.21) for

$$\lambda < -3\left(\frac{\gamma}{2}\right)^{2/3},$$

and the branching diagram looks like Figure 2.36 (solid lines). The point with a vertical tangent is a turning point. For $\gamma < 0$, the curves in Figure 2.36 are reflected in the λ-axis. As $\gamma \to 0$, the curves approach the pitchfork (dashed curve). For an arbitrary perturbation γ, the occurrence of a turning point is typical. Bifurcation is the rare exception because it occurs only for $\gamma = 0$.

Before stating a general rule, let us reconsider this example, focusing on the multiplicity of the solutions. The λ-value at the vertical tangent depends on γ:

$$0 = \left(\frac{\lambda}{3}\right)^3 + \left(\frac{\gamma}{2}\right)^2. \tag{2.22}$$

Now we consider Eq. (2.21) as a *two-parameter model,* with γ and λ having equal rights. Equation (2.22) establishes a curve in the (λ,γ)

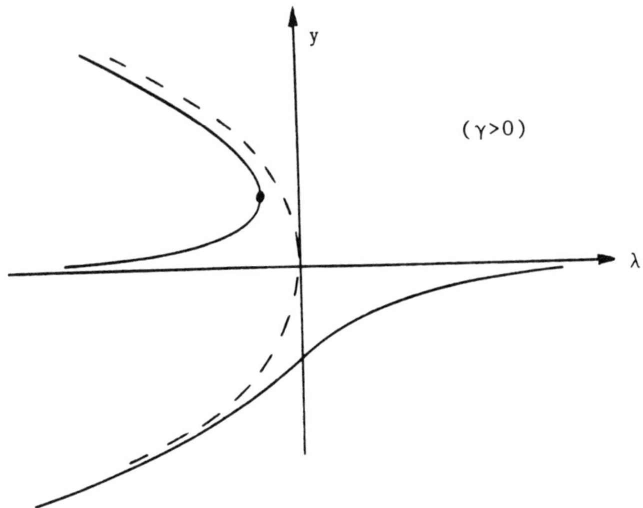

Figure 2.36

plane (Figure 2.37) with the form of a *cusp*. For all combinations of (λ,γ) values inside the cusp (hatched region), Eq. (2.21) has three solutions; for values outside the cusp, there is only one solution. This becomes clear when branching diagrams are drawn. Figure 2.36 for fixed $\gamma > 0$ depicts the situation on the dashed line (a) in Figure 2.37. Choosing $\lambda < 0$ leads to the branching diagram of Figure 2.38. The corresponding line in Figure 2.37 (dashed line (b)) crosses the cusp

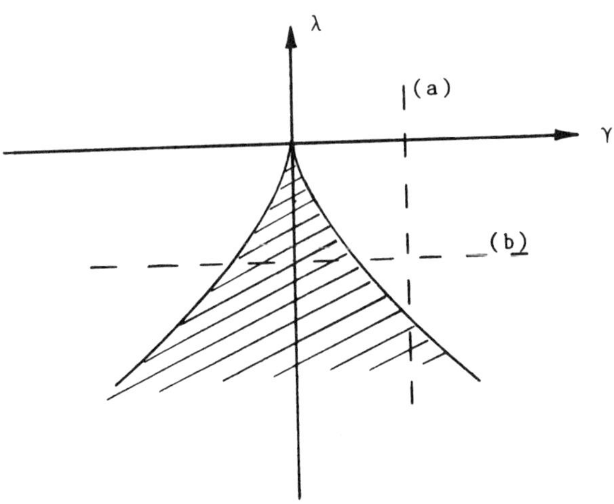

Figure 2.37

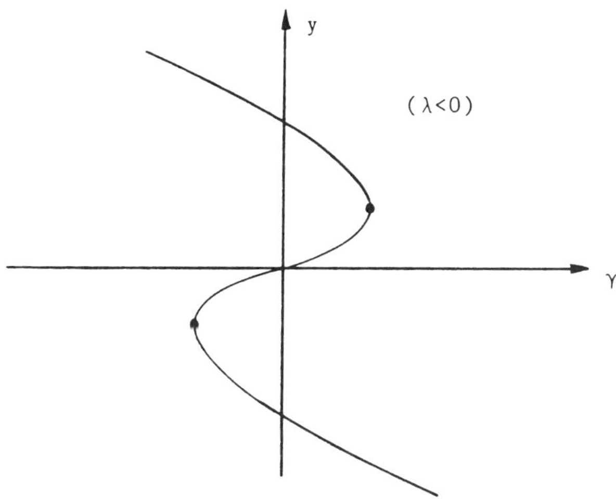

Figure 2.38

boundary twice. Accordingly, there are two turning points with hysteresis behavior. Clearly, when a line crosses the cusp curve in the (λ,γ) plane, a branching phenomenon takes place. Here it is a jump that occurs when the solution is on an appropriate part of the branch. Because Eq. (2.21) defines a two-dimensional surface in the three-dimensional (y,λ,γ) space, each branching diagram is a cross section (Exercise 2.13). Equation (2.21) will be analyzed in a more general context in Section 8.2.

Before proceeding to a discussion of what kind of branching is generic, we pause for some definitions. Suppose our system of equations depends on a number of parameters λ, γ, ζ, . . . , say ν parameters. These may be physical parameters such as Rayleigh number and Damköhler number, may stand for constants that can be varied, and so on. The parameters form a ν-dimensional vector

$$\Lambda = (\lambda,\gamma,\zeta, \ldots).$$

The *parameter space* of all feasible parameter values is called the *control space* because the values of the parameters control the solutions of the problem. In our previous example, the control space is the (λ,γ) parameter plane of Figure 2.37. The loci in a parameter space where a branching phenomenon occurs are called *critical boundaries*. In Eq. (2.21), the critical boundaries are formed by the solutions of Eq. (2.22)—that is, the critical boundaries are the curves that form the cusp in Figure 2.37. The subset of the parameter space formed by the critical boundaries is called the *branching set* or *catastrophe set*. A graphical representation of a part of the parameter space that displays critical boundaries or multiplicities of solutions will be called a *parameter chart*. Figure 2.37 is an example of a parameter chart. A parameter chart that displays regions of stable and unstable behavior will be called a *stability chart*. It sometimes makes sense to separate a certain branching parameter, say λ, from the remaining $(\nu - 1)$ control parameters (Exercise 2.14). This will be discussed in more detail in Section 8.2.

We mention in passing that so far we have developed four graphical tools for illustrating the behavior of the solutions of

$$\dot{\mathbf{y}} = \mathbf{f}(\mathbf{y},\Lambda),$$

namely,

(a) graphs of $(t,y_k(t))$ for various k ("time history")

(b) phase diagrams

(c) branching diagrams

(d) parameter charts (stability chart).

It is important that these different tools not be confused (see Exercise 2.15).

After these preliminaries, we address the question of what kind of branching is generic. As we shall see, the answer depends on the number of parameters involved. We start with the principal case of varying only one parameter λ. Any other parameters are kept fixed for the time being.

In one-parameter problems, only turning points and Hopf-type bifurcations are generic. If the underlying model undergoes small changes, turning points and Hopf points may vary, but they persist. To explain this situation for turning points, let us look again at Definition 2.4. Conditions (3) and (4) involve the relations "$\notin$" and "$\neq 0$," which represent a "normal state." If in condition (4), $d^2\lambda/d\sigma^2$ is continuous and $d^2\lambda(\sigma_0)/d\sigma^2 \neq 0$ for some σ_0, then $d^2\lambda/d\sigma^2 \neq 0$ in some neighborhood of σ_0 too. By a similar argument, it is clear that condition (3) is not affected by small perturbations. These perturbations do not cause the conditions of a turning point (Definition 2.4) to be violated, nor do they cause the conditions that create Hopf bifurcation to be violated. Consequently, turning points and Hopf bifurcations are abundant. Stability such as that of turning points and Hopf bifurcation with respect to small perturbations of the underlying problem is called *structural stability*.

Hysteresis points and stationary bifurcations are not generic; they are destroyed by perturbations. The *unfolding* of bifurcations and hysteresis points is illustrated in Figure 2.39. In the left column of Figure 2.39, we have the nongeneric situation, which is unfolded by a perturbation in one of two ways. The situations in the two right columns are generic because they involve only turning points. Although pitchfork bifurcations and transcritical bifurcations are not generic, we occasionally see them. As Arnol'd [16] states, these bifurcations are due to some peculiar symmetry or to the inadequacy of the idealization, in which some small effects are neglected. Simplifying assumptions often used in modeling can create artificial bifurcations in equations. The onset of a pitchfork bifurcation in a one-parameter model can be seen as an indication of limitations in our modeling capabilities and computers. For example, in considering Euler's buckling beam problem, it is convenient to assume perfect symmetry. It is this simplification that establishes a bifurcation (Exercise 2.14). If material imperfections are taken into consideration, the models are more complex and more difficult to solve. Because it will never be possible to avoid all simplifying assumptions, we cannot completely rule out the occurrence of stationary bifurcations in one-parameter problems. Simplifying assumptions such as symmetry play no role in hysteresis. For this reason, hys-

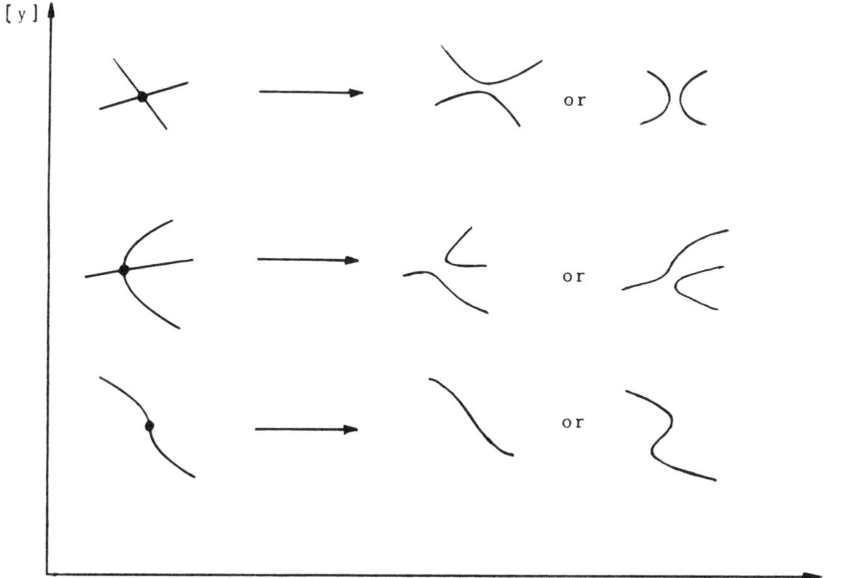

Figure 2.39

teresis points hardly ever occur for realistic one-parameter problems. The same holds true for multiple bifurcation points and isola centers.

The above discussion might suggest that one can forget about such nongeneric cases as hysteresis points. This may be justified for one-parameter problems, but not for multiparameter problems, where such points occur naturally. We illustrate this by studying the situation for a fictitious two-parameter model. Let us again denote the second parameter by γ. Consider a sequence of branching diagrams with respect to λ, say for $\gamma_1 < \gamma_2 < \gamma_3 < \ldots$. Such a sequence may look like Figure 2.40. For γ_1, we find two branches without connection; the upper branch is an isolated branch. Starting from point A, we come smoothly to point B when λ is increased. For γ_2 the situation is the same. Now consider the situation for γ_3, where we no longer encounter an isolated branch; the branches resemble a mushroom in shape. The consequences for the path from A to B are severe. There are two jumps (right figure, double arrows). Apparently there is a situation with a transcritical bifurcation for a value γ_0^B with $\gamma_2 < \gamma_0^B < \gamma_3$. One must go through this bifurcation when γ is increased from γ_2 to γ_3. Hence the bifurcation separates a continuous path A $\rightarrow$ B (Figure 2.40) from a path with jumps. This transcritical bifurcation can be seen as an *organizing center* separating a no-jump situation from a two-jump situation. In view of its structure in a two-parameter setting, a bifurcation point may be called *gap center* (compare Figure 2.39).

Increasing γ further, one may obtain Figure 2.41. Here it is assumed that, for increasing γ, the right hysteresis phenomenon disappears and only one jump remains (Figure 2.41, γ_5, right figure). Clearly, between γ_4 and γ_5 there is a value γ_0^{HY} where a hysteresis point occurs. The parameter combination $(\lambda_0^{HY}, \gamma_0^{HY})$ of a hysteresis point is another example of an organizing center and is therefore a *hysteresis center*.

Consider again the first picture of Figure 2.40 ($\gamma = \gamma_1$). We ask what might happen when smaller values of γ are taken, say γ_{-1}, γ_{-2}, and γ_{-3} with

$$\gamma_1 > \gamma_{-1} > \gamma_{-2} > \gamma_{-3}.$$

Possible branching behavior is depicted in Figure 2.42. Here, for decreasing γ, the isolated branch gets smaller. For some value γ_0^I between γ_{-2} and γ_{-3}, the two turning points coalesce in an isola center. (The unfolding of isola centers can be illustrated as in Figure 2.39.)

The situation depicted by Figures 2.40 through 2.42 is summarized in the parameter chart of Figure 2.43. The lines in Figure 2.43 show the curves where turning points are located. Two of these curves meet in the bifurcation point with parameter pair $(\lambda_0^B, \gamma_0^B)$, two curves meet

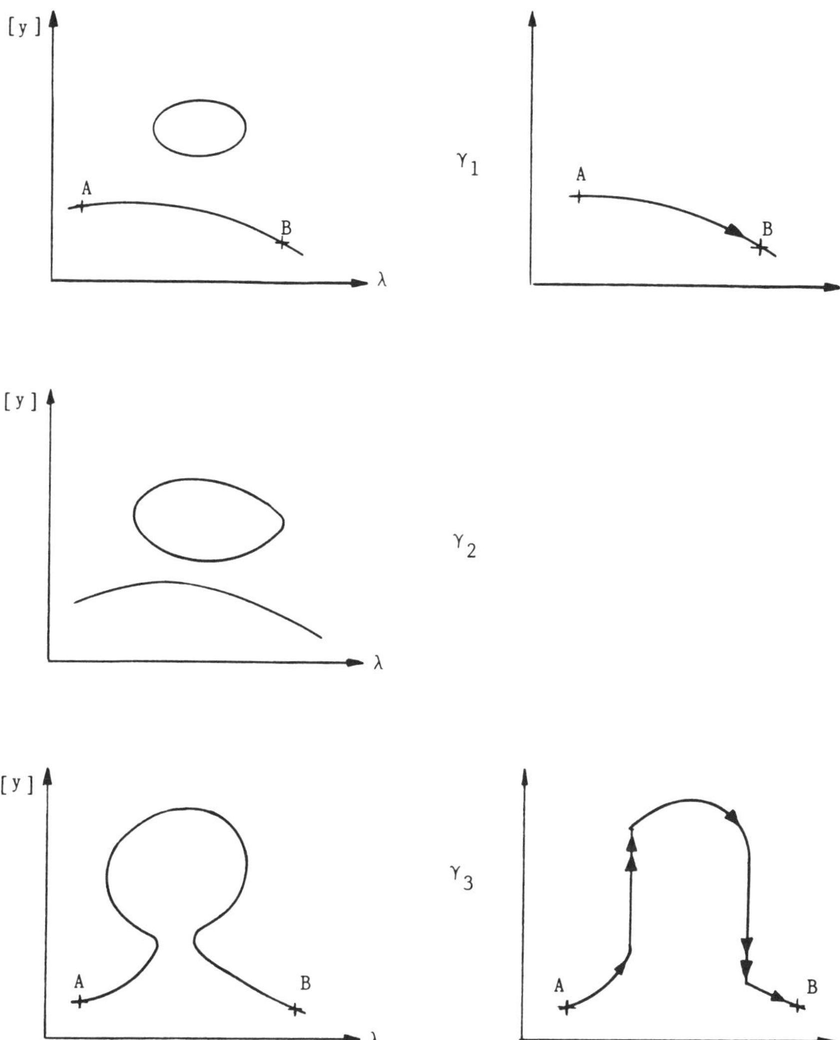

Figure 2.40

in $(\lambda_0^{HY},\gamma_0^{HY})$ (hysteresis point), and two curves merge at (λ_0^I,γ_0^I), which is the location of an isola center. At these three points a new dynamic behavior is organized (no multiplicity, then no jump, then two jumps, then one jump). This leads us to call these points "organizing centers." A similar interpretation can be found for the onset of a pitchfork bifurcation (Exercise 2.16). Let us note that examples with such phenomena as those depicted in Figures 2.40 through 2.42 occur in reality.

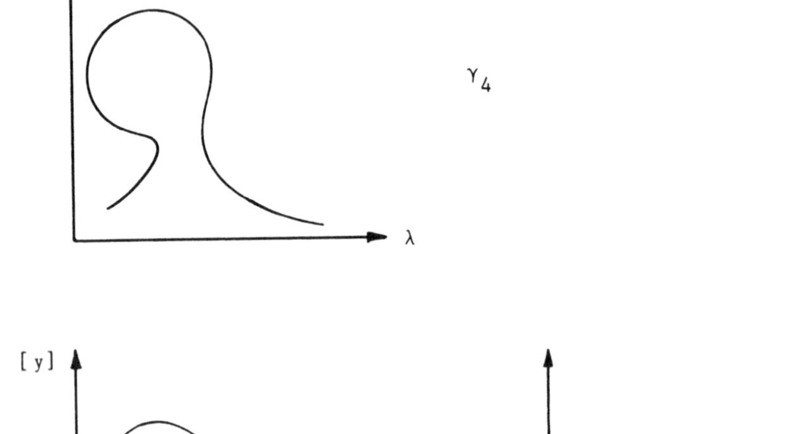

Figure 2.41

For example, the CSTR problem discussed in [148] exhibits these phenomena. The jumps to the upper level (Figure 2.40) are interpreted as ignition of the CSTR because [**y**] can be related to a temperature. Correspondingly, the jumps to the lower level are extinctions of the reactor. In this example, a turning point is either a point of ignition or a point of extinction. Phenomena such as those depicted in Figures 2.40 through 2.42 can be described by simple equations (Exercise 2.17).

In the above discussion and in Figures 2.40 through 2.43, the emphasis is on stationary bifurcations. Hopf bifurcations too vary with parameters and can coalesce or interact with stationary bifurcations. Interactions between Hopf points and stationary bifurcations can trigger interesting effects [209, 210]. How Hopf points can coalesce in *Hopf centers* will be illustrated by means of a model describing nerve activities.

Example 2.5. Potential of Ventricular Action.
According to Beeler and Reuter [29], the action potential of ventricular myocardial fibers can be modeled by eight differential equations. The membrane potential across the capacity, V, and the intracellular calcium ion concentration $[Ca]$ obey the differential

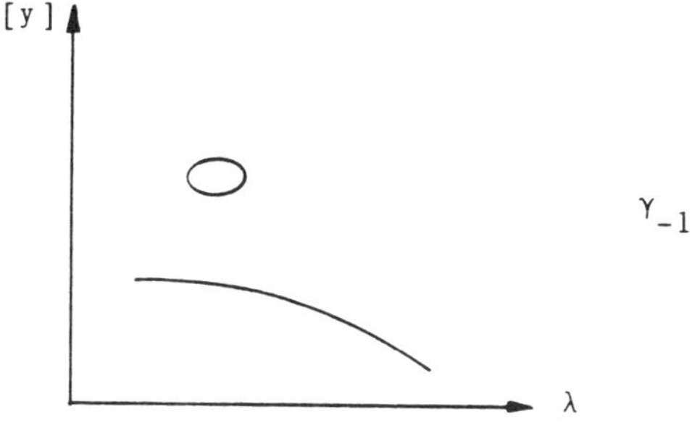

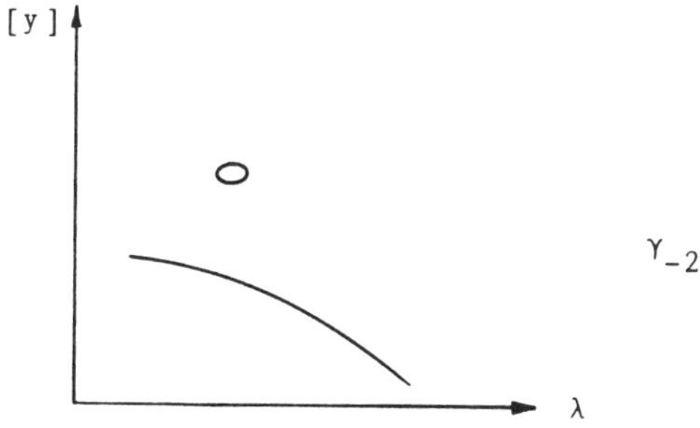

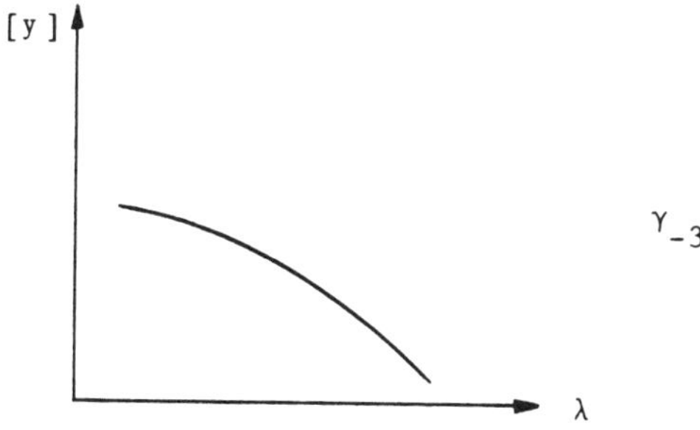

Figure 2.42

equations

$$\begin{aligned} \frac{dV}{dt} &= -i_K - i_x - i_{Na} - i_{Ca} + i_{\text{ext}} \\ \frac{d[Ca]}{dt} &= -\gamma i_{Ca} + 0.07(\gamma - [Ca]). \end{aligned} \tag{2.23}$$

Here, i_K, i_x, i_{Na}, and i_{Ca} are currents given by six further differential

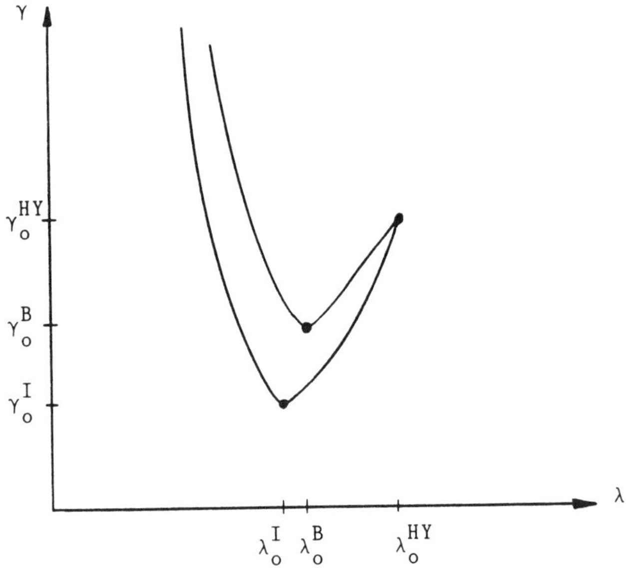

Figure 2.43

equations [29]. The external current serves as a branching parameter, $\lambda = i_{\text{ext}}$. The scale factor γ ($\gamma = 10^{-7}$ in [29]) can be varied. For γ in the interval

$$10^{-8} < \gamma < 10^{-3},$$

many Hopf points (H) and turning points (noses N) were calculated by means of BIFPACK (see Appendix 6). The results obtained for stationary solutions are summarized in Figure 2.44. Hysteresis behavior starts (or ends) at the hysteresis point HY; this corresponds to one part of Figure 2.43. At the Hopf center HC, two Hopf points coalesce (are born, respectively).

The difference between a one-parameter setting and a two-parameter setting is great. When varying λ for any one given γ, hitting a hysteresis point is unlikely. This is the situation of a one-parameter setting, which dominates practical calculations. Interpreting the same example as a two-parameter model with continuously varying γ, one sees how coalescing points such as isola centers, hysteresis centers, bifurcations, or Hopf centers occur naturally as organizing centers. This holds a fortiori for models involving more than two parameters. In such cases, organizing centers can be still more complicated, a topic that will be discussed in Chapter 8.

Exercise 2.13.
Draw two branching diagrams for Eq. (2.21):

(a) for a fixed $\lambda > 0$
(b) for $\lambda = \gamma + 1$.

Equation (2.21) defines a two-dimensional surface in the three-dimensional (y,λ,γ) space. Sketch this surface.

Exercise 2.14.
In a finite-element analog of Euler's beam problem, the scalar equation

$$0 = y - \alpha - 2\lambda \sin(y) + \beta \cos(y)$$

occurs [112] (see Figure 2.45). The dependent variable y and the control

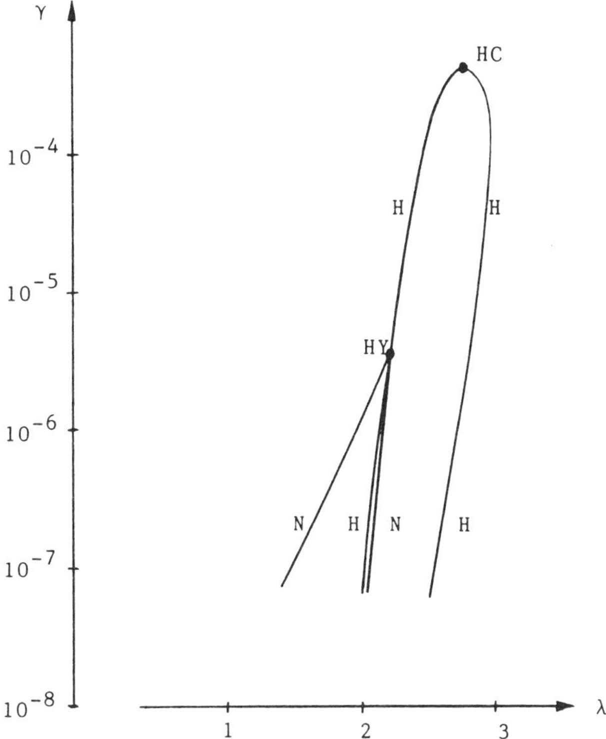

Figure 2.44

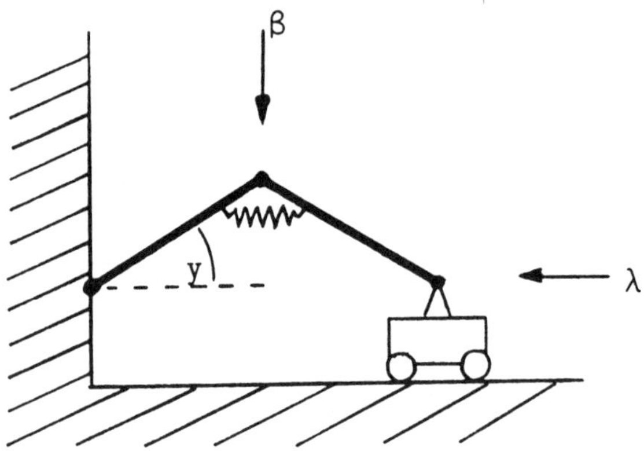

Figure 2.45

parameters are

y, the angle of displacement
λ, the compressive force
β, the central load
α, the angle at which spring exerts no torque.

(a) Assume $\alpha = 0$ (perfect symmetry, no initial curvature) and $\beta = 0$ (weightless rods). What solutions exist for $0 < \lambda < 5$? Sketch a branching diagram.
(b) What solutions exist for $0 < \lambda < 5$ when $\alpha = \beta \neq 0$?
(c) Which perturbations must be expected in the general case $\alpha \neq 0$, $\beta \neq 0$? (Hint: Calculate a relation between α and β that establishes a hysteresis point.)
(d) In the (α,β) plane, draw the critical boundaries of the bifurcation points and the hysteresis points—that is, draw a parameter chart. Shade the region for which every branching diagram with respect to λ has exactly three turning points.

Exercise 2.15.
Illustrate the behavior of the solutions of Eq. (2.17) by drawing four different diagrams: time dependence of y_1, phase diagram(s), a branching diagram, and a parameter chart. (One rarely draws a parameter chart for a one-parameter problem, but it is useful here as an educational tool.)

Exercise 2.16.
Draw a sequence of branching diagrams illustrating a pitchfork bifurcation of a two-parameter model (see Figures 2.40 through 2.42).

Exercise 2.17.
Consider Eq. (2.21). For the *straight* lines in the parameter chart of Figure 2.37, one has the branching diagrams of Figure 2.36 and Figure 2.38. Following appropriate *curved* paths in the parameter chart yields diagrams as in Figure 2.40 and Figure 2.41. Find paths that correspond to the situations γ_1, γ_3, and γ_5. (Hint: The meaning of parameters λ and γ in Figure 2.37 differs from that in Figures 2.40 and 2.41. Introduce new parameters λ' and γ'.)

3 Practical Problems

The diagrams in Chapters 1 and 2 are idealizations of the results one gets in practice. Numerical reality does not look quite as good as these diagrams. The smooth curves of the branching diagrams are drawn from a skeleton of computed points. It requires imagination and confidence to draw a good diagram from the few points the computer gives us; the many difficulties are seldom mentioned. In this chapter we discuss the practical tools that are required to solve a particular example. We also attempt to sharpen the eyes of the researcher who must interpret results.

3.1 READILY AVAILABLE TOOLS AND LIMITED RESULTS

Let us look at the readily available tools for solving equations. First of all, we need a solution procedure for the particular class of problems under investigation. Later in this book, we shall call related standard software of numerical analysis "SOLVER." For example, if we handle a model described by a system of algebraic equations, a Newton-like method (see Section 1.4) may be required. In the case of an ODE problem, we will need a solver for boundary-value problems. It will be assumed that standard software of numerical analysis is available and that an approximation of a solution of the equation

$$\mathbf{F}(\mathbf{Y}) = \mathbf{0} \tag{3.1}$$

or the boundary-value problem

$$\dot{\mathbf{Y}} = \mathbf{F}(\mathbf{Y}), \quad \mathbf{R}(\mathbf{Y}(a), \mathbf{Y}(b)) = \mathbf{0} \tag{3.2}$$

can be obtained. The notation **F** and **Y** rather than **f** and **y** indicates that **Y** includes **y** and that **F** is the result of a transformation of **f**. This includes the cases $\mathbf{Y} = \mathbf{y}$ and $\mathbf{F} = \mathbf{f}$. Equation (3.2) represents a two-point boundary-value problem for $a \leq t \leq b$ with boundary conditions defined by $\mathbf{R}$. Although it may be laborious to solve Eq. (3.1) or Eq. (3.2), we take the stand that a solution for such problems is available when needed.

The dependence of the problem on parameters increases the computational effort. One can solve the equations for only a finite number of parameter values. Suppose that, because of limitations in time and money, we can find approximate solutions for only 90 parameter values $\lambda_1, \lambda_2, \ldots, \lambda_{90}$. We do not know exactly what solutions look like for any parameter values other than these 90. The situation is comparable to replacing a movie by a set of snapshots. Plotting the results means drawing a chain of dots (Figure 3.1b). Each dot represents one solution. Connecting the dots by a smooth curve gives a nice branching diagram, but unfortunately no guarantee of correctness. One may be lucky, as in Figure 3.1, where the interpolating curve represents the true parameter dependence; apparently the calculation of further solutions for intermediate parameter values will not reveal more information. In the sections that follow, we shall see what can happen between two "consecutive" solutions.

3.2 PRINCIPAL TASKS

In a multiparameter problem we keep all parameters fixed except one (denoted, as usual, by λ). During the course of a parameter study, parameters may be switched, and another parameter might become the branching parameter. Suppose the parameter range of physical interest is

$$\lambda_a \leq \lambda \leq \lambda_b.$$

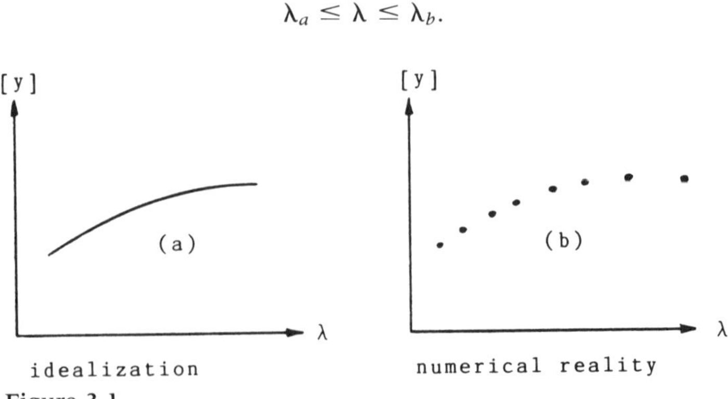

Figure 3.1

Let us calculate a first solution for $\lambda_1 = \lambda_a$. The calculation of this first solution is often the most difficult part. Initial guesses may be based on solutions to simplified versions of the underlying equations. Sometimes choosing special parameters (often = 0) leads to equations that are easily solved. In complicated cases a hierarchy of simplifications must be gradually relaxed step by step, the solution of each equation serving as an initial guess to the following more complicated equation. Solving a sequence of equations with diminishing degree of simplification until finally the "full" equation is solved is called *homotopy* (Appendix 5). Here we neglect possible trouble and assume that a solution to the full problem can be calculated for $\lambda = \lambda_a$.

After having succeeded with λ_1, the next question is how to choose λ_2. Assume again that available computer time allows us to compute no more than 90 solutions. We can try to get enough information by selecting an *increment* $\Delta\lambda$ (say $\Delta\lambda = (\lambda_b - \lambda_a)/89$) and taking

$$\lambda_{i+1} = \lambda_i + \Delta\lambda, \qquad i = 1, 2, \ldots .$$

Such a strategy of choosing equidistant values of the parameter might be successful—that is, we might get a solution for each λ_i. It is likely, however, that one or more of the following problems occurs (Figure 3.2):

A bifurcation (A) is overlooked.

A turning point (B) occurs, and SOLVER does not converge for some value of λ.

$\Delta\lambda$ is too large or too small, causing a waste of money.

Such situations are common. Figure 3.3 shows the numerical

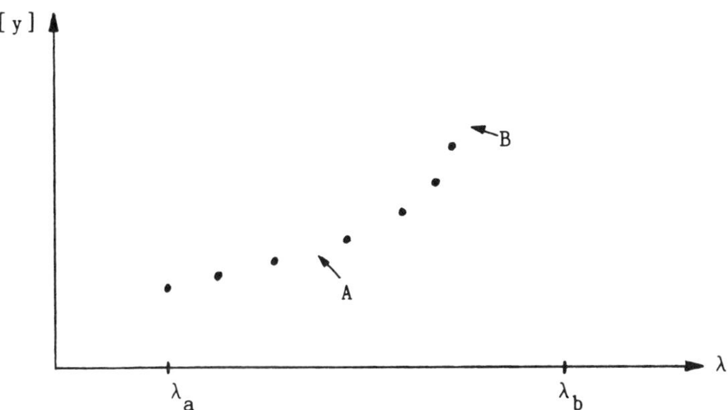

Figure 3.2

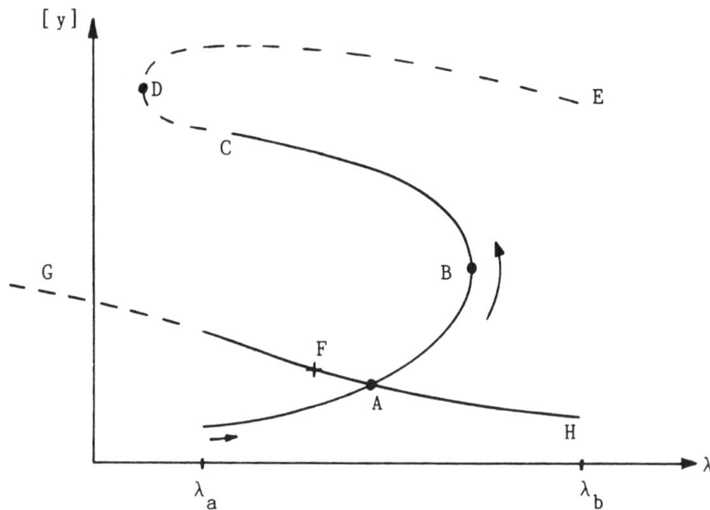

Figure 3.3

methods that are required. First we need a *continuation method.* A continuation method generates a chain of solutions. In a good continuation procedure, reasonable candidates for the increment $\Delta\lambda$ are calculated adaptively. Using a continuation method, one can track a branch around a turning point. A device that indicates whether a bifurcation point is close or has been passed is needed. In many problems, stability must be checked. Finally, we need a method for switching branches. It suffices to calculate one solution on the "new" branch (Figure 3.3, F); then, by means of continuation methods, the entire branch (G/H) can be generated. The real situation underlying Figure 3.2 may look like that in Figure 3.3, for example. Detecting the bifurcation point (A) offers a straightforward way to reach λ_b (via H). A device for passing the turning point (B) provides information about another part of the branch. Extending the path outside the range $\lambda_a \leq \lambda \leq \lambda_b$ may provide another way to reach λ_b (Figure 3.3, C, D, E). Following the new branch leftward from F for a sufficient distance may convince the investigator that this particular branch neither turns back (G) nor exhibits a bifurcation that leads back into the parameter range of interest.

Let us summarize the basic tools required for a parameter study. We need:

(a) continuation methods with devices for detecting bifurcation and checking stability

(b) methods for switching from one branch to another with or without the option of calculating the branch point itself.

The most time-consuming part of this is the use of the continuation method. Typically, 90 to 99 percent of the total effort consists in generating branches by the continuation method, that is, solving sequences of equations. In comparison, the handling of bifurcation points is inexpensive.

3.3 WHAT ELSE CAN HAPPEN

In the preceding section we briefly discussed some common situations. The interpretation was straightforward. The many other phenomena are difficult to interpret. In this section we give some examples that encourage the reader to double-check results carefully.

One frequent problem is undesired jumping between branches. Such jumps are easily overlooked. Figure 3.4 depicts a branch whose upper part grows to infinity for λ approaching zero. Suppose a continuation starts at point A with tracing of the branch to the right (see the arrows). After rounding the turning point, the upper part B can be traced using decreasing values of λ. At C the continuation procedure should stop because "infinite" values of **y** are not likely to make sense. Instead of stopping there, a jump to the lower part of the branch (D)

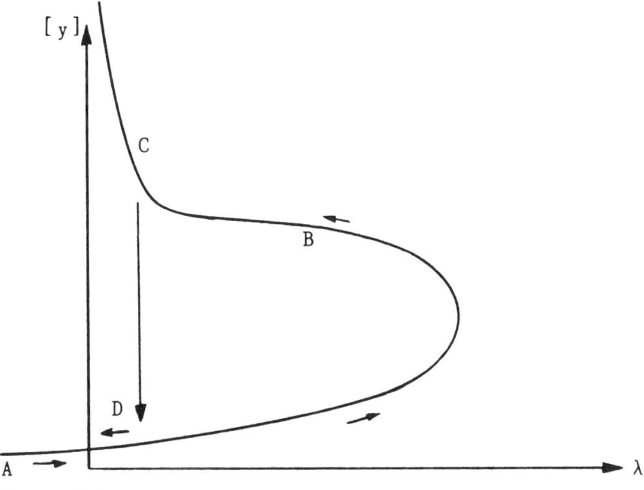

Figure 3.4

may take place. This jump may be interpreted as indicating a closed path and is in a sense a "failure." A similar situation is shown in Figure 3.5. Tracing a hysteresis for increasing values of λ, one can see results that behave like a physical experiment. The continuation may not detect the turning point at A quickly enough and may try to calculate a solution for $\lambda > \lambda_0$. In this situation, SOLVER may fail to converge, or the solution may jump to the other part of the branch (C). In the second case, the hysteresis phenomenon is often overlooked, especially if it is small.

In Figures 3.4 and 3.5 the question "Is the continuation procedure giving us the same branch or a different branch?" arises. This common question is illustrated in Figure 3.6. Suppose we are tracking a closed branch (left of Figure 3.6) in a counterclockwise direction starting from A (symbol "+"). At B we change the symbol to "O." Arriving again at parameter values in a neighborhood of A, we ask whether we are generating the same branch or a different branch (right of Figure 3.6). In practice it is often difficult to answer this question.

Let us give more examples of the same dilemma. In Figure 3.7 a situation where the continuation procedure jumps across a gap is depicted; the "+" symbols represent consecutive solutions. Another example is depicted in Figure 3.8. This figure shows a transcritical bifurcation with an exchange of stability. Three consecutive solutions obtained during the continuation are marked. The instability of the solution at λ_3 may cause the algorithm to search for a bifurcation in the wrong interval $\lambda_2 \leq \lambda \leq \lambda_3$.

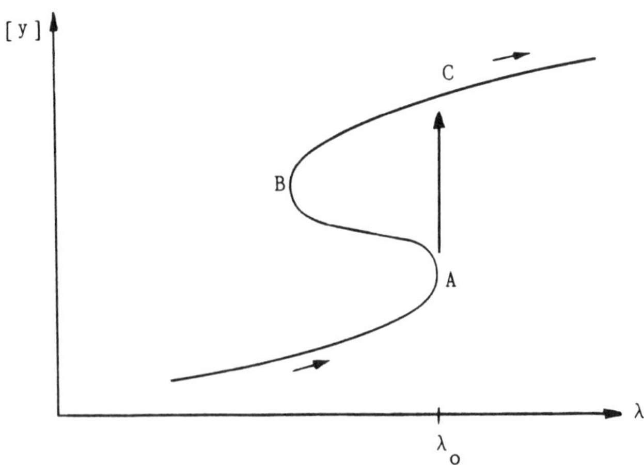

Figure 3.5

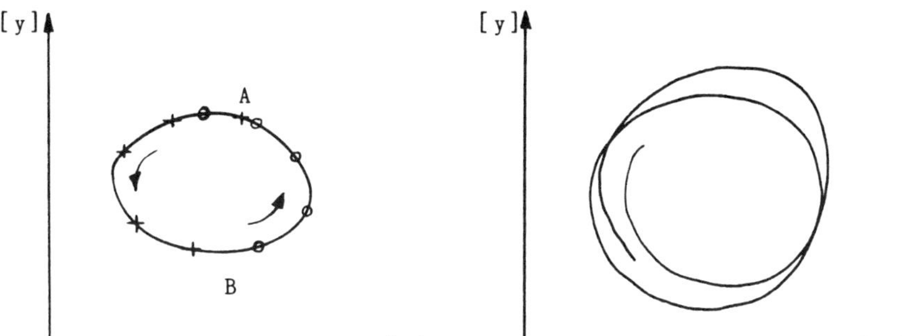

Figure 3.6

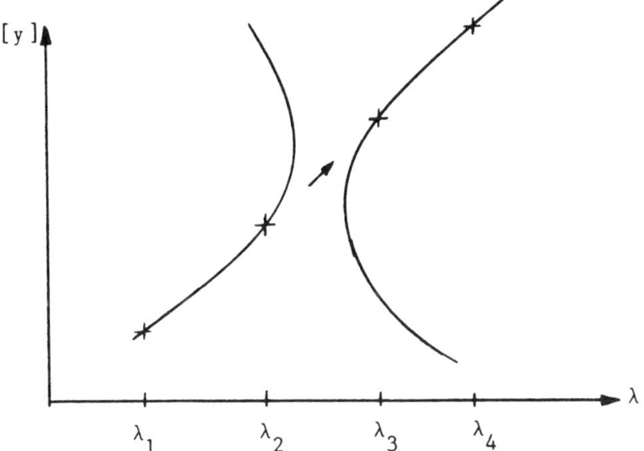

Figure 3.7

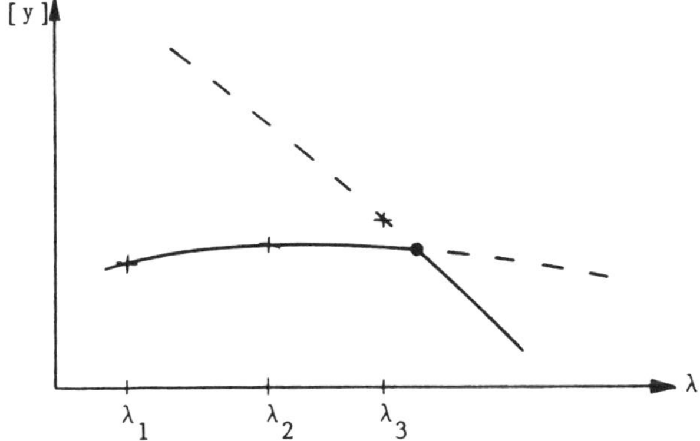

Figure 3.8

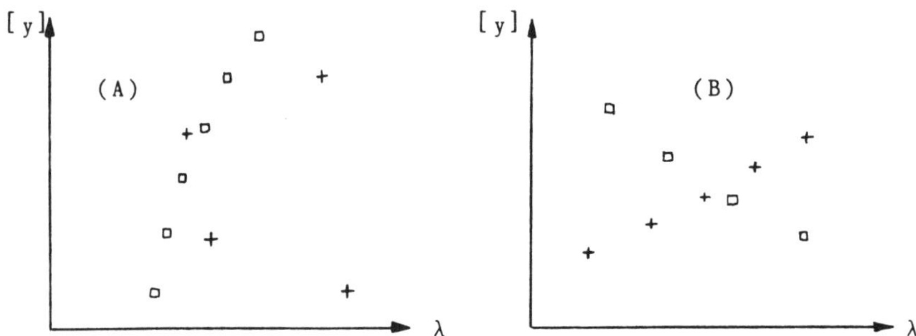

Figure 3.9

Another question that is sometimes not easy to answer is whether two branches intersect (in a bifurcation point). Figure 3.9 shows two such situations. The solutions of one branch are indicated by crosses, the solutions of the other branch by squares. In such situations, clarity is achieved by investigating the branches in more detail. If there are bifurcation points on either branch with the same value of λ_0, we can be sure that both branches intersect. An alternative is to interpolate the values of neighboring solutions on each branch and compare the results. Close to a bifurcation, the solution values of both branches must match. Situation A in Figure 3.9 may be a pitchfork bifurcation, and B might be no bifurcation at all.

Finally, we give an example of how bifurcation can be overlooked (Figure 3.10). In a first run with coarse step sizes, a turning point is passed. All the calculated stationary solutions (+ in the left figure) are unstable. The same branch is then recalculated with smaller step sizes. During this second run, a short branch of stable equilibria bounded by a Hopf bifurcation (H) and the turning point (T) is discovered. There is an emanating branch of stable periodic solutions. The results suddenly look completely different. This example shows how results are affected

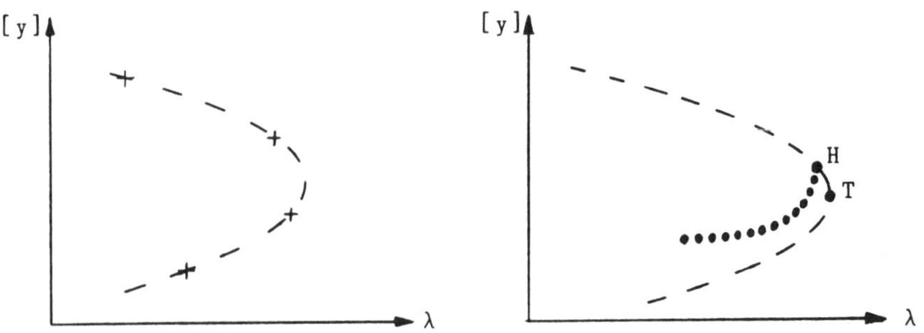

Figure 3.10

by step sizes in continuation methods. Unfortunately, one rarely knows beforehand how small a "good" step size must be.

In this section we discussed diagrams without reference to specific examples. It should be stressed that the difficulties adduced are common. In the following section, an example where all the above-mentioned difficulties come about is presented.

3.4 MARANGONI CONVECTION

Section 2.7 discussed the Rayleigh-Bénard problem. Heating of a fluid lowers its density. Gravity (buoyancy) will cause convection if the heating is from below and the density gradient is strong enough. In addition to gravity, another phenomenon induces convection when a liquid is heated. When temperature is increased, surface tension decreases. Higher surface tension drags a liquid. If the pull is strong enough, convection appears. Convection driven by surface tension is called convective thermocapillarity, or the Marangoni effect [263, 289, 327, 372].

The Marangoni effects are widespread [304]. For instance, they are responsible for irregular patterns that form on drying paint films. Here temperature gradients are caused by evaporation of solvents. Frequently, both the Rayleigh-type convection (driven by gravity) and the Marangoni-type convection (driven by surface tension) act simultaneously; it can be difficult to distinguish between them because these two effects reinforce each other. Apparently Rayleigh's explanation of Bénard's experiments was misleading [252]. Bénard ran his experiments with free surface and thin layers, a situation where Marangoni instabilities are likely to dominate over Rayleigh-type convection. Related convection phenomena have been called Rayleigh-Bénard-Marangoni instabilities.

Thermocapillary forces play an important role in zone refining, especially in purifying semiconductor material. Figure 3.11 shows a typical configuration. A rod of semiconductor material is moved through a ring heater, where it melts. Upon cooling, the material solidifies, forming a single crystal. Because of the nonuniform temperature distribution outside the melted part, thermocapillary forces act on the liquid. Under appropriate conditions steady convection occurs, leading to a uniform distribution of contamination. For certain temperature gradients, the convection becomes unstable, leading to a flow that causes undesired striations in the semiconductor crystal. The related loss of stability can be explained by bifurcations.

The Navier-Stokes equations, which describe this process, have been investigated repeatedly (see, for instance, [63, 190, 289, 327]).

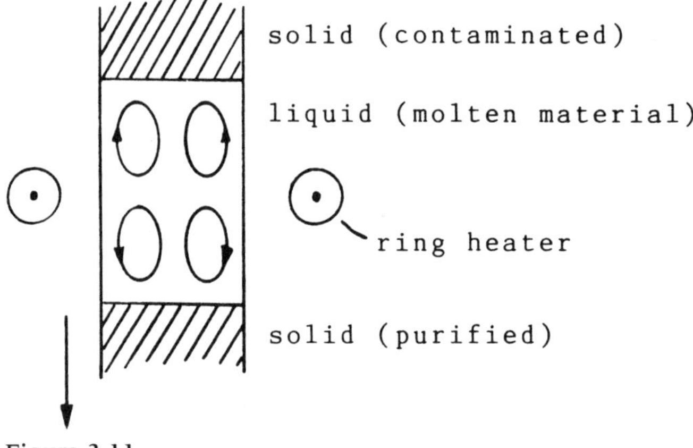

Figure 3.11

In this section we follow a model developed by Gill [104] that is based on the two-dimensional time-dependent momentum equations, energy equation, and continuity equation for a circular cylinder of radius R and length L. The approach of [104] is mixed: A Galerkin approach is used to approximate the dependence on one of the space variables; the other space variable is discretized by finite differences. Without presenting details, we list the equations that are the outcome of the Galerkin-type first phase of the analysis. Here, $\mathbf{y}(\rho,t)$ is an artificial variable related to the flow. With the variables and constants

$\dot{\mathbf{y}} = \partial\mathbf{y}/\partial t$, $\mathbf{y}' = \partial\mathbf{y}/\partial\rho$, ($\rho$ normalized radius)

$A = R/L$: aspect ratio

Re: Reynolds number

Pr: Prandtl number

$Q = 2A^3 Re$

Nu: Nusselt number

Θ_0, Θ_2, Θ_4 coefficients modeling the outer temperature distribution

$\xi = 1/\rho$ (abbreviation)

the set of partial differential equations is

$$\dot{y}_1 = y_1'' + (\xi + Qy_2)y_1' - Qy_1y_2' - \xi^2 y_1 - 2\xi Qy_1y_2 + 3y_3/Re$$

$$0 = y_2'' + \xi y_2' - \xi^2 y_2 - y_1$$

$$\dot{y}_3 = y_3'' + (\xi + Qy_2)y_3' - \xi^2 y_3 - Qy_3(4\xi y_2 + 3y_2')$$
$$+ Qy_4(3y_1' - 4\xi y_1) - Qy_1y_4'$$

$$
\begin{aligned}
0 &= y_4'' + \xi y_4' - \xi^2 y_4 - y_3 \\
Pr\,\dot{y}_5 &= y_5'' + \xi y_5' + Pr\cdot Q y_2 y_5' - y_6 \\
Pr\,\dot{y}_6 &= y_6'' + \xi y_6' - Pr\cdot Q(2\xi y_6[y_2 + \xi y_2'] - y_2 y_6') \\
&\quad + 6A^3(y_7/Q - Pr\cdot y_4 y_5') \\
Pr\,\dot{y}_7 &= y_7'' + \xi y_7' - Pr\cdot Q(4\xi y_7[y_4 + \xi y_4'] \\
&\quad + 4\xi y_7[y_2 + \xi y_2'] - y_2 y_7' - 6 y_4 y_6').
\end{aligned}
$$

The boundary conditions for $\rho = 0$ and $\rho = 1$ are

$$
\begin{aligned}
y_1(0) &= y_2(0) = y_3(0) = y_4(0) = 0 \\
y_2(1) &= y_4(1) = 0 \\
y_1(1) &= y_6(1),\ y_3(1) = y_7(1) \\
y_5'(0) &= y_6'(0) = y_7'(0) = 0 \\
y_5'(1) &= Nu[\Theta_0/(2A^2) - y_5(1)] \\
y_6'(1) &= Nu[\Theta_2 - y_6(1)] \\
y_7'(1) &= -Nu[2Q\Theta_4/A + y_7(1)].
\end{aligned}
$$

The above system of seven PDEs can be reduced to a set of five equations by eliminating y_2 and y_4 [104]. Using *N lines* in a discretization of the derivatives with respect to ρ, we come up with a set of $5N$ scalar ODEs modeling a time-dependent 2-D Navier-Stokes problem. Solutions of the resulting system vary with the seven parameters,

$$Q,\ Pr,\ A,\ Nu,\ \Theta_0,\ \Theta_2,\ \Theta_4.$$

In what follows, we discuss some numerical results for

$$Pr = 0.014, \quad \Theta_0 = \Theta_2 = A = 1. \tag{3.3}$$

The branching parameter will be $\lambda = Q$. Let us investigate the branching diagrams for stationary solutions with $N = 9$ lines for varying Nusselt numbers Nu. After many experiments (all of them run with the continuation method of [315], it turned out that the value $\Theta_4 = 0$ plays a critical role. For this value some interesting bifurcations occur. Figure 3.12 shows a branching diagram for $Nu = 100$ and $\Theta_4 = 0$. The vertical axis displays the radial velocity component of the flow, measured at the half-radius. We see two branches A and C, with turning points at $Q = 32$ and $Q = 2328$. The branches A and C are connected via two bifurcations and branch B, which represents the onset of a second concentric convection cell. There is an exchange of stability at these two bifurcation points. The stability of branches A, B, and C is indicated in Figure 3.12 in the usual manner (dashed means unstable).

Figure 3.12 shows a bifurcation point that connects branches C and C′. This bifurcation and branch C′ appear to be *spurious*. Spurious solutions are those that solve the discretized model but have no counterpart in the original continuous PDE. Hence, spurious solutions have no physical significance. Relations between a continuous model and its discretized analogs have been studied repeatedly. In the context of bifurcation theory, we refer to [35, 38, 39, 254].

We now look at how the branching diagram varies as the Nusselt number Nu is decreased from the $Nu = 100$ used for Figure 3.12. The bifurcation that connects branches A and B occurs at small values of Q and is affected only slightly. We turn our attention to high values of Q, where the situation is different (Figures 3.13 and 3.14). For $Nu = 0.5$ (Figure 3.13), branch B bends back on itself and there is a new branch D connecting with C. There is a stable section on branch C, with two short stable extensions on branches B and D. These extensions lose their stability at the Hopf bifurcation points H. Reducing the Nusselt number further, the bistability vanishes as the two Hopf points join.

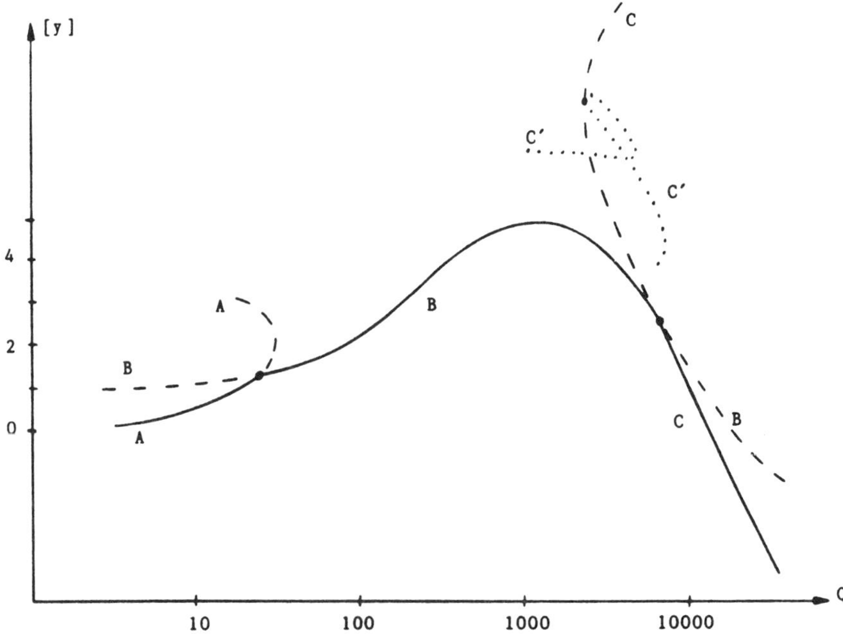

Figure 3.12

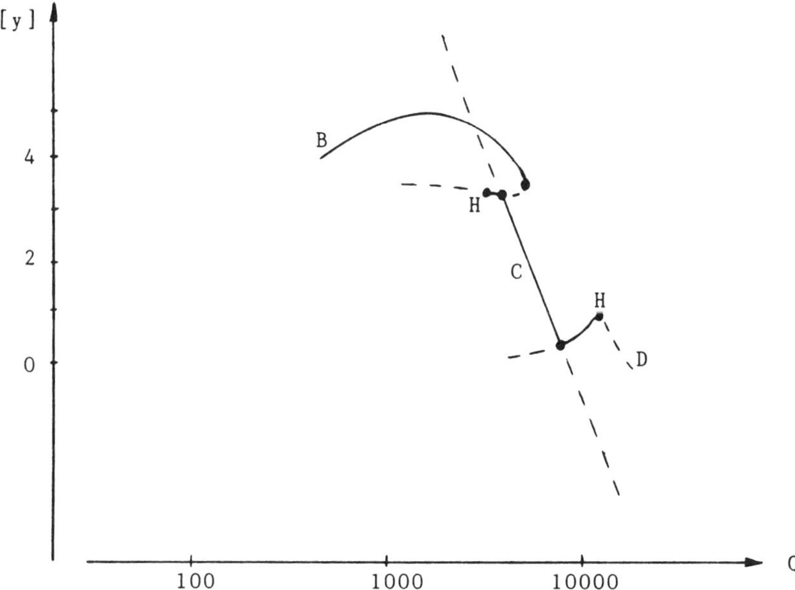

Figure 3.13

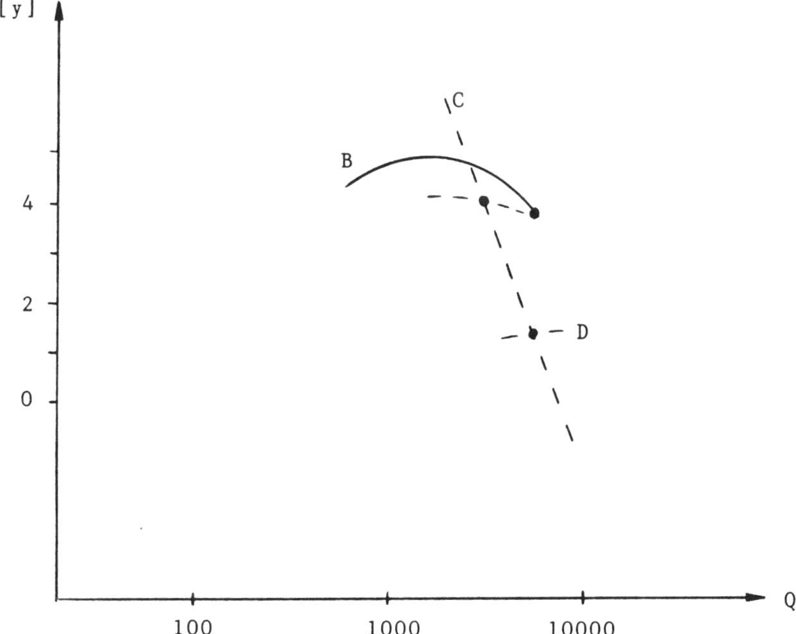

Figure 3.14

For $Nu = 0.43$ (Figure 3.14) the connections between branches B, C, and D are still present, but the stability has been lost. Reducing Nu even further, branch C becomes disconnected from B and D; the two bifurcation points of Figure 3.14 disappear. Finally, for $Nu \approx 0.3$, branch C vanishes.

The bifurcation structure in the Marangoni convection equations is rich, and the numerical procedures used to solve these equations are complex. Tracing branches requires careful attention. Small but critical details are too easily overlooked. We shall return to the example of this section later in Section 5.2 and in Section 5.4.7.

3.5 THE ART AND SCIENCE OF PARAMETER STUDY

In the following chapters we shall discuss numerical methods in greater depth. So far, we have mentioned the principal tools of continuation and branch switching. We learned about various practical problems that have a strong impact on the use of the numerical methods. In the present section, we summarize how one goes about doing a parameter study.

The scenarios described in Section 3.3 show that it is not always sufficient to rely on numerical methods only. To find all the branches, good luck is also needed. Rashness in interpreting computer output must be avoided. Using a plotter that traces smooth branching curves through points calculated by the continuation method may produce misleading results. Too often points belonging to different branches are connected. It is generally advisable first to plot only the points and to study them carefully, checking for any irregularities. An experienced researcher with knowledge and a good imagination may find an interpretation that is different from that of a machine. Although it is in principle possible to construct a "black box" that takes care of almost everything, this would have some drawbacks. First, a fully automated parameter study carries no proof of correct interpretation. Second, it is difficult to write algorithms that decide, for instance, whether "it is the same branch" or not. A machine can easily waste time in closed-loop branches or regions that are of no interest. To make a decision, a fully automated black box would require accuracy greater than that needed when the human mind is doing the interpreting. For the present, automatic procedures should be confined to calculating a short continuation or carrying out a branch switching. The investigator then interprets the results and gives orders for another automatic subroutine to be carried out. This is done in an interactive fashion, with the help of graphical display and output of representative numbers.

> After a pilot study is conducted and interpreted in an interactive fashion, subsequent runs with slightly modified parameters might be delegated to a black box. For this it is advisable to store the "history" of the initial semiautomated procedure to the extent possible.

Generally, parameter studies are expensive. Each additional parameter to be varied increases the cost by an order of magnitude. This situation calls for reduced accuracy requirements for the bulk of a continuation process. For instance, calculating the branches of the thermocapillary problem of the previous section, a variable grid was used—that is, a great part of the continuation was done with a few lines only (N small). Intermittently, results were confirmed by recalculating selected solutions with higher accuracy (N larger). In this way the overall cost of branch tracing can be reduced significantly.

High sensitivity to changes in parameters can occasionally create problems. Recall that in the Marangoni convection problem $\Theta_4 = 0$ is a kind of organizing center, because bifurcations occur as illustrated in Figure 3.12. These bifurcations disappear for $\Theta_4 \neq 0$. As was discussed in Section 2.9 (see Figure 2.39), small perturbations can lead to completely different branching diagrams; see Figure 3.15. Imagine that a continuation starts with small values of Q. For $\Theta_4 = 0.01$, the branch

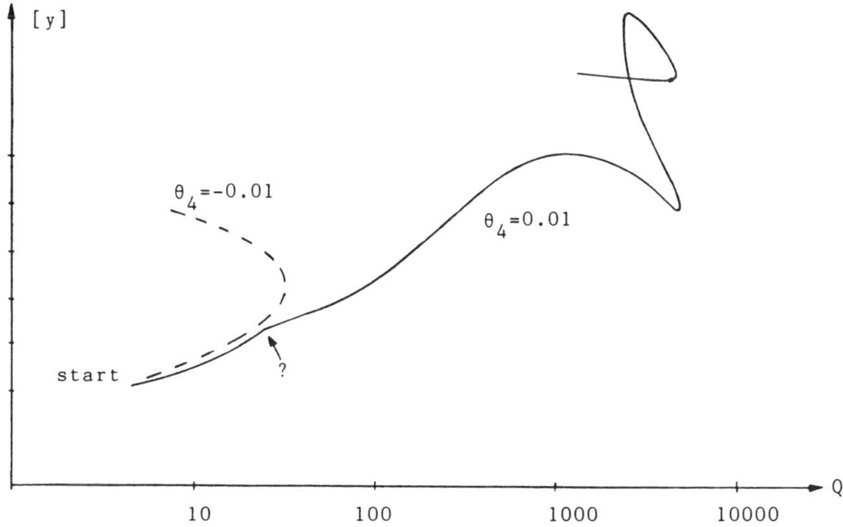

Figure 3.15

followed by the continuation process differs drastically from the branch followed for $\Theta_4 = -0.01$. This can be expected because of the role of $\Theta_4 = 0$. Before the role of $\Theta_4 = 0$ is known, a tiny maximum of the lower curve is the only hint suggesting a nearby bifurcation (see question mark in Figure 3.15).

The reader may now have the impression that a parameter study is an art, one that requires some creativity and imagination. This is partly justified, but the methods described in the following chapters show that a parameter study is also a science.

4 Principles of Continuation

In the preceding chapter the continuation procedure was used for the important task of *branch tracing* or *path following*. In this chapter we discuss in depth the principles underlying this procedure. The system of nonlinear "algebraic" equations

$$\mathbf{0} = \mathbf{f}(\mathbf{y},\lambda) \tag{4.1}$$

serves as a basis for the discussion. Here, $\mathbf{y}$ denotes an n-dimensional vector. No generality is lost by restricting attention to Eq. (4.1). Both ODEs and PDEs are "solved" by approximating them by such systems of nonlinear equations (see Appendix 5). Principles similar to those discussed here can be applied directly to the differential equations.

Papers introducing basic ideas of continuation were published in the 1960s by Haselgrove [133], Klopfenstein [189], and Deist and Sefor [76]. In the same period, continuation procedures were introduced in engineering and scientific applications, including civil engineering [108], flow research [290, 244, 323], chemical reactions [196, 197], solidification [59], and combustion [223]. Today continuation is in wide use. In what follows we confine ourselves to "classical" continuation methods that have proved successful in applications. An introduction to simplicial continuation methods is beyond the scope of this book; for an exposition of this class of methods, see [8].

Assume that at least one solution of Eq. (4.1) has been calculated. Let us denote this "first" solution by $(\mathbf{y}^1,\lambda_1)$. The continuation problem is to calculate further solutions,

$$(\mathbf{y}^2,\lambda_2),\ (\mathbf{y}^3,\lambda_3),\ \ldots,$$

until one reaches a target point, say at $\lambda = \lambda_b$. (The superscripts are not to be confused with exponents.)

The j-th continuation step starts from (an approximation of) a solution $(\mathbf{y}^j,\lambda_j)$ of Eq. (4.1) and attempts to calculate the solution

$(\mathbf{y}^{j+1},\lambda_{j+1})$ for the "next" λ, namely, λ_{j+1},

$$(\mathbf{y}^j,\lambda_j) \to (\mathbf{y}^{j+1},\lambda_{j+1}).$$

With *predictor-corrector* methods, the step $j \to j + 1$ is split into two steps:

$$\underset{\text{predictor step}}{(\mathbf{y}^j,\lambda_j) \to (\bar{\mathbf{y}}^{j+1},\bar{\lambda}_{j+1})} \underset{\text{corrector step}}{\to (\mathbf{y}^{j+1},\lambda_{j+1})}$$

(see Figure 4.1). In general, the predictor $(\bar{\mathbf{y}},\bar{\lambda})$ is not a solution of Eq. (4.1). The predictor merely provides an initial guess for corrector iterations that home in on a solution of Eq. (4.1). The major portion of the work is either on the predictor step (resulting in an approximation $(\bar{\mathbf{y}},\bar{\lambda})$ close to the branch) or on the corrector step (if a "cheaper" predictor has produced a guess far from the branch). The distance between two consecutive solutions $(\mathbf{y}^j,\lambda_j)$ and $(\mathbf{y}^{j+1},\lambda_{j+1})$ is called the *step length* or *step size*. In addition to Eq. (4.1), a relation that identifies the location of a solution on the branch is needed. As we shall discuss in Section 4.2, this identification is related to the kind of *parameterization* strategy chosen to trace the branch.

Continuation methods differ, among other things, in the following:

(a) predictor
(b) parameterization strategy
(c) corrector
(d) step length control.

The first three of these four items can be chosen independently of each other. The step length control must correspond to the predictor, the

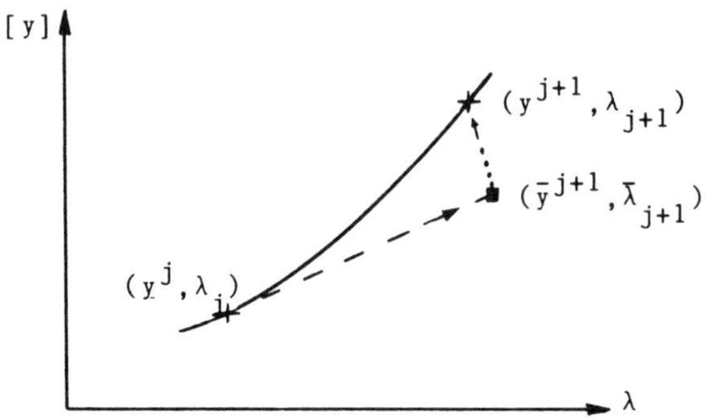

Figure 4.1

corrector, and the underlying parameterization. In the next three sections, the first three items will be explained in detail and brief remarks will be made on step controls.

4.1 PREDICTORS

Predictors can be divided into two classes:

(a) ODE-methods, which are based on $\mathbf{f}(\mathbf{y},\lambda)$ and its derivatives
(b) polynomial extrapolation, which uses only solutions $(\mathbf{y},\lambda)$ of Eq. (4.1).

4.1.1 ODE-Methods; Tangent Predictor

Taking the differential of both sides of Eq. (4.1), one obtains

$$\mathbf{0} = d\mathbf{f} = \mathbf{f_y} d\mathbf{y} + \mathbf{f}_\lambda d\lambda \tag{4.2}$$

and hence

$$\frac{d\mathbf{y}}{d\lambda} = -(\mathbf{f_y})^{-1}\mathbf{f}_\lambda.$$

Integrating this system starting from the initial value $(\mathbf{y}^1,\lambda_1)$, one obtains the branch on which $(\mathbf{y}^1,\lambda_1)$ lies. This procedure, proposed by Davidenko [73, 1, 100], fails at turning points because of the singularity of $\mathbf{f_y}$. In multiparameter problems this difficulty can be overcome by switching to an alternate parameter [13].

Another way of overcoming the failure of the method at turning points is to change the parameter to the arclength s. Both $\mathbf{y}$ and λ are considered to be functions of the arclength parameter s: $\mathbf{y} = \mathbf{y}(s)$, $\lambda = \lambda(s)$. From Eq. (4.2) one obtains

$$0 = \mathbf{f_y}\frac{d\mathbf{y}}{ds} + \mathbf{f}_\lambda \frac{d\lambda}{ds}. \tag{4.3a}$$

The arclength s satisfies the relation

$$(dy_1/ds)^2 + \cdots + (dy_n/ds)^2 + (d\lambda/ds)^2 = 1. \tag{4.3b}$$

Equations (4.3a) and (4.3b) form an implicit system of $n + 1$ differential equations for the $n + 1$ unknowns

$$dy_1/ds, \ldots, dy_n/ds, d\lambda/ds.$$

A specific method for solving Eq. (4.3) will be described later.

An important ODE predictor is the tangent predictor. With the notation z_i = i-th component of $\mathbf{z}$ and

$$z_i = dy_i (1 \leq i \leq n), \; z_{n+1} = d\lambda,$$

one derives formally from Eq. (4.2) the equation

$$(\mathbf{f_y} \mid \mathbf{f}_\lambda)\mathbf{z} = \mathbf{0}$$

for a tangent $\mathbf{z}$. A normalization of $\mathbf{z}$ must be imposed to give this equation a unique solution. One can use, for example,

$$\mathbf{e}_k^{\text{tr}}\mathbf{z} = z_k = 1,$$

where $\mathbf{e}_k$ is the $(n + 1)$-dimensional vector with all elements equal to zero except the k-th, which equals unity (recall that tr denotes transpose). The tangent $\mathbf{z}$ is the solution of the linear system

$$\begin{pmatrix} \mathbf{f_y} \mid \mathbf{f}_\lambda \\ \hline \mathbf{e}_k \end{pmatrix} \mathbf{z} = \mathbf{e}_{n+1}. \tag{4.4}$$

Provided the full-rank condition $\text{rank}(\mathbf{f_y} \mid \mathbf{f}_\lambda) = n$ holds along the whole branch, Eq. (4.4) has a unique solution at any point on the branch (k may have to be changed). In particular, the tangent can be calculated at turning points.

Several continuation methods—for example, those in [178, 235, 272, 274]—are based on tangent predictors. In these methods the predictor point (initial approximation) for $(\mathbf{y}^{j+1}, \lambda_{j+1})$ is

$$(\bar{\mathbf{y}}^{j+1}, \bar{\lambda}_{j+1}) = (\mathbf{y}^j, \lambda_j) + \sigma_j \mathbf{z} \tag{4.5}$$

(see Figure 4.1). Here σ_j is an appropriate step length. This tangent predictor, Eq. (4.5), can be considered a step of the Euler method for solving a differential equation that describes the branch.

We now return to differential equation (4.3), which implicitly defines the $n + 1$ derivatives of $y_1, \ldots, y_n, y_{n+1} = \lambda$ with respect to arclength s. A solution of Eq. (4.3) can be obtained by the procedure proposed by Kubicek [195]. Note that system (4.3a) is linear in its $n + 1$ unknowns, while the arclength equation (4.3b) is nonlinear. Temporarily, we consider one of the components to be prescribed:

$$dy_m/ds \neq 0 \text{ for some fixed } m, \quad 1 \leq m \leq n + 1.$$

Then Eq. (4.3a) consists of n linear equations in n unknowns, which can be readily solved. The solutions of these equations depend linearly on dy_m/ds:

$$dy_i/ds = c_i \, dy_m/ds, \qquad i = 1, 2, \ldots, m - 1, m + 1, \ldots, n + 1.$$

(The coefficients c_i vary with $(\mathbf{y}, \lambda)$.) Substituting these expressions into

Eq. (4.3b) yields dy_m/ds:

$$(dy_m/ds)^2 = (1 + c_1^2 + \cdots + c_{m-1}^2 + c_{m+1}^2 + \cdots + c_{n+1}^2)^{-1}.$$

This constitutes for each $(\mathbf{y},\lambda)$ explicit expressions for the derivatives with respect to s. The numerical integration of the above equations is carried out by a multistep method, which yields better predictors than a one-step method, such as the tangent predictor discussed above.

4.1.2 Polynomial Extrapolation; Secant Predictor

A polynomial in λ of degree ν that passes through the $\nu + 1$ points

$$(\mathbf{y}^j,\lambda_j),\ (\mathbf{y}^{j-1},\lambda_{j-1}),\ \ldots,\ (\mathbf{y}^{j-\nu},\lambda_{j-\nu})$$

can be constructed. One assumes that this polynomial provides an approximation to the branch in the region to be calculated next ($\lambda \approx \lambda_{j+1}$). The predictor consists in evaluating the polynomial at $\lambda = \lambda_{j+1}$.

The trivial predictor is the zeroth-order polynomial ($\nu = 0$) when the predictor point is the previous solution:

$$(\bar{\mathbf{y}}^{j+1},\bar{\lambda}_{j+1}) = (\mathbf{y}^j,\lambda_j). \tag{4.6a}$$

The slightly modified version

$$(\bar{\mathbf{y}}^{j+1},\bar{\lambda}_{j+1}) = (\mathbf{y}^j,\lambda_{j+1}) \tag{4.6b}$$

is also used.

A polynomial of the first order ($\nu = 1$), the secant, can also be used:

$$(\bar{\mathbf{y}}^{j+1},\bar{\lambda}_{j+1}) = (\mathbf{y}^j,\lambda_j) + \sigma_j(\mathbf{y}^j - \mathbf{y}^{j-1},\lambda_j - \lambda_{j-1}) \tag{4.7}$$

(see Figure 4.2). The quantity σ_j is an appropriate step length. Higher-

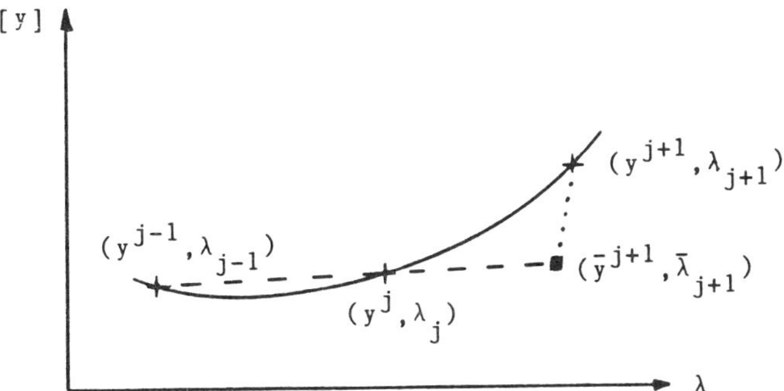

Figure 4.2

order formulas ($\nu \geq 2$), some of which use values of previous tangents, are given in [133]. Frequently, however, lower-order predictors (tangent, secant) are in the long run less costly and are therefore preferred.

4.2 PARAMETERIZATIONS

A branch is a connected curve consisting of points in $(\mathbf{y},\lambda)$ space that are solutions of $\mathbf{f}(\mathbf{y},\lambda) = \mathbf{0}$. This curve must be "parameterized." A parameterization is a kind of measure along the branch, a mathematical way of identifying each solution on the branch. Terms like "next" or "previous" solution are quantified by introducing a parameterization.

Not every (part of a) branch allows any arbitrary parameterization. To see this, imagine a particle moving along the branch. Imagine also a rope attached to the particle. Different parameterizations correspond to the different directions in which the rope can be pulled. Clearly, not every direction works. Pulling normal to the branch is the direction that does not work.

The most obvious parameter is the control variable λ. While this parameter has the advantage of having physical significance, it encounters difficulties at turning points, where the pulling direction is normal to the branch.

With little extra effort, however, a parameterization by λ can be sustained even at turning points. This can be accomplished by using the predictor $\bar{\mathbf{h}}$ of [307, 312] (cf. Eq. (5.21)). This predictor is perpendicular to the λ-axis and provides a way of jumping over the turning point and calculating a solution for the same value of λ on the opposite side of the branch (Figure 4.3). In the remainder of this section, parameterizations that do not need special provisions at turning points will be discussed.

4.2.1 Parameterization by Adding an Equation

If another variable, say γ, is chosen as the parameter, solutions of $\mathbf{f}(\mathbf{y},\lambda) = \mathbf{0}$ depend on γ:

$$\mathbf{y} = \mathbf{y}(\gamma),\ \lambda = \lambda(\gamma).$$

For a particular value of γ, the system $\mathbf{f}(\mathbf{y},\lambda) = \mathbf{0}$ consists of n equations for the $n + 1$ unknowns $(\mathbf{y},\lambda)$. If the parameterization is established by one additional scalar equation,

$$p(\mathbf{y},\lambda,\gamma) = 0,$$

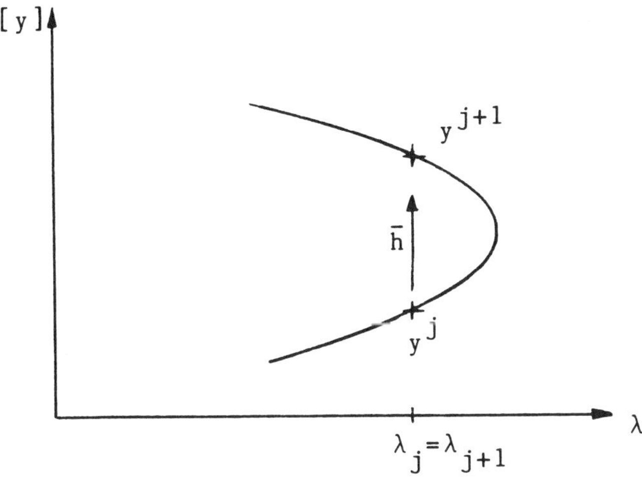

Figure 4.3

one can formulate an extended system

$$\mathbf{F}(\mathbf{Y},\gamma) = \begin{pmatrix} \mathbf{f}(\mathbf{y},\lambda) \\ p(\mathbf{y},\lambda,\gamma) \end{pmatrix} = \mathbf{0}, \tag{4.8}$$

which consists of $n + 1$ scalar equations for the $n + 1$ unknowns $\mathbf{Y} = (\mathbf{y},\lambda)$. The general setting Eq. (4.8) includes all types of parameterization. For example, the option of taking λ as continuation parameter means choosing $\gamma = \lambda$ and

$$p(\mathbf{y},\lambda) = \lambda - \lambda_{j+1}.$$

Other examples will follow soon. Solving Eq. (4.8) for specified values of γ yields information on the dependence $\mathbf{y}(\gamma),\lambda(\gamma)$. Attaching a normalizing or parameterizing equation is a standard device in numerical analysis; in the context of branch tracing, refer to [276, 178, 272, 308].

In Eq. (4.8) the parameterizing equation $p(\mathbf{y},\lambda,\gamma) = 0$ is attached to the given system (4.1). This allows one to apply a SOLVER such as a Newton method to Eq. (4.8); in this way the desired parameterization is imposed on Eq. (4.1) automatically. Alternatively, one can apply a Newton method to Eq. (4.1) and impose an appropriate side condition on the iteration [133, 235]. This latter approach will be described in Section 4.3. The formulation (4.8) has the advantage of not being designed to be used with a special version of a Newton method. Any modification or any other iteration method for solving equations can be applied to solve Eq. (4.8).

4.2.2 Arclength and Pseudo Arclength

As outlined earlier, arclength can be used for the purpose of parameterizing a branch—that is, $\gamma = s$. This means pulling the imaginary particle in the direction of the branch; turning points do not pose problems. A corresponding equation $p(\mathbf{y},\lambda,s) = 0$ is obtained by multiplying the arclength Eq. (4.3b) by $(ds)^2$. The resulting parameterizing equation is

$$0 = p_1(\mathbf{y},\lambda,s) = \sum_{i=1}^{n} (y_i - y_i(s_j))^2 + (\lambda - \lambda(s_j))^2 - (s - s_j)^2. \tag{4.9}$$

If $(\mathbf{y}^j,\lambda_j) = (\mathbf{y}(s_j),\lambda(s_j))$ is the solution previously calculated during continuation, Eq. (4.9) together with Eq. (4.1) fix the solution $(\mathbf{y}(s),\lambda(s))$ at arclength distance $\Delta s = s - s_j$. Arclength parameterization has been used repeatedly; in particular, it was advocated by [189, 195, 178, 58, 74]. In [178] "pseudo arclength" was proposed —that is, for $0 < \zeta < 1$ either

$$0 = p_2(\mathbf{y},\lambda,s) = \zeta \sum_{i=1}^{n} (y_i - y_i(s_j))^2 + (1 - \zeta)(\lambda - \lambda(s_j))^2 - (s - s_j)^2$$

or

$$0 = p_3(\mathbf{y},\lambda,s) = \zeta \sum_{i=1}^{n} (y_i - y_i(s_j))\, dy_i(s_j)/ds$$
$$+ (1 - \zeta)(\lambda - \lambda(s_j))\, d\lambda(s_j)/ds - (s - s_j).$$

The tune factor ζ allows us to place different emphasis on $\mathbf{y}$ or λ. The latter parameterization p_3 is motivated by the Taylor expansion,

$$\mathbf{y}(s) - \mathbf{y}(s_j) = (s - s_j)\, d\mathbf{y}(s_j)/ds + O(|s - s_j|^2).$$

The derivatives are evaluated at the previously calculated solution. The parameterization p_3 has the advantage of being linear in the increments $\Delta\mathbf{y}$, $\Delta\lambda$, and Δs. Among the problems solved by arclength continuation are driven cavity flows between rotating disks [181] and equations describing premixed laminar flames [140].

4.2.3 Local Parameterization

All the components y_i $(i = 1, \ldots, n)$ can be admitted as parameters, including $y_{n+1} = \lambda$. This leads to the parameterizing equation

$$p(\mathbf{y},\eta) = y_k - \eta,$$

with an index k, $1 \leq k \leq n + 1$, and a suitable value of η. The index k and the parameter $\eta = y_k$ are locally determined at each continuation step $(\mathbf{y}^j,\lambda_j)$ in order to keep the continuation flexible. This kind of parameterization has been called *local parameterization* [274]. For $k \leq n$, local parameterization means pulling the mentioned particle in a direction perpendicular to the λ-axis. Hence, an index k for local parameterization exists also at turning points. With $\mathbf{Y} = (\mathbf{y},\lambda)$, Eq. (4.8) is written as

$$\mathbf{F}(\mathbf{Y},\eta,k) = \begin{pmatrix} \mathbf{f}(\mathbf{Y}) \\ y_k - \eta \end{pmatrix} = \mathbf{0}. \qquad (4.10)$$

Such systems have been used in the context of branch tracing—for example, in [2, 77, 235, 272, 308]. At least two codes that implement a local parameterization strategy have been developed [274, 315, 316].

It is easy to find a suitable index k and parameter value η. A continuation algorithm based on tangent predictors Eq. (4.4) and Eq. (4.5) determines k such that

$$| z_k | = \max\{| z_1 |, \ldots, | z_n |, | z_{n+1} |\}. \qquad (4.11)$$

This choice picks the component of the tangent $\mathbf{z}$ that is maximal. In the algorithm of [315], which is based on the secant predictor, Eq. (4.7), index k is chosen such that the relative changes $\Delta_r y_k$ are maximal in y_k,

$$\Delta_r y_k = \max\{\Delta_r y_1, \ldots, \Delta_r y_n, \Delta_r y_{n+1}\}, \qquad \text{with } \Delta_r y_i = | y_i^j - y_i^{j-1} | / | y_i^j |. \qquad (4.12)$$

Criteria (4.11) and (4.12) establish indices k that can be expected to work effectively. These choices, however, do not guarantee that the corresponding continuation is especially fast; rather, the emphasis is on reliability. It may happen that a choice of k from Eq. (4.11) or Eq. (4.12) fails. A failure is likely when a turning point "arrives" suddenly. Such a scenario is depicted in Figure 4.4. Assume that a continuation step with local parameterization has been carried out, producing the solution $(\mathbf{y}^j,\lambda_j)$ close to the turning point. (Figure 4.4 shows a turning point with respect to y_k.) Assume further that the step control does not notice the change and still maintains y_k as parameter. In Figure 4.4, in both case (a) and case (b) the continuation fails, because it seeks for a solution in the same direction characterized by $y_k^{j+1} > y_k^j$. In situation (a) the step size must be reduced too drastically in order to obtain a solution. A continuation procedure trying to avoid unnecessary cost interprets a reduction in step size below a certain limit as failure. In situation (b) of Figure 4.4, a successful calculation of $\mathbf{y}^{j+1}$ means a backward step of the continuation. As a result, the orientation of the branch tracing is reversed.

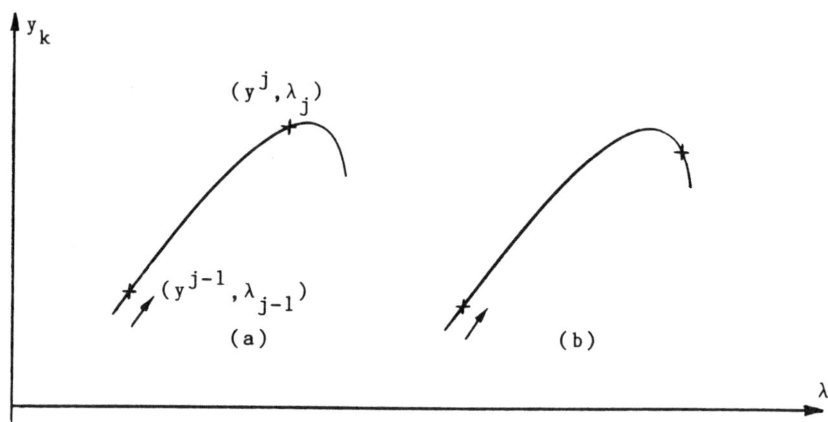

Figure 4.4

In order to overcome a failure of a chosen k, it makes sense to have an alternative k_1 available. The choices Eq. (4.11) and Eq. (4.12) suggest choosing for k_1 the index that corresponds to the second largest value of the right-hand sides in Eqs. (4.11) and (4.12). In case Eq. (4.10) with k from Eqs. (4.11) and (4.12) does not lead to a solution, Eq. (4.10) can be tried again with k_1.

After an index k is fixed, a suitable parameter value η must be determined. A value of η depends on the index k, on the location of the current solution on the branch (that is, on j), and on the desired step size σ,

$$\eta = \eta(k,j,\sigma).$$

A simple strategy relates the current step size $\eta - y_k^j$ to the previous step size $y_k^j - y_k^{j-1}$ by

$$\eta = y_k^j + (y_k^j - y_k^{j-1})\xi. \tag{4.13}$$

There is an example for the adjustable factor ξ in Section 4.4. Other strategies have been proposed—for instance, by [272, 274].

The local parameterization concept can be generalized to a continuation that involves qualitative aspects. For example, a continuation by asymmetry is furnished by

$$p(\mathbf{y},\lambda,\gamma) = y_k - y_m - \gamma \qquad (\gamma \neq 0).$$

For this parameterization we refer to Section 5.4.6.

4.3 CORRECTORS

In this section, we briefly discuss methods that are typically used to solve Eq. (4.1) or the extended system Eqs. (4.8) or (4.10). Because the latter systems are ready to be solved by the standard software SOLVER, we characterize this type of corrector only briefly. In the remainder, we shall concentrate on the previously mentioned alternative that establishes a parameterization by imposing a side condition on the iteration method.

First, we characterize correctors that result from solving Eq. (4.8); see Figure 4.5. The final limit $(\mathbf{y}^{j+1},\lambda_{j+1})$ of the corrector iteration is determined by the parameterizing equation $p = 0$ in Eq. (4.8). The *path* of the iteration is not prescribed (dotted line in Figure 4.5). This path depends on the particular parameterization and on the kind of method implemented in SOLVER; the user does not care about the path. In principle, the same limit $(\mathbf{y}^{j+1},\lambda_{j+1})$ can be produced by any parameterization if a proper step length is found. In practice, however, nothing is known about relations among y_i, λ, and arclength s. Consequently, different parameterizations usually lead to a different selection of solutions on the branch. Figure 4.6 illustrates correctors of the three parameterizations

(a) by branching parameter λ (step length $\Delta\lambda$)

(b) local parameterization (step length Δy_k)

(c) arclength parameterization (step length Δs).

Imagine that each dot in Figure 4.5 and Figure 4.6 stands for one iterate

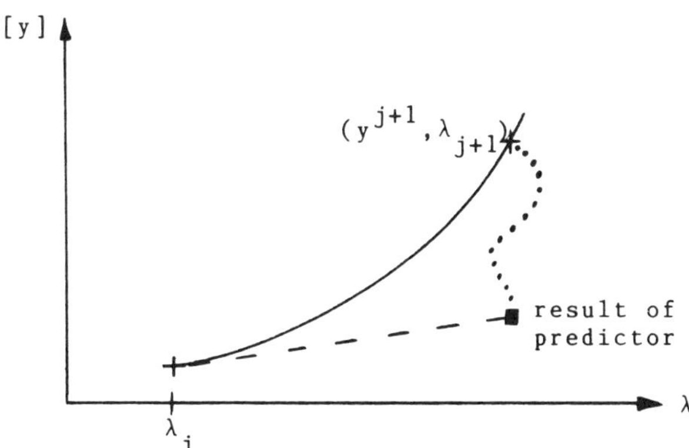

Figure 4.5

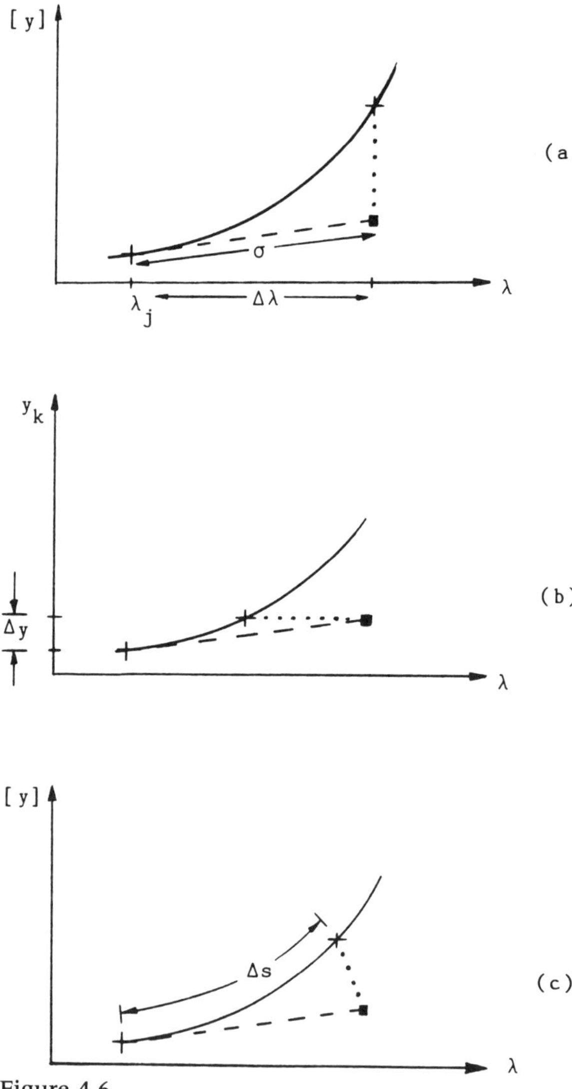

Figure 4.6

of the corrector iteration. The upper and middle figures in Figure 4.6 reflect a choice of η in Eq. (4.10) (and hence of $\Delta\lambda$ or $\Delta \mathbf{y}$) that matches the predictor $\bar{\mathbf{y}}$. In other words, η and the step lengths of the chosen parameter are derived from the predictor step length σ. In this case the corrector iterations in Figure 4.6(a) and (b) can be expected to move more or less parallel to the axes. A corresponding identification of η

with $\bar{y}_k$ is comfortable but not required. In general, relations between the step lengths σ and Δy_k are more involved.

Now we turn to the parameterization of the corrector iteration. To this end, consider one step of the Newton iteration, formulated for the $(n + 1)$-dimensional vector $(\mathbf{y},\lambda)$,

$$\begin{aligned} \mathbf{f}_\mathbf{y}\Delta\mathbf{y} + \mathbf{f}_\lambda\Delta\lambda &= -\mathbf{f}(\mathbf{y}^{(\nu)},\lambda^{(\nu)}), \\ \mathbf{y}^{(\nu+1)} = \mathbf{y}^{(\nu)} + \Delta\mathbf{y}, &\quad \lambda^{(\nu+1)} = \lambda^{(\nu)} + \Delta\lambda. \end{aligned} \tag{4.14}$$

In Eq. (4.14) the first-order partial derivatives are evaluated at $(\mathbf{y}^{(\nu)},\lambda^{(\nu)})$. The system of linear equations in Eq. (4.14) consists of n equations in $n + 1$ unknowns $\Delta\mathbf{y}$, $\Delta\lambda$. The system can be completed by attaching one further equation that parameterizes the course of the corrector iteration. Following [133, 235], one requests that the iteration be perpendicular to the tangent $\mathbf{z}$ from Eq. (4.4). Using the $(n + 1)$ scalar product, this is written as

$$(\Delta\mathbf{y}^{\mathrm{tr}},\Delta\lambda)\mathbf{z} = 0. \tag{4.15}$$

Starting from the predictor,

$$(\mathbf{y}^{(0)},\lambda^{(0)}) = (\bar{\mathbf{y}}^{j+1},\bar{\lambda}_{j+1}),$$

Eq. (4.14) and Eq. (4.15) define a sequence of corrector iterations

$$(\mathbf{y}^{(\nu)},\lambda^{(\nu)}), \qquad \nu = 1, 2, \ldots$$

that is to converge to a solution $(\mathbf{y}^{j+1},\lambda_{j+1})$ (see Figure 4.7). This sequence eventually hits the branch in the solution that is the intersection

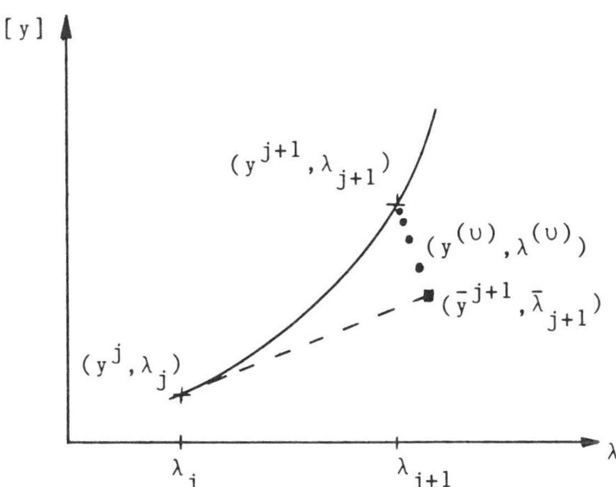

Figure 4.7

of the curve defined by $\mathbf{f}(\mathbf{y},\lambda) = \mathbf{0}$ and the *hyperplane* defined by Eq. (4.15). Hence the parameterization of the branch is established by Eq. (4.15). In the parameterizing equation (4.15), the tangent $\mathbf{z}$ can be replaced by another suitable vector.

Now that the parameterization of Eq. (4.15) is characterized by a hyperplane in the $(n + 1)$-dimensional $(\mathbf{y},\lambda)$-space, we realize that also the local parameterization concept can be seen as defined by hyperplanes (see Figure 4.6(a) and (b)). A straightforward analysis (Exercise 4.1) reveals that conceptual differences among various parameterizations are not significant. In a practical implementation, however, differences become substantial. Note that Eqs. (4.8) and (4.10) can be solved without requiring specific software.

Exercise 4.1.

The $(n + 1)^2$ Jacobian $\mathbf{J}$ associated with the Newton method that corresponds to Eq. (4.8) or Eqs. (4.14) and (4.15) is composed of matrix $\mathbf{f_y}$, vector $\mathbf{f}_\lambda$, a vector $\mathbf{c}$, and a scalar ξ,

$$\mathbf{J} = \begin{pmatrix} \mathbf{f_y} & \mathbf{f}_\lambda \\ \mathbf{c}^{tr} & \xi \end{pmatrix} .$$

What are the vectors $\mathbf{c}$ and the scalars ξ in the cases of local parameterization, Eq. (4.10), pseudo arclength parameterization p_3, and the parameterization Eq. (4.15)?

Exercise 4.2.

Programming a continuation algorithm based on the secant predictor Eq. (4.7) is easy when three arrays are used. Design a continuation algorithm requiring only two arrays.

4.4 STEP CONTROLS

In simple problems the principles introduced above may work effectively without a step length control. That is, constant step lengths are taken throughout, say with $\Delta s = 0.1$ in case of parameterization by arclength. If the step size is small enough, such a step strategy may be successful for a wide range of problems. But such results are often obtained in an inefficient manner, involving too many steps along "flat" branches. The step length should be adapted to the actual convergence behavior. Ultimately, it is the flexibility of a step control that decides whether a continuation algorithm works well. Because step controls must meet the features of the underlying combination predictor/parameterization/corrector, step-length algorithms are specifically designed. In the following, some general aspects are summarized briefly.

Step controls can be based on estimates of the convergence quality of the corrector iteration. Corresponding measures of convergence are not easy to develop. In [77] it was shown that an extrapolation of convergence radii of previous continuation steps into future is hardly possible. In [77, 274] the branch was locally replaced by a quadratic polynomial. This enables a derivation of convergence measures involving the curvature of the branch.

As pointed out in [306, 315], step-length algorithms can be based on statistical arguments. Such a strategy is based on the observation that the total amount of work involved in a continuation depends on the average step length in a somewhat convex manner: the continuation is expensive both for very short steps (too many steps) and for very large steps (slow or no convergence of correctors). The costs of a continuation are moderate for a certain medium step length, which is related to an optimal number N_{opt} of iterations of a corrector. This number depends on the type of corrector and on the prescribed error tolerance ϵ. For example, with quasi-Newton correctors and $\epsilon = 10^{-4}$ the optimal number has been shown to be about $N_{\text{opt}} \approx 6$. The aim is to adjust the step size so that each continuation step needs about N_{opt} corrector iterations. Let N_j denote the number of iterations needed to approximate the previous continuation step. Then a simple strategy is to reduce the step size in case $N_j > N_{\text{opt}}$ and to increase the step size in case $N_j < N_{\text{opt}}$. This can be done, for example, by updating the previous step size by multiplication with the factor

$$\xi = N_{\text{opt}}/N_j;$$

compare Eq. (4.13). In case of local parameterization, this leads to the value

$$\eta = y_k^j + (y_k^j - y_k^{j-1})N_{\text{opt}}/N_j \tag{4.16}$$

to be prescribed in Eq. (4.10) for the current parameter y_k. This step control is easy to implement (Exercise 4.4).

Exercise 4.3.

In designing a continuation algorithm, one does not allow the step length to become arbitrarily large. Hence, the absolute value of a maximal step size should be prescribed as input data. The following questions may stimulate one to think about a practical realization.

(a) Does a limitation in the step length affect the choice of k in the local parameterization criteria Eq. (4.11) or Eq. (4.12)?
(b) Design a step length control that relates the bound on the step length to the λ-component only.

Exercise 4.4.
Design a continuation algorithm implementing elements of your choice. The following should be implemented: a predictor, a parameterization strategy, a corrector ("CALL SOLVER" suffices), and a step length control. (Hint: A simple but successful algorithm is obtained by combining Eqs. (4.7), (4.10), and (4.12) for k and k_1, and Eq. (4.16).)

4.5 CONTINUATION SUBJECT TO CONSTRAINTS

So far we have discussed an idealized continuation carried out in an environment where enough computing time is available and no interference is expected. In many applications, however, a step control is affected by certain side conditions or restrictions. Let us illustrate this widespread situation by a typical example.

Assume that one is investigating a problem of periodic solutions and is concentrating on the stability. A main concern is to find a critical value of λ at which the stability is lost. One usually wants to study the situation close to a loss of stability (bifurcation) in more detail. This calls for substantially smaller step sizes. Therefore a step control needs to incorporate a "constraint."

The practical implications of this situation are significant. We may like to trace a branch close to the point of the loss of stability by prescribing a small step length in the physical parameter λ, not affected by any step control. The desire to choose equidistant values of λ temporarily renders superfluous any step-length algorithm and calls for a semiautomated and interactive way of tracing a branch.

Compromises between fully automating the continuation and taking into account the physical significance of λ are possible. For example, the variables y_i ($i = 1, \ldots, n$) can be treated differently from λ. This can be accomplished by making use of a weighted local parameterization [315]. The choice of the index k (see Eqs. (4.11) and (4.12)) is subjected to a suitable weighting factor strongly preferring λ. This strategy still enables us to pass turning points. Such a weighting, however, may slow down the turning around noses, because it tends to stick too long to a parameterization by λ. Allowing more algorithmic complexity, step-length algorithms can be modified so that the values of λ chosen are as even as possible.

There is another limitation that affects automatic continuation procedures too often, a limitation that is not scientific but annoying. Many computing centers restrict the available computing time to time slices. This severely affects a continuation, because it is not known beforehand how many steps can be completed before the next break-off. It is not easy to design or modify a continuation strategy when only a limited amount of computing time is available. Sophisticated continuation algorithms may suffer under severe restrictions.

4.6 MORE PRACTICAL ASPECTS

A common situation in running continuations is a halt followed by a restart. The solution calculated last (say, $(\mathbf{y}^j,\lambda_j)$) must be kept available for a subsequent restart of the continuation. In case the continuation is based on a tangent predictor, Eqs. (4.4) and (4.5), a guess $(\bar{\mathbf{y}}^{j+1},\bar{\lambda}_{j+1})$ for the next solution is readily obtained, provided the tangent $\mathbf{z}$ and information on the last predictor length σ have been saved. For a reasonable restart, some information on the history of the completed part of the continuation must be available. Apart from $(\mathbf{y}^j,\lambda_j)$, the data of σ_j, the index k in case of local parameterization and the last actual step length must be saved. Then the new continuation can benefit from what was calculated before. Whether the tangent $\mathbf{z}$ should be saved or recalculated upon restart will depend on the amount of available storage.

We discuss the matter of storage for the secant predictor Eq. (4.7), which is based on two previous solutions. Sufficient data for generating a new predictor is available when two previous solutions are stored, but this strategy may encounter difficulties. Typically, a parameter study must be split up into sequences of various continuation runs. Storing all terminating solutions can easily produce a flood of data. Storing two solutions of each termination leads to double the amount of data, which might exceed a particular allowance. One alternative is to restart a continuation with the trivial predictor Eq. (4.6). This choice usually requires a starting step length that is drastically smaller than the terminating step size that may have taken advantage of secant predictors. It is worth starting as follows: In a pre-iteration, a zeroth step is calculated with a small step length of, say, $\Delta\lambda = \lambda 10^{-3}$. The resulting auxiliary solution only serves to generate a secant, now enabling us to proceed with the terminating step length of the previous part of a continuation run.

The reader might have the impression that a nontrivial predictor (such as a tangent or a secant) always leads to better results than just starting from the previous solution (trivial predictor, Eq. (4.6)). In fact, nontrivial predictors generally result in faster convergence of the subsequent corrector. But this convergence is not always toward the desired solution. As an example, we reconsider the Duffing equation, Eq. (2.9). A continuation along the upper branch with decreasing values of the exciting frequency ω produced results as depicted in Figure 4.8. In this example, a secant predictor provided a guess close to the emanating branch. As a result, the corrector converged to this branch. This example shows that a predictor such as Eq. (4.5) or Eq. (4.7) may lead to an undesired branch switching. The difficulty did not arise with a weighting of the branching parameter λ [315].

The extrapolation provided by tangent and secant predictors speeds up the continuation because it makes larger steps possible. But larger steps reduce the sensitivity. In branching problems it is not uncommon to have two different continuation runs tracing the same part of the branch reveal a "different" branching behavior. For example, in Figure 4.7 a run with extrapolation based on a secant or tangent predictor obtained the indicated results, whereas the trivial predictor Eq.

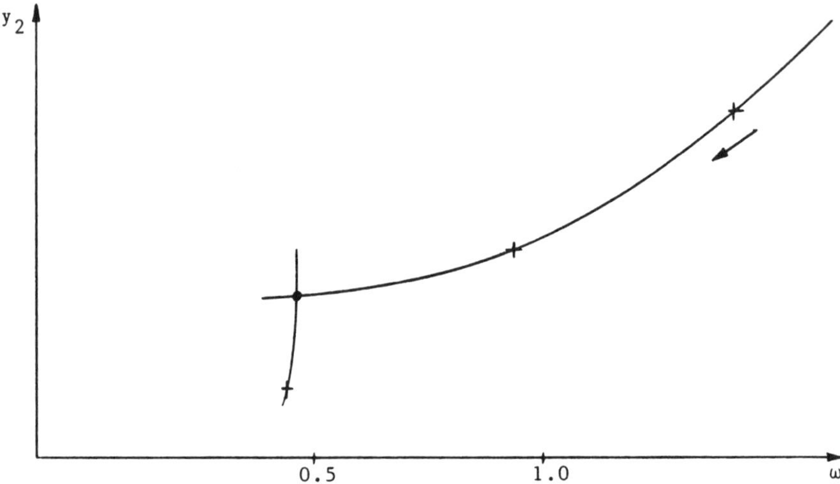

Figure 4.8

(4.6) probably would have traced the left branch with the turning point. Hence, in branching problems no predictor is generally "better" than others. Instead, in many models it might be recommended that continuations be run with another predictor strategy. We conclude that a continuation algorithm should offer the option of switching between extrapolation yes or no—that is, between an optimistic view and a more pessimistic view.

The continuation approach does not guarantee that all possible solutions can be located in a specific example. In particular, chances to detect isolated branches are small. For some problems it might be worthwhile to solve the governing equations for a random choice of parameters and to check whether an obtained solution is part of the already constructed branching diagram. In case a "new" solution is found, the branch tracing algorithm will follow the new branch, which might turn out to be disconnected (isolated) from the already calculated branches. To some extent, a parameter study of a difficult problem is a venturous exploration.

At the present stage, no particular continuation method can be recommended exclusively. The preceding sections pointed out that different alternatives for combining the basic elements (predictor, parameterization, corrector, step control) lead to a vast number of possible continuation methods. This explains why, so far, a numerical comparison of some of the major branch tracing methods is still missing. Certainly all the continuation methods have their merits. For an engineer or scientist working on a particular problem, it may be encouraging that relatively simple continuation principles work satisfactorily for complicated problems.

5 Calculation of the Branching Behavior of Nonlinear Equations

Equipped now with some knowledge about continuation, we assume that we are able to trace branches. We take for granted that the entire branch can be traced, provided one solution on that branch can be found. In this chapter we address problems of locating branch points and branch switching. Essential ideas and methods needed for practical bifurcation and stability analysis are presented.

The basic principles discussed here apply both to ODE boundary-value problems and to systems of algebraic equations. Although some of the ideas were first introduced for the more complicated case of boundary-value problems, we begin with the simpler situation of systems of "algebraic" equations:

$$\mathbf{f}(\mathbf{y},\lambda) = \mathbf{0}. \tag{5.1}$$

First we concentrate on varying one branching parameter only. The results can be used directly in analyzing multiparameter models. In Section 5.7, models with two branching parameters are treated. Methods for boundary-value problems are deferred to Chapter 6.

5.1 CALCULATING STABILITY

In this section we interpret Eq. (5.1) as an equation that defines the equilibria of the ODE system $\dot{\mathbf{y}} = \mathbf{f}(\mathbf{y},\lambda)$. Consequently, gain and loss of stability are crucial aspects of branching. The principles of local stability were introduced in Chapter 1. Recall that the n^2 Jacobian matrix $\mathbf{J} = \mathbf{f_y}$ is evaluated at the solution $(\mathbf{y},\lambda)$, the stability of which is to be investigated. The n eigenvalues

$$\mu_j(\lambda) = \alpha_j(\lambda) + i\beta_j(\lambda) \qquad (j = 1, \ldots, n) \tag{5.2}$$

of the Jacobian vary with λ. Linearized or local stability is ensured, provided all real parts α_j are negative. The basic principles are straightforward. Carrying them out by calculating the Jacobian $\mathbf{J}$ and the eigenvalues μ is not as simple.

Let us discuss first how the eigenvalues are affected by the accuracy in evaluating the Jacobian $\mathbf{J}$. This problem splits into two subproblems. The first is *how* the first-order partial derivatives are evaluated; the second is *where* they are evaluated—that is, for which argument(s) $(\mathbf{y},\lambda)$. The question of how to evaluate $\mathbf{J}$ was discussed in Section 1.4. Recall that $\mathbf{J}$ can be approximated by numerical differentiation, updated by rank-one approximations or evaluated analytically. The latter produces the most accurate results but is usually limited to simple examples. Often the solution to Eq. (5.1) is calculated by a quasi-Newton method. Because an approximate Jacobian is calculated on each step of a quasi-Newton method, we have an approximation $\bar{\mathbf{J}}$ available on the last step. The quality of this Jacobian is strongly affected by the history of the iteration. The initial guess and the rank-one update strategy both have significant influence on $\bar{\mathbf{J}}$. For instance, inaccuracies in $\bar{\mathbf{J}}$ are caused by the following factors:

SOLVER needs only one iteration. $\bar{\mathbf{J}}$ is then evaluated at the initial guess, which is usually not close to a solution of $\mathbf{f}(\mathbf{y},\lambda) = \mathbf{0}$.

The continuation step length is "large." Rank-one updates to $\bar{\mathbf{J}}$ converge more slowly to the true Jacobian than the iterates $(\mathbf{y}^{(\nu)},\lambda^{(\nu)})$ converge to the solution $(\mathbf{y},\lambda)$. In general, the quality of $\bar{\mathbf{J}}$ deteriorates with increasing step length.

Predictor Eq. (4.6b) is used instead of Eq. (4.6a). Hence the initial approximation for the Newton method "leaves" the branch. The resulting Jacobians $\bar{\mathbf{J}}$ are frequently worse than when Eq. (4.6a) is used.

The question is how useful eigenvalues of a poor approximation $\bar{\mathbf{J}}$ of $\mathbf{J} = \mathbf{f}_{\mathbf{y}}$ can be. Experience shows that the eigenvalues based on $\bar{\mathbf{J}}$ are often good enough. Close to critical situations such as loss of stability, however, the eigenvalues and thus $\bar{\mathbf{J}}$ must be calculated more accurately. Here the real part of a critical eigenvalue is close to zero, and the relative error may be unacceptable even though the absolute error might be tolerable. Occasionally, smaller continuation steps improve the accuracy of $\bar{\mathbf{J}}$. Higher demands for accuracy call for an additional approximation of the Jacobian by numerical differentiation at the final Newton iterate. The advanage in efficiency gained by approximating $\mathbf{J}$ by rank-one updates can be lost when accurate eigenvalues are required.

Software mentioned in Section 1.4 and Appendix 6 is available for calculating the eigenvalues. The widely used QR method is relatively expensive. In fact, the cost of carrying out a Newton step and of evaluating the eigenvalues by the QR method are of the same order of magnitude ($O(n^3)$). Hence, for large systems (n large) alternatives are sought [163]. One alternative that often provides information about stability is simulation of the transient behavior by numerically integrating the ODE system associated with Eq. (5.1). This simulation was described in Chapter 1. For the special case of symmetric Jacobians, an inexpensive method for approximating the eigenvalue with real part closest to zero has been proposed in [322].

We conclude this section by showing how eigenvalues monitored during a continuation procedure can be used to draw practical conclusions about bifurcation behavior. Consider the following table for $n = 4$ equations, in which the four eigenvalues μ_1, μ_2, μ_3, μ_4, are listed for consecutive values of the branching parameter λ:

Parameter	μ_1	μ_2	μ_3	μ_4
$\lambda_1 = \dots$	0.82	$-5.1 + 0.4i$	$-5.1 - 0.4i$	-17.3
$\lambda_2 = \dots$	0.02	$-5.9 + 0.43i$	$-5.9 - 0.43i$	-17.1
$\lambda_3 = \dots$	-0.51	$-6.5 + 0.45i$	$-6.5 - 0.45i$	-16.8
$\vdots$		$\vdots$		

For clarity, the eigenvalues in the table are ordered in the same way for all λ, although this situation cannot be expected from the QR method in "raw" output data. The QR method frequently changes the positions of the eigenvalues, causing difficulties in tracing "the same" eigenvalue. As the table shows, one eigenvalue changes the sign of its real part for a value of λ between λ_2 and λ_3 (first column). The question is whether the crossing of the imaginary axis is real or complex. That is, the investigator must decide whether a stationary bifurcation or a Hopf bifurcation has been passed. Although the three calculated values of μ_1 are all real, one might expect a tiny subinterval between λ_2 and λ_3 with complex values of μ_1. This situation is difficult to survey when there are many eigenvalues (n large). In case of uncertainty, it is advisable to retrace the branch in the critical subinterval, using a small step length. The specific situation illustrated by the table above is easy to resolve. In order to become complex, the eigenvalue μ_1 needs a real-valued partner to coalesce with. The numbers taken by μ_2 and μ_3 obviously rule out μ_2 and μ_3 as candidates; they are partners of each other. The remaining candidate μ_4 does not show any desire to join with μ_1. Thus the example clearly indicates that a real crossing by μ_1 of the imaginary axis is as certain as numerical results can be.

5.2 BRANCHING TEST FUNCTIONS

During branch tracing a bifurcation must not be overlooked. Suppose that a chain of solutions

$$(\mathbf{y}^1,\lambda_1),\ (\mathbf{y}^2,\lambda_2),\ \ldots$$

to $\mathbf{f}(\mathbf{y},\lambda) = \mathbf{0}$ has been approximated and that a bifurcation point is straddled by $(\mathbf{y}^{j-1},\lambda_{j-1})$ and $(\mathbf{y}^j,\lambda_j)$. How can we recognize from the data from these solutions whether a bifurcation $(\mathbf{y}_0,\lambda_0)$ is close? The required information is provided by a *test function* $\tau(\mathbf{y},\lambda)$, which is evaluated during the branch tracing. A bifurcation is indicated by a zero of τ—that is, a branching test function satisfies the property

$$\tau(\mathbf{y}_0,\lambda_0) = 0 \tag{5.3}$$

and is continuous in a sufficiently large interval. The application of τ is illustrated in Figure 5.1. The upper part of the figure shows several solutions calculated on one branch. During these computations, the values of a branching test function are monitored (lower part of figure). Because it is unlikely to hit the zero (or the bifurcation, respectively) exactly, we must check for a change of sign of τ. A bifurcation is most likely to be passed if the criterion

$$\tau(\mathbf{y}^{j-1},\lambda_{j-1})\tau(\mathbf{y}^j,\lambda_j) < 0 \tag{5.4}$$

is satisfied. Figure 5.1 depicts an ideal situation with clearly isolated zeros of τ; step length $|\lambda_{j-1} - \lambda_j|$ and accuracy of τ together ensure a smooth interpolating curve. It is difficult to verify a bifurcation via criterion (5.4) when step lengths are too large or when the computed τ exhibits artificial oscillations that are the result of inaccurate evaluation.

Some examples of test functions are immediately obvious, others are defined in an artificial way. We delay a discussion of the latter until later. A natural choice for τ is the maximum of all real parts of the eigenvalues $\alpha_j + i\beta_j$ of the Jacobian $\mathbf{J}$,

$$\tau = \max\{\alpha_1, \ldots, \alpha_n\}. \tag{5.5}$$

This choice has the advantage of being physically meaningful because $\tau < 0$ guarantees local stability. τ is continuous, provided $\mathbf{f}(\mathbf{y},\lambda)$ is con-

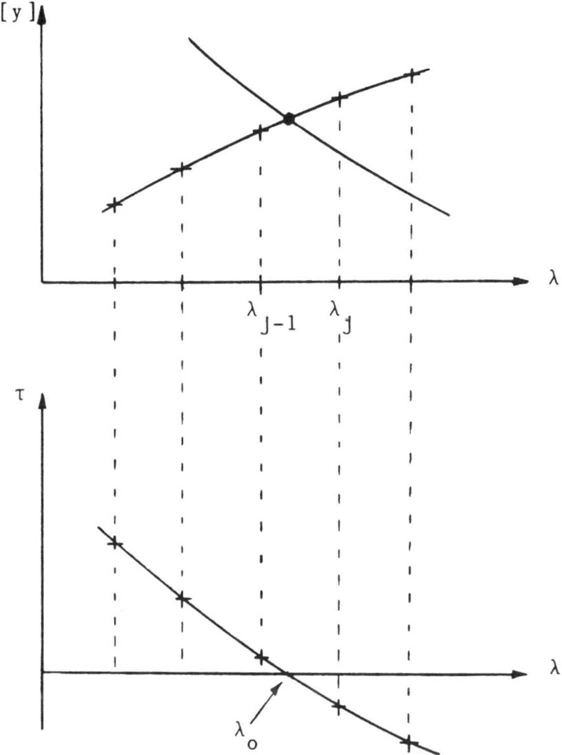

Figure 5.1

tinuously differentiable. However, τ from Eq. (5.5) does not need to be smooth.

Example 5.1.
We consider again the system of ODEs that models Marangoni instabilities (Section 3.4), with the values Eq. (3.3) and $Nu = 100$. Recall that a bifurcation occurs for $\Theta_4 = 0$ and $Q_0 = \lambda_0 \approx 24.5$. The branch A (Figure 3.12) that is traced is stable for $Q < Q_0$ and unstable for larger Q. The corresponding test function τ (in Figure 5.2) clearly indicates the bifurcation by a smooth shape and a zero at Q_0. Varying Θ_4, the bifurcation gets unfolded, which affects the test function. Figure 5.3 depicts the results for $\Theta_4 = 0.002$. For $Q \approx 26$ the test function fails to be smooth. This phenomenon can be explained when the second largest real part also is drawn (Figure 5.4). At $Q \approx 26$ two real eigenvalues coalesce, forming a pair of complex-conjugate eigenvalues for $Q > 26$. The test func-

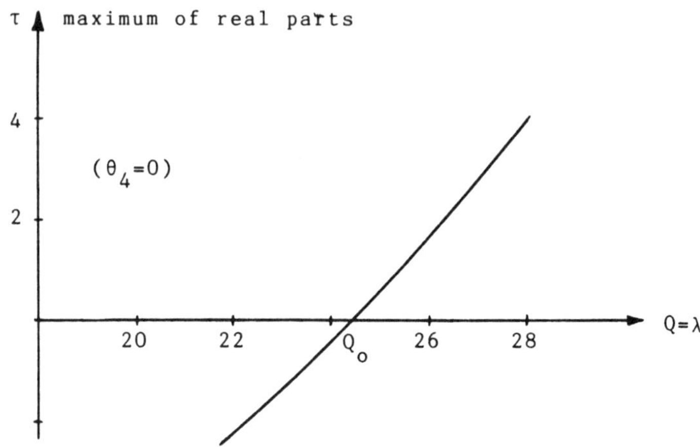

Figure 5.2

tion for $\Theta_4 = 0.002$ approaches the Q-axis closely, thereby indicating that a zero (bifurcation) is close for a nearby parameter constellation. The test function (5.5) also indicates turning points that exhibit a loss of stability. This is illustrated in Figure 5.5 for $\Theta_4 = -0.01$. The two turning points with the gap between are reflected by the behavior of the test function.

The test function (5.5) does not signal turning points in which two unstable half-branches coalesce. The same holds for bifurcations that connect unstable branches only. At such branch points, all real parts are negative, and hence τ from Eq. (5.5) is nonzero. This minor

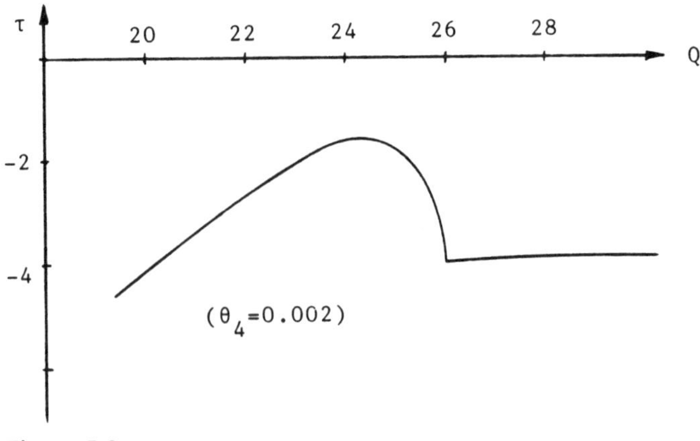

Figure 5.3

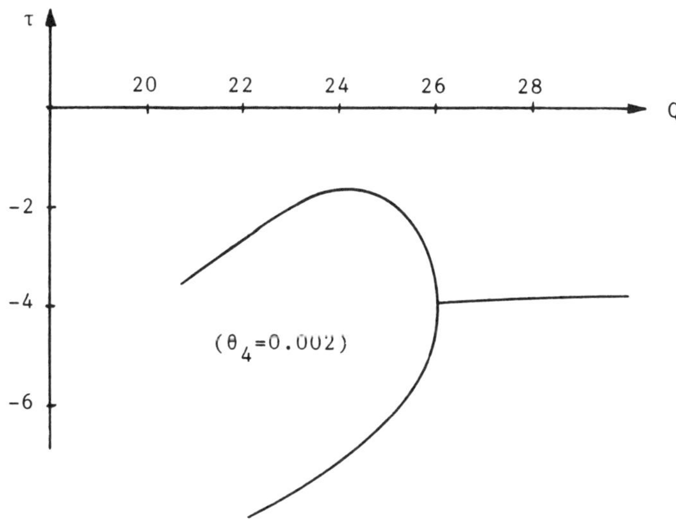

Figure 5.4

problem can be overcome by modifying Eq. (5.5) to

$$\tau = \alpha_k \quad \text{with } |\alpha_k| = \min\{|\alpha_1|, \ldots, |\alpha_n|\}; \tag{5.6}$$

this was used in [322]. A strong advantage of Eq. (5.5), compared with the test function to be discussed next, is that it also indicates Hopf

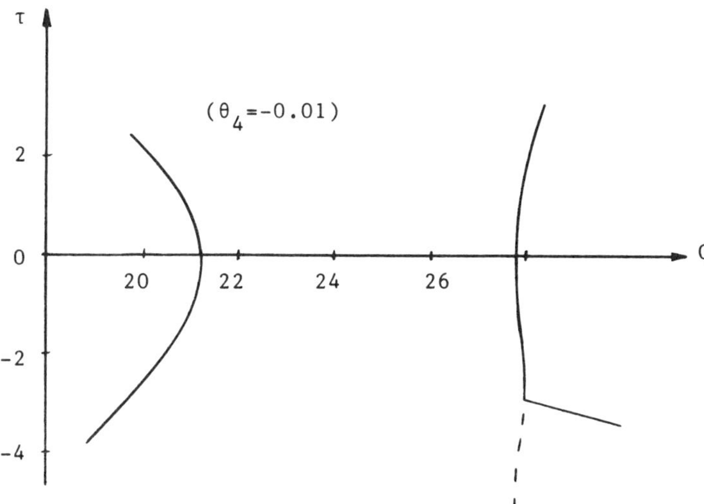

Figure 5.5

bifurcations. Problems concerning accuracy and cost in evaluating τ were investigated in the preceding section.

Another test function can be derived from the characterization of bifurcation points (Definition 2.3) and turning points (Definition 2.4). Since the Jacobian evaluated at $(\mathbf{y}_0,\lambda_0)$ is singular,

$$\det \mathbf{f_y}(\mathbf{y}_0,\lambda_0) = 0.$$

This suggests monitoring the test function

$$\tau(\mathbf{y},\lambda) = \det \mathbf{f_y}(\mathbf{y},\lambda) \tag{5.7}$$

during continuation. A most attractive feature of Eq. (5.7) is its availability as by-product of the Newton method. Because the Newton method makes use of the Jacobian, this matrix (or an approximation) is available along with the final iterate. Typically, for solving the linear system (1.8a) the Jacobian matrix is decomposed into

$$\mathbf{Pf_y} = \mathbf{LU}.$$

$\mathbf{P}$ is a permutation matrix (see Section 1.4). Taking the determinant of these matrix products yields

$$(\pm 1)\cdot\det \mathbf{f_y} = 1\cdot\det \mathbf{U}$$

and hence

$$\tau = \pm\det \mathbf{U}.$$

Because the determinant of a triangular matrix equals the product of the diagonal elements, $\det(\mathbf{U})$ and thus $|\tau|$ are easily available. Care must be taken to obtain the correct sign of τ, which depends on the permutations required for the $\mathbf{LU}$ decomposition. As with all branching test functions, the accuracy of τ depends on how accurately the Jacobian is evaluated. There are two known drawbacks for Eq. (5.7). First, evaluating Eq. (5.7) gets expensive for large systems when $\mathbf{LU}$ decompositions are not used. Second, as reported in [2], the order of magnitude of Eq. (5.7) is occasionally inconvenient. That is, a loss of significance due to scaling problems may occur. Alternative test functions, which apparently do not suffer from these disadvantages, were introduced in [2] and [307]; the latter will be discussed in Section 5.3.4. Apart from some remarks in [2], not much attention has been paid to such questions as which drawbacks are significant or which test function is recommended for which purpose.

Exercise 5.1.

Study again Figures 5.2 through 5.5, which depict test functions that correspond to an unfolded bifurcation (depending on Θ_4). Complete these figures by sketching one figure showing the unfolded τ for several values of Θ_4 with $|\Theta_4|$ small.

5.3 GENERAL METHODS FOR CALCULATING BRANCH POINTS

The general concept of branch points, which include both bifurcation points and turning points, is met by methods that are able to approximate either of them. We report on such general methods in this section, deferring special methods for special purposes to Section 5.5. For reasons that will become apparent later, we distinguish between indirect methods and direct methods.

5.3.1 Indirect Methods

Standard methods for calculating branch points benefit from the data obtained during the continuation procedure, completed by values of a branching test function. With τ_j denoting the value of a test function evaluated at $(\mathbf{y}^j, \lambda_j)$, the data

$$(\mathbf{y}^j, \lambda_j, \tau_j), \qquad j = 1, 2, \ldots$$

are available. Assume that a bifurcation has been located in the subinterval $\lambda_{j-1} < \lambda < \lambda_j$ by noticing a change of the sign of τ (see criterion (5.4)). It is a straightforward calculation to interpolate the data in order to obtain an approximation of the zero of τ (see Figure 5.6). To this end, assume that τ has exactly one zero λ_0 in the subinterval under consideration. Then an approximation $\bar{\lambda}_0$ is obtained by calculating the zero of the interpolating straight line (dashed in Figure 5.6). The

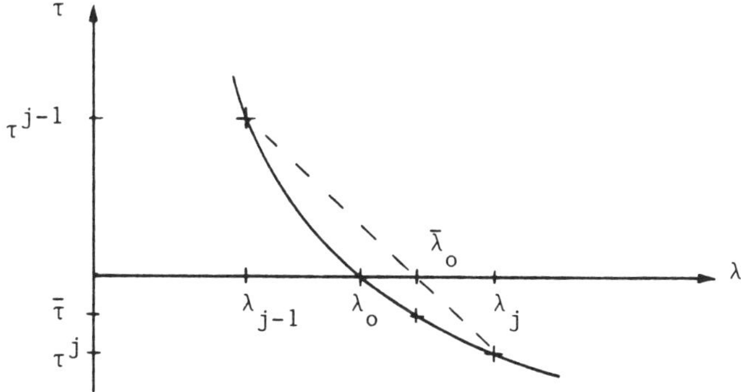

Figure 5.6

formula is

$$\bar{\lambda}_0 = \lambda_{j-1} + (\lambda_j - \lambda_{j-1})\tau_{j-1}/(\tau_{j-1} - \tau_j). \tag{5.8}$$

In many applications this approximation is close enough to λ_0, because the continuation step length $|\lambda_j - \lambda_{j-1}|$ is typically small when bifurcations are investigated. Formula (5.8) is well defined, because $\tau_{j-1}\tau_j < 0$ holds. In some cases it is necessary to improve the approximation or to get information on the error $|\bar{\lambda}_0 - \lambda_0|$. In order to achieve this objective, the data of at least one additional solution are required. Three solutions with three values of τ enable the approximation of τ by an interpolating parabola. In order to save work, one may take the data obtained at λ_{j-2}. The results will become better when an additional solution for a value of λ close to the expected bifurcation is approximated. The natural candidate is $\lambda = \bar{\lambda}_0$. Solving $\mathbf{f}(\mathbf{y},\lambda) = \mathbf{0}$ for this parameter value provides a third value $\bar{\tau}$ of the test function. We illuminate this with Figure 5.6, now interpreting the continuous curve as the interpolating parabola and λ_0 as its zero λ_0^*. The formulas are (see Exercise 5.2):

$$\begin{aligned}\lambda_0^* - \bar{\lambda}_0 = \{&-\zeta - \gamma(\bar{\lambda}_0 - \lambda_j)\\ &\pm ([\zeta + \gamma(\bar{\lambda}_0 - \lambda_j)]^2 - 4\gamma\bar{\tau})^{1/2}\}/(2\gamma)\end{aligned} \tag{5.9}$$

with

$$\zeta = (\tau_j - \bar{\tau})/(\lambda_j - \bar{\lambda}_0),$$

$$\gamma = [(\tau_{j-1} - \bar{\tau})/(\lambda_{j-1} - \bar{\lambda}_0) - \zeta]/(\lambda_{j-1} - \lambda_j).$$

The sign in Eq. (5.9) is uniquely determined because there is one and only one value of λ_0^* with

$$\lambda_{j-1} < \lambda_0^* < \lambda_j.$$

The relation (5.9) is used in two ways. First, this formula gives a corrector to be added to $\bar{\lambda}_0$ in order to obtain a better approximation. Second, assuming that the second approximation λ_0^* is by far better than the first approximation $\bar{\lambda}_0$,

$$|\lambda_0^* - \lambda_0| \ll |\bar{\lambda}_0 - \lambda_0|,$$

Eq. (5.9) offers an estimate of the error of the first approximation $\bar{\lambda}_0$. To this end, substitute λ_0 for λ_0^*. In this way, no error estimate for $\lambda_0 - \lambda_0^*$ is obtained. But finding a small value of $|\lambda_0^* - \bar{\lambda}_0|$ by Eq. (5.9) suggests that $|\lambda_0 - \lambda_0^*|$ is much smaller. On the other hand, a large value of Eq. (5.9) ("large" compared with the interval length $|\lambda_j - \lambda_{j-1}|$) serves as a hint that the accuracy is not sufficient. In such a case the continuation steps usually have been too large. The procedure can be repeated, then resembling the Muller method for calculating zeros [334]. Convergence problems due to an almost singular Jacobian may occur when $\mathbf{f}(\mathbf{y},\lambda) = \mathbf{0}$ is solved for a λ too close to λ_0.

After an approximation $\bar{\lambda}_0$ (or λ_0^* respectively) has been obtained, the next step is to approximate the corresponding solution $\mathbf{y}$ of $\mathbf{f}(\mathbf{y},\lambda) = \mathbf{0}$. This can be done most economically by applying linear interpolation also to $\mathbf{y}$, rather than by solving the equation $\mathbf{f}(\mathbf{y},\lambda) = \mathbf{0}$. The approximation is accomplished by

$$\bar{\mathbf{y}}_0 = \mathbf{y}^{j-1} + (\mathbf{y}^j - \mathbf{y}^{j-1})\tau_{j-1}/(\tau_{j-1} - \tau_j). \tag{5.10}$$

The approximation $(\bar{\mathbf{y}}_0,\bar{\lambda}_0)$ obtained in this way is accurate enough in most practical applications. The cost of the interpolation formulas is negligible, compared with the cost of solving $\mathbf{f}(\mathbf{y},\lambda) = \mathbf{0}$. Note that Eqs. (5.8) and (5.10) are also valid for extrapolation: a guess on the next bifurcation is obtained when Eq. (5.8) is evaluated even before τ changes sign. This use of Eq. (5.8) is illustrated in Figure 5.7. The information that a $\bar{\lambda}_0$ is close can be used in advance for deciding whether the continuation step length is to be reduced. In general, the accuracy of the guess $\bar{\lambda}_0$ improves as the bifurcation is approached. There are, however, exceptions to this rule: note that the criterion $\tau_{j-1}\tau_j < 0$ is not met when Eq. (5.8) is used for extrapolation. Hence the guess $\bar{\lambda}_0$ may suffer severely from inaccurate τ values. In particular, an evaluation of Eq. (5.8) does not make sense when $\tau_{j-1} \approx \tau_j$.

We have thus far obtained approximations of a bifurcation point. In a similar fashion, approximations of turning points can be based on τ. Before going into detail, we stress that there are more economical methods that do not require test functions. We postpone this for a moment and briefly outline the application of test functions.

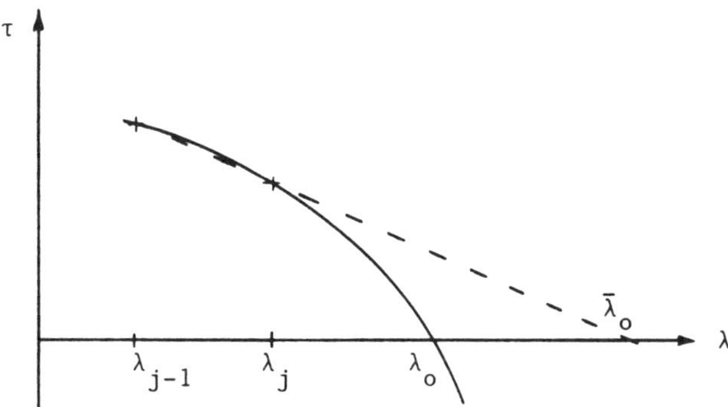

Figure 5.7

Because at a turning point a branching test function reflects the behavior of the branch, τ crosses the λ-axis at λ_0 perpendicularly, compare the example of Figure 5.5. This justifies modeling the shape of τ locally by the parabola

$$\lambda - \lambda_0 = \xi\tau^2.$$

This ansatz function makes it possible to compute an approximation $\bar{\lambda}_0$ based on only two solutions $(\mathbf{y}^{j-1},\lambda_{j-1})$, $(\mathbf{y}^j,\lambda_j)$ with associated τ_{j-1} and τ_j. As we can easily see (Exercise 5.6), the relation

$$\bar{\lambda}_0 = (\lambda_j\tau_{j-1}^2 - \lambda_{j-1}\tau_j^2)/(\tau_{j-1}^2 - \tau_j^2) \tag{5.11}$$

establishes an approximation $\bar{\lambda}_0$ of the value λ_0 of a turning point.

Note that an evaluation of Eq. (5.11) should be avoided if $|\tau_{j-1}| \approx |\tau_j|$ holds. For such values of τ_{j-1} and τ_j, Eq. (5.11) is not well defined; $\bar{\lambda}_0$ then reacts sensitively to inherent errors in τ. If $|\tau_{j-1}|$ and $|\tau_j|$ are close, one set of data (say $\mathbf{y}^j,\lambda_j,\tau_j$) may be replaced by corresponding values of another solution, aiming at $|\tau_j| < \frac{1}{2}|\tau_{j-1}|$. Based on the approximation of Eq. (5.11), approximate values for $\bar{\mathbf{y}}_0$ can be obtained (Exercise 5.6).

Applying the above approximation formula is easy, provided the bifurcation manifests itself by a change of sign of the branching test function. But there are bifurcations that do not do us this favor. Consider the pitchfork bifurcation depicted in Figure 5.8. The upper part of the figure shows the branching diagram, the lower part shows corresponding test functions. In both we have dotted the results that correspond to the branch that can be parameterized by λ. Here we focus on the branch that turns back at λ_0 (solid line). In contrast to what we have seen with turning points, τ does not change sign. This behavior of τ has two consequences. To begin with the advantageous aspect, such a feature of τ makes it possible to distinguish between turning point and pitchfork bifurcation without a closer investigation of this branch point. The information is obtained as by-product of a branch tracing that evaluates a test function. Recall that results of a plain continuation generally do not give any hint that discriminates between turning point and pitchfork. (Consider the dots in Figure 5.8 as unknown.) The unfavorable aspect is that the absence of a change of sign makes it difficult to decide on which side of the branch point a particular solution is situated. The approximation of Eq. (5.11) is not valid here. One must instead prefer Eq. (5.8).

The approximation formulas introduced so far are convenient in that they involve only two or three solutions $(\mathbf{y}^j,\lambda_j)$ together with τ_j. We now turn to approximating branch points when more solutions are involved. One either is content with the solutions available from con-

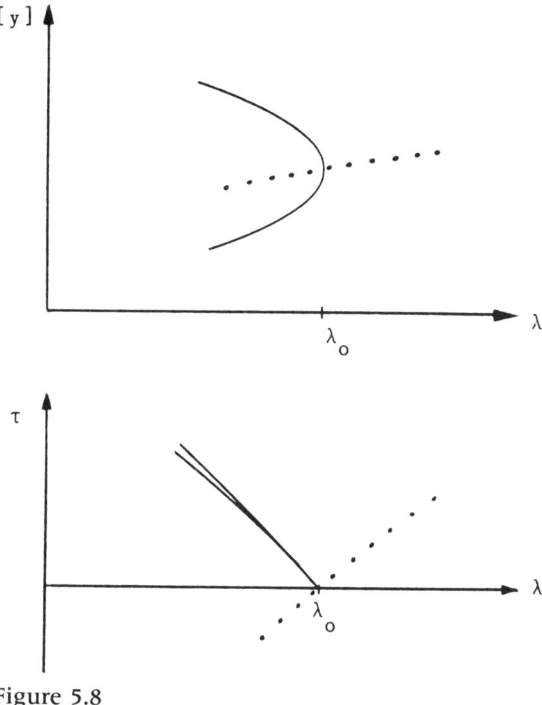

Figure 5.8

tinuation, or one calculates more solutions close to the branch point. A powerful tool that can be applied is *inverse interpolation.* Assume we have four to six solutions to $\mathbf{f}(\mathbf{y},\lambda) = \mathbf{0}$ available with corresponding values of τ in a range where τ changes sign and where its derivative with respect to λ does not vanish. For good results, make sure the solutions are chosen in such a way that about half have positive values of τ and the remaining have negative values of τ. Then consider the interpolating polynomial $\lambda = p(\tau)$ with

$$\lambda_j = p(\tau_j)$$

for the chosen data λ_j,τ_j. Evaluating this polynomial for $\tau = 0$ gives an approximation for λ_0 that is generally extremely accurate, provided the values of τ_j are accurate enough and the step lengths are not too large. The practical evaluation of $p(0)$ is done by applying Neville's method. (This recursion is described in any textbook of numerical analysis.) It is worthwhile to append a corresponding algorithm to any continuation method that provides values of a branching test function (see Exercise 5.7). The simple interpolations Eqs. (5.8) and (5.9) can be regarded as special cases of inverse interpolation.

Finally, let us point out that turning points can be approximated very cheaply without evaluating a particular test function. Instead, one can exploit the relation $d\lambda/d\mathbf{y} = 0$ that characterizes a turning point as a maximum or minimum in λ. Transferring this feature to an interpolating inverse polynomial leads to simple formulas based on $(\mathbf{y},\lambda)$ values only (see Exercise 5.8).

It has become customary to call methods that approximate λ_0 by interpolation "indirect methods." Indirect methods are based mostly on data generated by continuation. Before addressing "direct" methods, let us emphasize that indirect methods are to be recommended in practical computations. With small overhead, they are appropriate for standard demands of accuracy.

Exercise 5.2.
Verify Eq. (5.9) using

$$p(\lambda) = \bar{\tau} + \zeta(\lambda - \bar{\lambda}_0) + \gamma(\lambda - \bar{\lambda}_0)(\lambda - \lambda_j)$$

for the interpolating polynomial.

Exercise 5.3.
Assume that criterion (5.4) holds and that the subinterval $\lambda_{j-1} < \lambda < \lambda_j$ has exactly one zero of a continuous $\tau(\lambda)$. Show that there is exactly one of the two values of λ_0^* (furnished by Eq. (5.9)) in this subinterval.

Exercise 5.4.
Equation (5.9) may suffer from *cancellation* (loss of significant digits due to subtraction of nearly equal numbers). Establish a formula that does not suffer from cancellation.

Exercise 5.5.
Design a flowchart of an algorithm that exploits data $(\mathbf{y}^j,\lambda_j,\tau_j)$ provided by continuation. The objective is to locate bifurcations and to calculate the approximation $\bar{\lambda}_0$. You are encouraged to program your algorithm on a computer and to extend it by making use of Eq. (5.9).

Exercise 5.6 Approximation of turning points $(\mathbf{y}_0,\lambda_0)$ by means of test functions. Verify Eq. (5.11).
Establish an approximation to $\mathbf{y}_0$ based on the assumption that the components of $\mathbf{y}$ locally behave like parabolas.

Exercise 5.7 (project).
Construct an algorithm that approximates λ_0 by inverse interpolation. Design the algorithm so that it can be easily attached to any continuation procedure that also provides values of a branching test function. Let $p(\tau)$ be the interpolating polynomial explained in the text.

(a) A first version of the algorithm evaluates $p(0)$ after each continuation step, with an interpolation based upon four (say) pairs λ_j,τ_j. When no change of sign is involved, output $p(0)$ as a tentative guess of a possible λ_0. When a change of sign is straddled, output $p(0)$ as an approximation to λ_0.
(b) You may like to work on a sophisticated version, using a variable number of pairs λ_j,τ_j and a feedback of relevant information on $p(0)$, causing appropriate changes in the continuation step length. You may also like to extend the Neville scheme to calculate $dp(0)/d\tau$. Passing turning points, this value serves as hint about the accuracy of the approximation (why?).

Exercise 5.8 (project).
As mentioned above, turning points can be approximated most effectively without making use of any test function. This approach requires three or more solutions $(\mathbf{y}^j,\lambda_j)$, $j = 1, 2, \ldots$, calculated in the neighborhood of a current turning point $(\mathbf{y}_0,\lambda_0)$. For convenience we restrict ourselves to three solutions, interpreting y and y^j as any component (the first, say) of these vectors.

(a) Use

$$\lambda = p(y) = \lambda_1 + \zeta(y - y^1) + \gamma(y - y^1)(y - y^2)$$

and the relation

$$dp(y_0)/dy = 0$$

to establish an approximation $(\bar{y}_0,\bar{\lambda}_0)$ to (y_0,λ_0). (Hint: $2\bar{y}_0 = y^1 + y^2 - \zeta/\gamma$.)
(b) Incorporate these formulas into a continuation algorithm. Design a check that tells whether the turning point is straddled by the last three solutions. Output $(\bar{\mathbf{y}}_0,\bar{\lambda}_0)$ after each continuation step, with a text depending on whether $(\mathbf{y}_0,\lambda_0)$ is straddled or not.

Exercise 5.9 (project).
Design an algorithm that interrupts the continuation in order to approximate a turning point more accurately. To this end, use the formulas

of Exercise 5.8 and calculate an additional solution, using Eq. (4.10) with a value of η obtained from $\bar{\mathbf{y}}_0$. After a solution is calculated, a better approximation to the turning point can be obtained in one of two ways. For one, apply the formula in Exercise 5.8 based on three solutions. Replace the solution that is furthest from the current guess by the new approximation. As an alternative, use a polynomial of the next higher degree. After the approximation is obtained, the process can be repeated in an iterative way. One or two iterations of this simple scheme are usually sufficient to produce an accurate approximation to a turning point.

5.3.2 Direct Methods

A *direct method* for calculating branch points formulates a suitable equation

$$\mathbf{F}(\mathbf{Y}) = \mathbf{0} \tag{5.12}$$

in such a way that its solution is a branch point. To this end, the original system $\mathbf{f}(\mathbf{y},\lambda) = \mathbf{0}$ is enlarged by additional equations that characterize a branch point. The vector $\mathbf{Y}$ includes $(\mathbf{y},\lambda)$ as subvector and the function $\mathbf{F}$ includes $\mathbf{f}$. A direct method solves the single equation (5.12), preferably by just calling standard software of numerical analysis ("CALL SOLVER"). This might give one hope that eventually a researcher will first be able to calculate all branch points and then decide whether to connect the branch points with branches. Although this will be shown to be an illusion, it illustrates well the principal difference between an indirect method and a direct method for approximating branch points. By concept, a direct method is an iterative process because it exploits the iterative approach of the numerical method in SOLVER.

By the definition of a test function $\tau(\mathbf{y},\lambda)$, the enlarged system of equations

$$\mathbf{F}(\mathbf{Y}) = \mathbf{F}(\mathbf{y},\lambda) = \begin{pmatrix} \mathbf{f}(\mathbf{y},\lambda) \\ \tau(\mathbf{y},\lambda) \end{pmatrix} = \mathbf{0} \tag{15.13a}$$

is suitable for calculating branch points. Approaches based on Eq. (5.13a) were proposed by [276, 194, 2, 201, 260]. As an example, we apply the determinant Eq. (5.7). This leads to the enlarged system

$$\begin{pmatrix} \mathbf{f}(\mathbf{y},\lambda) \\ \det \mathbf{f}_{\mathbf{y}}(\mathbf{y},\lambda) \end{pmatrix} = \mathbf{0}. \tag{5.13b}$$

Both turning points and stationary bifurcation points are among the solutions. Solving Eq. (5.13b) by a Newton method is restricted to small n, because the determinant is required analytically, which is difficult to obtain.

Efficient methods based on Eqs. (5.13a) and (5.13b) have been proposed by Abbott [1, 2]. His approach is based on the method of Brown and Brent [43], which turns out to be suitable for equations (see Eq. (5.13a)) that have the feature that $\mathbf{f_y}$ is available analytically whereas a derivative of τ is not available. In [2], three choices of the scalar equation $\tau(\mathbf{y},\lambda) = 0$ are presented; one of them is Eq. (5.13b).

Another prototype of direct methods was introduced in [306]. We describe the basic method in some detail for two reasons. First, a specific aspect of the method will give rise to a class of branch switching methods and to new test functions. Second, this method can be modified in various ways to adapt its performance to special applications. Special versions for special purposes will be discussed in Section 5.5; the present section is confined to the original version of [306, 309], which is able to calculate different types of branch points without requiring modifications.

The linearization of $\mathbf{f}(\mathbf{y},\lambda) = \mathbf{0}$ about $\mathbf{y}$ is the system of n equations

$$\mathbf{f_y}(\mathbf{y},\lambda)\mathbf{h} = \mathbf{0}. \tag{5.14}$$

The vector $\mathbf{h}$ consists of n components. We know (see Section 2.5) that the Jacobian matrix $\mathbf{J}(\mathbf{y},\lambda) = \mathbf{f_y}(\mathbf{y},\lambda)$ is singular if evaluated at a branch point $(\mathbf{y}_0,\lambda_0)$. An equivalent criterion for $\mathbf{f_y}$ being singular is that the linearization Eq. (5.14) has a nontrivial solution $\mathbf{h}_0 \neq 0$. The argument $(\mathbf{y},\lambda)$ in Eq. (5.14) must solve $\mathbf{f}(\mathbf{y},\lambda) = 0$, which equation can be attached to Eq. (5.14). Adding a relation that forces $\mathbf{h}$ to be nonzero leads to a large system with as many equations as unknowns:

$$\begin{pmatrix} \mathbf{f}(\mathbf{y},\lambda) \\ \mathbf{f_y}(\mathbf{y},\lambda)\mathbf{h} \\ \varphi(\mathbf{h}) - \xi \end{pmatrix} = \mathbf{0}. \tag{5.15}$$

In Eq. (5.15), φ is a functional that imposes a nonzero value ξ on $\mathbf{h}$, $\varphi(\mathbf{h}) = \xi \neq 0$. Branch points can be calculated by solving Eq. (5.15). The practical choice of φ is to require the k-th component of $\mathbf{h}$ to take a certain prescribed value ξ,

$$h_k = \xi. \tag{5.16}$$

Because Eq. (5.14) can be multiplied by any nonzero number without changing the quality of the linearization, we are free to impose any nonzero value of ξ. For the sake of a reasonable scaling of variables, we choose $\xi = 1$. To transform the extended system (5.15) into the form of Eq. (5.12), define the vector $\mathbf{Y}$ of $2n + 1$ components by

$$\mathbf{Y} = \begin{pmatrix} \mathbf{y} \\ \lambda \\ \mathbf{h} \end{pmatrix}. \tag{5.17}$$

For convenience we shall write such vectors as rows. The system $\mathbf{F}$ of $2n + 1$ scalar functions is given by

$$\mathbf{F}(\mathbf{Y}) = \begin{pmatrix} \mathbf{f}(\mathbf{y},\lambda) \\ \mathbf{f}_\mathbf{y}(\mathbf{y},\lambda)\mathbf{h} \\ h_k - 1 \end{pmatrix} = \mathbf{0}. \tag{5.18}$$

Because for both turning points and bifurcation points $(\mathbf{y}_0,\lambda_0)$ the linearization has a nontrivial solution $\mathbf{h}_0$, it is evident that these combined vectors $(\mathbf{y}_0,\lambda_0,\mathbf{h}_0) = \mathbf{Y}_0$ solve $\mathbf{F}(\mathbf{Y}) = 0$. Referring to its general applicability, the system (5.18) has been called a *branching system.*

Solving Eq. (5.18) by such standard methods as Newton methods requires a reasonable initial guess $(\bar{\mathbf{y}},\bar{\lambda},\bar{\mathbf{h}})$ to start the iterative process. That is, the iteration that generates a sequence of vectors $\mathbf{Y}^{(1)}$, $\mathbf{Y}^{(2)}, \ldots$ must be supplied by an initial approximation

$$\mathbf{Y}^{(0)} = \bar{\mathbf{Y}} = (\bar{\mathbf{y}},\bar{\lambda},\bar{\mathbf{h}})$$

that is close enough to the expected solution $\mathbf{Y}_0$ to $\mathbf{F}(\mathbf{Y}) = \mathbf{0}$. Without a close initial approximation, success of the Newton method here is doubtful; it will converge slowly if it converges at all. In view of this, a direct method based on Eq. (5.18) requires a precursive phase that generates $\bar{\mathbf{Y}}$. The easy part is to find $(\bar{\mathbf{y}},\bar{\lambda})$, provided one has located the branch point. One possibility is to take the current solution of $\mathbf{f}(\mathbf{y},\lambda) = \mathbf{0}$ at which a change of sign for a branching test function was detected. This choice of $(\bar{\mathbf{y}},\bar{\lambda})$ can be replaced by results of an approximation $(\bar{\mathbf{y}}_0,\bar{\lambda}_0)$ that is obtainable by the interpolation outlined in the preceding subsection. The problem remains to calculate a vector $\bar{\mathbf{h}}$, depending on $(\bar{\mathbf{y}},\bar{\lambda})$, such that $\bar{\mathbf{h}} \approx \mathbf{h}_0$.

This problem is solved in [306] based on the following principle: The linearized system

$$\mathbf{f}_\mathbf{y}(\bar{\mathbf{y}},\bar{\lambda})\mathbf{h} = \mathbf{0} \tag{5.19}$$

has no nontrivial solution, because the Jacobian $\mathbf{J}(\bar{\mathbf{y}},\bar{\lambda}) = \mathbf{f}_\mathbf{y}(\bar{\mathbf{y}},\bar{\lambda})$ evaluated at $(\bar{\mathbf{y}},\bar{\lambda}) \neq (\mathbf{y}_0,\lambda_0)$ is nonsingular. Removing one of the n scalar equations of Eq. (5.19) produces a solvable system. It is convenient to replace the removed equation by the normalizing equation $h_k = 1$, which enforces $\mathbf{h} \neq \mathbf{0}$. The process can be described formally as follows:

Let us denote by l the index of the removed equation and define the n^2 matrix

$$\bar{\mathbf{J}}_{lk} = \mathbf{J}_{lk}(\bar{\mathbf{y}},\bar{\lambda}) = (\mathbf{I} - \mathbf{e}_l\mathbf{e}_l^{tr})\mathbf{J}(\bar{\mathbf{y}},\bar{\lambda}) + \mathbf{e}_l\mathbf{e}_k^{tr}. \tag{5.20}$$

As above, the n-vector $\mathbf{e}_i$ denotes the i-th unit vector. The matrix $\bar{\mathbf{J}}_{lk}$ essentially consists of $\mathbf{J}(\bar{\mathbf{y}},\bar{\lambda})$, except that the l-th row is replaced by $\mathbf{e}_k$,

$$\bar{\mathbf{J}}_{lk} = \begin{pmatrix} \mathbf{J}(\bar{\mathbf{y}},\bar{\lambda}) \\ \ldots 0\;1\;0 \ldots \\ \phantom{\mathbf{J}} \end{pmatrix} \leftarrow l.$$

$$\uparrow$$
$$k$$

With these notations, the method of producing a solvable linear system is equivalent to solving the equation

$$\bar{\mathbf{J}}_{lk}\mathbf{h} = \mathbf{e}_l. \tag{5.21}$$

Definition 5.1. $\bar{h}$ *is the solution of Eq.* (5.21).

Modifications are possible, working with the matrix $\mathbf{f}_{\mathbf{y}}(\bar{\mathbf{y}},\bar{\lambda})$ instead of $\bar{\mathbf{J}}_{lk}$ (Exercise 5.11). The vector $\bar{\mathbf{h}}$ varies with $(\bar{\mathbf{y}},\bar{\lambda})$ and changes with the choice of the indices l and k. Remarks on the choices of l and k will follow below.

Let us make sure that the $\bar{\mathbf{h}}$ defined as the solution of Eq. (5.21) meets the requirement of being close to $\mathbf{h}_0$. The vector $\bar{\mathbf{h}}$ qualifies as approximation if

$$\text{rank } \mathbf{J}(\mathbf{y}_0,\lambda_0) = n - 1,$$

which is satisfied by turning points and (simple) bifurcation points. In this case of rank deficiency one, indices l and k can be found such that $\mathbf{J}_{lk}(\mathbf{y}_0,\lambda_0)$ is nonsingular. The following continuity property has been established (for a proof, see [309]):

Theorem 5.1. Assume that $(\mathbf{y}_0,\lambda_0,\mathbf{h}_0)$ *solves Eqs.* (5.15) *or* (5.18), *that* $\mathbf{f}$ *is two times continuously differentiable, and that* $\mathbf{J}_{lk}(\mathbf{y}_0,\lambda_0)$ *is nonsingular. Then* $\bar{\mathbf{h}}$ *from Eq.* (5.21) *approaches* $\mathbf{h}_0$ *if* $(\bar{\mathbf{y}},\bar{\lambda})$ *approaches* $(\mathbf{y}_0,\lambda_0)$.

The direct method can be summarized as follows:

Algorithm 5.1. Method for Calculating Branch Points. Let a branch point be located between the continuation steps $(\mathbf{y}^{j-1},\lambda_{j-1})$ and $(\mathbf{y}^j,\lambda_j)$ with values τ_{j-1} and τ_j of the branching test function.

Evaluate Eq. (5.8) and Eq. (5.10) and choose $(\bar{\mathbf{y}},\bar{\lambda}) = (\bar{\mathbf{y}}_0,\bar{\lambda}_0)$ as initial approximation. Establish matrix $\bar{\mathbf{J}}_{lk}$ from Eq. (5.20) and solve Eq. (5.21) yielding $\bar{\mathbf{h}}$. Call SOLVER to solve Eq. (5.18), starting from $(\bar{\mathbf{y}},\bar{\lambda},\bar{\mathbf{h}})$.

The main disadvantage of this direct method is that the number $2n + 1$ of scalar equations to be solved in Eq. (5.18) is essentially doubled. Because solvers work with matrices (as Jacobians) of this size, the amount of storage required to solve the branching system with standard software is about four times as high as in performing an indirect method. This restricts the use of Algorithm 5.1 to small or moderate values of n. Another typical disadvantage of direct methods is that analytical expressions for the first-order derivatives in $\mathbf{f}_\mathbf{y}$ must be calculated and implemented in an appropriate procedure. The disadvantage of the high storage requirements can occasionally be removed by exploiting the special block structure that underlies Eq. (5.18) (discussed in Section 5.5), but the need to establish $\mathbf{f}_\mathbf{y}$ is a severe handicap. Also, one must be aware that in solving Eq. (5.18) by Newton-type methods, second-order derivatives of $\mathbf{f}$ are involved in the Jacobian of $\mathbf{F}$. Although this adds to the complexity, the convenience of solving Eq. (5.18) is hardly affected because the numerical differentiation device of a Newton method takes care of the derivatives. In spite of storage problems with high dimensions n, the method in Algorithm 5.1 has been applied also to PDE examples [288, 309].

So far we have introduced basic versions of indirect and direct methods; more specific direct methods will follow later. Faced with various methods for similar purposes, we need criteria that help judge the effectiveness of the methods. We list some criteria here and illustrate them by means of the direct method of Algorithm 5.1. Essential criteria are

(C1) analytical preparations to be made by the user (such as establishing derivatives)

(C2) storage requirements

(C3) overhead (how much software is required)

(C4) computing time

(C5) applicability

(C6) reliability (difficult to estimate at the present state of the art and hence not discussed in the following).

The criteria influence one another. For example, for a specific problem a specific method may be constructed with very good features in C2 and C4. This efficiency must often be paid for by large overhead, reduced applicability, or many analytical preparations. The above list can be extended, involving such criteria as comfort, but this would lead too far.

The direct method in Algorithm 5.1 is advantagous in C3 and C5 (small overhead and large applicability) and poor with regard to C1 and C2 (the Jacobian must be established in a routine, and the storage increases with order $O(4n^2)$). The first-order derivatives may be obtained through a symbolic manipulator, and the storage can be reduced by means of a specific Newton algorithm. However, such improvements in C1 and C2 require additional overhead, criterion C3.

The answer to the question of how well criterion C4 is satisfied by the method in Algorithm 5.1 is complicated. Let us assume that Eq. (5.18) is solved by Newton-like methods. Because the method in Algorithm 5.1 takes care of a reasonable initial guess, global convergence can be expected to be fast. That is, it is likely that $\bar{\mathbf{Y}}$ is already in the domain of attraction. Concerning the local convergence, one must distinguish between turning point and bifurcation point. At a turning point the Jacobian of $\mathbf{F}$ is nonsingular, which guarantees fast local convergence. For a bifurcation point the Jacobian of $\mathbf{F}$ is singular (since its upper part consists of $(\mathbf{f_y} \mid \mathbf{f}_\lambda)$ with $\mathbf{f}_\lambda \in \text{range}(\mathbf{f_y})$). Hence, one may expect locally slow (linear) convergence. Surprisingly, convergence has shown to be good enough for most practical applications. That is, for average demands of accuracy (four to five significant digits), Newton-like methods show almost the same convergence behavior for the nonsingular case of a turning point and the singular case of a bifurcation point. In particular, the convergence in the λ-component is satisfactory. Extensive experiments on a computer with about 14 decimal digits have shown that the iteration switches from fast to slow convergence after an average of eight digits accuracy is obtained (in λ). Therefore, concerning criterion C4 we may say that Algorithm 5.1 shows fast convergence to turning points for all accuracies and to bifurcation points for average accuracies.

Further remarks on the singularity: Convergence behavior of Newton or Newton-like methods at solutions with singular Jacobians has been investigated repeatedly; see [75, 117, 184, 268] and the references therein. Although in general the convergence to such singular solutions must be expected to be slow, there are certain exceptions to that "rule." Werner [365] found an explanation for the surprisingly good convergence behavior of Newton-like methods that iterate on the branching system (5.18) in order to approximate a bifurcation point: if one reduces the domain of $\mathbf{y}$ and $\mathbf{h}$ to certain subdomains (symmetric and antisymmetric vectors, to be introduced in Section

5.4.4), then the Jacobian of **F** is guaranteed to be nonsingular. Apparently the method for constructing an initial approximation (Algorithm 5.1) establishes an iteration that sticks closely enough to these subdomains. This may explain why the convergence behavior of the general direct method in Algorithm 5.1 has been satisfactory.

5.3.3 An Electrical Circuit

In [260] an electrical circuit is presented (as in Figure 5.9). Several resistors are connected with an operational amplifier in which a voltage U_A is induced, modeled by

$$U_A(U) = 7.65 \arctan(1962U).$$

Two diodes are modeled by

$$I(U) = 5.6 \cdot 10^{-8}(e^{25U} - 1).$$

The voltages at the nodes are denoted by y_i, $i = 1, \ldots, 6$. The input voltage is λ, the output voltage is y_6. The currents satisfy the first law

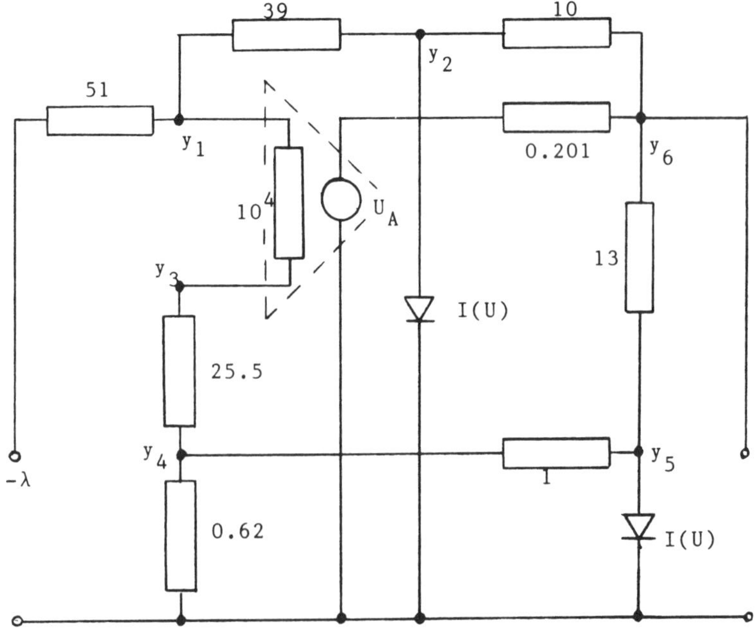

Figure 5.9

of Kirchhoff. Hence, at the six nodes the following six equations hold:

$$
\begin{aligned}
f_1 &= (y_1 - y_3)/10000 + (y_1 - y_2)/39 + (y_1 + \lambda)/51 = 0 \\
f_2 &= (y_2 - y_6)/10 + I(y_2) + (y_2 - y_1)/39 = 0 \\
f_3 &= (y_3 - y_1)/10000 + (y_3 - y_4)/25.5 = 0 \\
f_4 &= (y_4 - y_3)/25.5 + y_4/0.62 + y_4 - y_5 = 0 \\
f_5 &= (y_5 - y_6)/13 + y_5 - y_4 + I(y_5) = 0 \\
f_6 &= (y_6 - y_2)/10 - [U_A(y_3 - y_1) - y_6]/0.201 \\
&\quad + (y_6 - y_5)/13 = 0.
\end{aligned}
\tag{5.22}
$$

Upon varying the input voltage λ, one observes jumps of the output voltage. As expected, the jumps can be explained by hysteresis (compare Figure 5.10). A level of high voltage ($y_6 > 11$) and a low-voltage level are connected by a branch of unstable solutions. In short, the circuit exhibits *trigger* behavior.

We demonstrate the branching system with this example. To this end, the first-order partial derivatives of $\mathbf{f}(\mathbf{y},\lambda)$ are required. Because most expressions are linear, the calculation of the Jacobian is easy. The only nonlinear terms arise in f_2, f_5, and f_6. The nonzero entries c_{ij} of

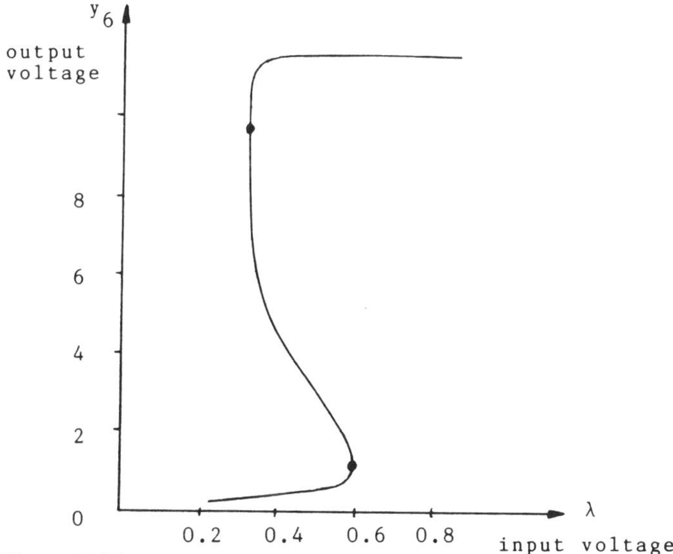

Figure 5.10

the Jacobian are

$$c_{11} = 10^{-4} + \frac{1}{39} + \frac{1}{51}$$

$$c_{12} = c_{21} = -\frac{1}{39}$$

$$c_{13} = -c_{31} = 10^{-4}$$

$$c_{22} = \frac{1}{10} + \frac{1}{39} + 1.4 \cdot 10^{-6}[\exp(25y_2) - 1]$$

$$c_{26} = c_{62} = -\frac{1}{10}$$

$$c_{33} = 10^{-4} + \frac{1}{25.5}$$

$$c_{34} = c_{43} = -\frac{1}{25.5}$$

$$c_{44} = 1 + \frac{1}{25.5} + \frac{1}{0.62}$$

$$c_{45} = c_{54} = -1$$

$$c_{55} = 1 + \frac{1}{13} + 1.4 \cdot 10^{-6}[\exp(25y_5) - 1]$$

$$c_{56} = c_{65} = -\frac{1}{13}$$

$$c_{61} = -c_{63} = 150009.3/[1 + 3849444(y_3 - y_1)^2]$$

$$c_{66} = \frac{1}{10} + \frac{1}{0.201} + \frac{1}{13}.$$

With the notations

$$y_7 = \lambda,\ y_8 = h_1, \ldots, y_{13} = h_6,$$

the six equations of the linearized problem

$$\mathbf{f}_\mathbf{y}(\mathbf{y},\lambda)\mathbf{h} = \mathbf{0}$$

can be written as

$$\sum_{j=1}^{6} c_{ij}h_j = 0 \qquad (i = 1, \ldots, 6).$$

For instance, the seventh (= $(n + 1)$st) component of the branching system (5.18) is

$$f_7 = \sum_{j=1}^{6} c_{ij} y_{j+7} = 0.$$

It is possible but tedious to write the equations

$$f_7 = 0, \ldots, f_{12} = 0$$

down explicitly. We omit that, emphasizing that this kind of formula is best organized by carrying out multiplications matrix*vector (Exercise 5.12). This convenient strategy is only slightly more expensive than an explicit formulation of each equation. In order to apply a direct method, the Jacobian is implemented in a separate procedure (with the name COEFF, say, and output c_{ij}). As is typical for a wide range of examples, the Jacobian is *sparse*—that is, mostly zero. In the above example of a specific electric circuit, one evaluation of COEFF takes even fewer arithmetic operations than evaluating the function **f** of Eq. (5.22) itself, provided the constants are implemented as decimals. But notice how much effort is needed before COEFF is constituted.

The equations of the linearization are

$$f_{6+i} = \sum_{j=1}^{6} c_{ij} y_{6+1+j} = 0, \qquad i = 1, \ldots, 6.$$

The normalizing equation $h_k = 1$ with $k = 1$ is

$$f_{13} = y_8 - 1 = 0.$$

This completes the branching system. This system of 13 equations requires a good initial approximation, as proposed earlier in this section. We omit further details and merely give the two obtained solutions to the branching system in the table below. The **h**-components ($y_8, \ldots, y_{13}$) are not listed.

y_1	0.049367	0.235778
y_2	0.547358	0.662969
y_3	0.049447	0.237598
y_4	0.049447	0.237602
y_5	0.129201	0.620832
y_6	1.166020	9.608997
λ	0.601853	0.322866.

The results are rounded to six digits behind the floating point. There is no value of the input voltage λ such that a value of the output voltage y_6 exists in the range

$$1.16602 < y_6 < 9.608997.$$

The critical input voltages where the bistability manifests itself by jumps are

$$0.601853 \quad \text{and} \quad 0.322866$$

(compare Figure 5.10).

5.3.4 A Family of Test Functions

We go back to the description of the direct method in Algorithm 5.1 and in particular review the calculation of the initial approximation $\overline{\mathbf{h}}$. Remember that the l-th equation of the linearized system

$$\mathbf{f_y}(\overline{\mathbf{y}},\overline{\lambda})\mathbf{h} = \mathbf{0}$$

has been removed in order to produce a solvable system. After having calculated $\overline{\mathbf{h}}$ by solving Eq. (5.21), one can calculate the value of the removed component:

$$\mathbf{e}_l^{tr}\mathbf{f_y}(\overline{\mathbf{y}},\overline{\lambda})\overline{\mathbf{h}}.$$

This number measures the distance to a branch point $(\mathbf{y}_0,\lambda_0)$ because it is zero for $(\overline{\mathbf{y}},\overline{\lambda}) = (\mathbf{y}_0,\lambda_0)$ and nonzero for $(\overline{\mathbf{y}},\overline{\lambda}) \neq (\mathbf{y}_0,\lambda_0)$. Thus the variable

$$\tau_{lk}(\mathbf{y},\lambda) = \mathbf{e}_l^{tr}\mathbf{f_y}(\mathbf{y},\lambda)\mathbf{h} \tag{5.23}$$

with

$$[(\mathbf{I} - \mathbf{e}_l\mathbf{e}_l^{tr})\mathbf{f_y}(\mathbf{y},\lambda) + \mathbf{e}_l\mathbf{e}_k^{tr}]\mathbf{h} = \mathbf{e}_l$$

is a branching test function. The matrix in brackets is $\mathbf{J}_{lk}$, which was shown to be easily established. Hence, formally, the test function (5.23) can be written

$$\tau_{lk} = \mathbf{e}_l^{tr}\mathbf{J}\mathbf{J}_{lk}^{-1}\mathbf{e}_l.$$

An evaluation of τ_{lk} consists of the two steps

(a) solve the linear system $\mathbf{J}_{lk}\mathbf{h} = \mathbf{e}_l$

(b) evaluate the scalar product of $\mathbf{h}$ with the l-th row of the Jacobian $\mathbf{J}$.

(An alternative is offered by the representation in Exercise 5.11c.)

Because **h** depends on the choice of the indices l and k, Eq. (5.23) defines quite a number of test functions τ_{lk}. The requirement that $\mathbf{J}_{lk}(\mathbf{y}_0,\lambda_0)$ must be nonsingular hardly restricts the choice of l and k. One may think of picking l and k in such a way that the nonsingularity of $\mathbf{J}_{lk}(\mathbf{y}_0,\lambda_0)$ becomes most pronounced. A related criterion (proposed in [171]) is based on the Householder decomposition of this matrix. Such a strategy faces the inherent difficulty that $\mathbf{J}_{lk}$ is evaluated at $(\overline{\mathbf{y}},\overline{\lambda}) \neq (\mathbf{y}_0,\lambda_0)$. A nonsingularity of $\mathbf{J}_{lk}(\overline{\mathbf{y}},\overline{\lambda})$ does not ensure the nonsingularity of $\mathbf{J}_{lk}(\mathbf{y}_0,\lambda_0)$. Fortunately, in most practical applications an arbitrarily chosen pair of indices works well, say, with $l = k = 1$. In examples with very large values of n with sparse or banded Jacobians, l and k should be chosen so that $\mathbf{J}_{lk}$ retains its structure. Then the special iterative solvers that solve the linear system of a Newton step can also be applied for calculating **h**. An alternative has already been mentioned (Exercise 5.11c).

We conclude the introduction of the test function τ_{lk} by giving two practical hints:

(a) $\mathbf{J}_{lk}(\mathbf{y},\lambda)$ can be singular for some $(\overline{\mathbf{y}},\overline{\lambda})$, although $\mathbf{J}_{lk}(\mathbf{y}_0,\lambda_0)$ is nonsingular. This singularity of $\mathbf{J}_{lk}$ causes a pole of the test function τ_{lk}. For sufficiently small continuation step size, a pole can be discovered by incorporating a third value of τ in Eq. (5.4): Require

$$|\ \tau(\mathbf{y}^{j-2},\lambda_{j-2})\ | > |\ \tau(\mathbf{y}^{j-1},\lambda_{j-1})\ |$$

to hold in addition to the change of sign.

(b) Formulas establishing a calculation of the derivative $d\tau/d\lambda$ of τ were derived. These formulas are costly because they require numerical approximation of the second-order derivatives of $\mathbf{f}(\mathbf{y},\lambda)$ with respect to $\mathbf{y}$ [309].

Example 5.2.

We reconsider the example of the electrical circuit in Figure 5.9, with voltages satisfying the system of equations (5.22). Upon tracing the lower branch (Figure 5.10, $y_6 < 1.16$), test functions τ_{11} and τ_{42} are calculated. The choice of these indices was random. The discrete values of the test functions are connected by smooth interpolating curves. The result is shown in Figure 5.11. Both test functions signal the presence of the turning point. The figure also illustrates that the characteristic behavior of various test functions may differ even though the roots are identical.

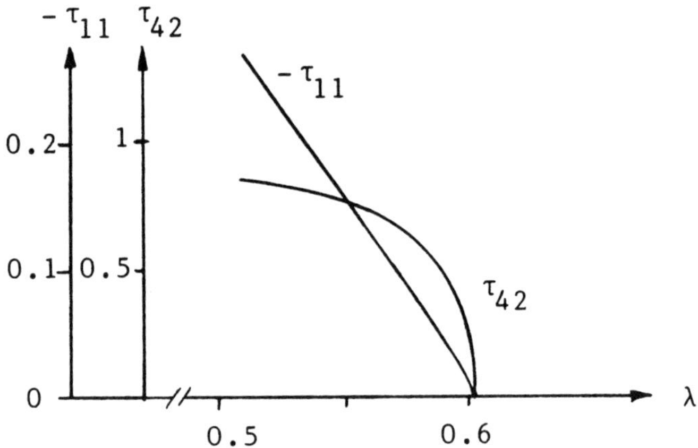

Figure 5.11

Exercise 5.10 (an academic example, following [65]).
Consider the equations

$$f_1(y_1,y_2,\lambda) = y_1 + \lambda y_1(y_1^2 + y_2^2 - 1) = 0$$

$$f_2(y_1,y_2,\lambda) = 10y_2 - \lambda y_2(1 + 2y_1^2 + y_2^2) = 0.$$

(a) Establish the branching system for this example.
(b) For the branch $y_1 = ((\lambda - 1)/\lambda)^{1/2}$, $y_2 = 0$, calculate the matrices $\mathbf{J}_{11}$ and $\mathbf{J}_{22}$.
(c) Show that τ_{11} and τ_{22} signal distinct branch points.

Exercise 5.11.
Consider the system

$$\begin{pmatrix} \mathbf{f}(\mathbf{y},\lambda) \\ (\mathbf{I} - \mathbf{e}_l\mathbf{e}_l^{tr})\mathbf{f}_\mathbf{y}(\mathbf{y},\lambda)\mathbf{h} \\ h_k - 1 \\ p(\mathbf{y},\lambda,\mathbf{h}) \end{pmatrix} = \mathbf{0}.$$

Solutions are parameterized by the scalar equation $p(\mathbf{y},\lambda,\mathbf{h}) = 0$ in one of two ways:

(1) $p_1(\mathbf{y},\lambda,\mathbf{h}) = \lambda - \eta$
(2) $p_2(\mathbf{y},\lambda,\mathbf{h}) = \mathbf{e}_l^{tr}\mathbf{f}_\mathbf{y}(\mathbf{y},\lambda)\mathbf{h} - \eta.$

The solutions depend on η, $\mathbf{y}(\eta)$, $\lambda(\eta)$, and $\mathbf{h}(\eta)$. The parameterization p_2 establishes an *embedding* of the branching system [306]. This idea gives rise to modifications of the calculation of an initial approximation.

(a) For which choices of parameterization does one obtain the initial approximation $\bar{\mathbf{h}}$ from Eq. (5.21) and the branching system (5.18)?
(b) How can this be used to construct a homotopy (continuation) for solving the branching system?
(c) Assume that the inverse of the Jacobian exists. Show that $\bar{\mathbf{h}}$ from Eq. (5.21) can be defined as solution of the inhomogeneous equation

$$\mathbf{f_y}(\bar{\mathbf{y}},\bar{\lambda})\mathbf{h} = \mathbf{z}, \quad \mathbf{z} = \mathbf{e}_l(\mathbf{e}_k^{tr}\mathbf{f_y}^{-1}\mathbf{e}_l)^{-1}.$$

(d) Is it possible to dispense with indices l and k?
(e) Assume that an **LU** decomposition of the Jacobian is calculated. Show that for $l = n$ the calculation of $\mathbf{h}$ requires only one back substitution.

Exercise 5.12.

(a) Design an algorithm that constitutes the function $\mathbf{F}$ of Eq. (5.18). Input: n, $\mathbf{Y}$; assume that procedures FX (calculates $\mathbf{f}(\mathbf{y},\lambda)$) and COEFF (calculates $\mathbf{f_y}(\mathbf{y},\lambda)$) are available.
(b) Modify the algorithm so that the known component h_k is saved and not recalculated as part of the branching system.

5.3.5 Temperature Distribution in a Reacting Material

In [26] the partial differential equation

$$\frac{\partial \Theta}{\partial t} = \nabla^2\Theta + \lambda \exp \frac{\Theta}{1 + \epsilon\Theta} \tag{5.24}$$

is investigated. The variable

$$\Theta = (T - T_a)E/(RT_a^2)$$

is a normalized temperature, with

T: actual temperature
T_a: surface temperature

R: gas constant

E: Arrhenius activation energy

$\epsilon = \mathrm{R}T_a/E$

λ: nondimensional lumped parameter

$\nabla^2\Theta$: Laplacian (with respect to dimensionless space variable).

Equation (5.24) models the temperature within a material with exothermic reaction. We confine ourselves to the steady-state situation in a unit square $0 \leq x_1 \leq 1$, $0 \leq x_2 \leq 1$,

$$\begin{aligned} -\nabla^2\Theta &= \lambda \exp \frac{\Theta}{1 + \epsilon\Theta} && \text{inside the square,} \\ \Theta &= 0 && \text{on the boundary.} \end{aligned} \tag{5.25}$$

Employing the standard five-point star for discretizing the Laplacian on a uniform square mesh with N^2 interior grid points

$$\left(\frac{i}{N+1}, \frac{j}{N+1}\right), \qquad i,j = 1, 2, \ldots, N$$

yields a system $\mathbf{f}(\mathbf{y},\lambda) = \mathbf{0}$ of N^2 equations (Exercise 5.13).

Calculating solutions for various values of λ and ϵ produces branching diagrams as depicted in Figure 5.12. For $\epsilon = 0$ there is one

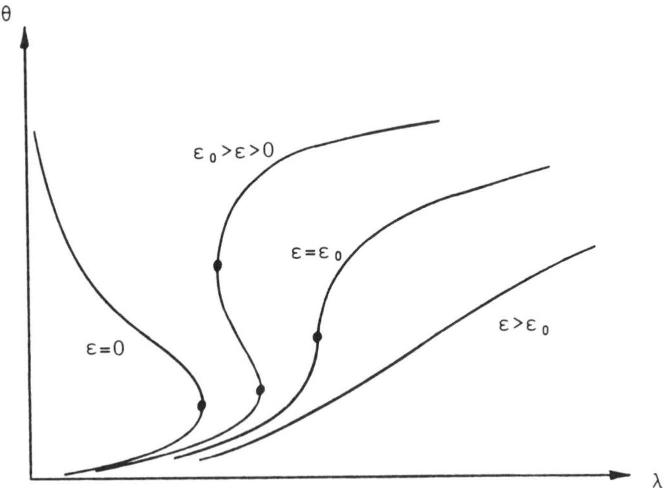

Figure 5.12

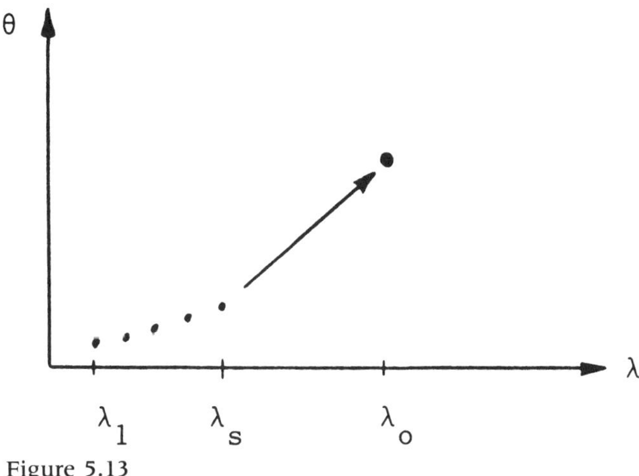

Figure 5.13

turning point only; for $\epsilon > 0$ the branch turns back in a second turning point. Eventually, if ϵ is increased to ϵ_0, the turning points collapse in a hysteresis point. The critical value λ_0 of the lower-temperature turning point has a physical meaning: if λ is increased past λ_0, a jump to higher temperature occurs and the material ignites. For $\epsilon > \epsilon_0$ there is no such jump. The value of ϵ_0 depends on the geometry [26, 27].

This example illustrates characteristic differences between direct and indirect methods. The two classes of methods will be discussed and compared at the end of this section. We confine ourselves to the case $\epsilon = 0$, which has been treated repeatedly in the literature (for example, [2, 236, 242, 322]).

For the numerical experiment to be described next, see Figure 5.13. We run a continuation starting with the (small) value of $\lambda = \lambda_1$. At λ_s we stop tracing the branch and change to a direct method in order to calculate the turning point at λ_0. This model calculation is purely for the purpose of demonstration and is run with the coarse discretization $N = 3$. The direct method of solving Eq. (5.18) yields the value of $\lambda_0 = 6.6905$. We measure the overall computing time needed from λ_1 to λ_0. The time depends on the strategy of choosing λ_s. The following table reports on the computing times (run on a CDC Cyber).

λ_s	6.67	6.6	6.2	5.2	4.05	3.07	2.07	1.07	0.27
sec	0.94	0.8	0.67	0.65	0.59	0.65	0.73	0.39	0.32

As the above table shows, a minimum of computing time was attained with a strategy that left the continuation at $\lambda_s \approx 4$ in order to jump directly to λ_0. If the jump is too large, a direct method has convergence problems as a rule. The surprisingly short computing times for $\lambda_s \approx 1$ reported in this particular experiment are an exception due to the simple structure of the almost linear equations. Based on these results, one might think of switching to a direct method early enough in order to save time. But this kind of strategy is not realistic. First, one does not know which value of λ_s is "fast." Second, nothing is known about the solution structure in the gap between λ_s and λ_0. This lack of information cannot usually be tolerated. Hence, a direct method should not replace part of the continuation. Instead, a direct method can be applied *after* at least rudimentary results have been obtained by an indirect method.

Another aspect that distinguishes between indirect and direct methods is the effort required to obtain branch points in high accuracy. If the iteration of a direct method converges, then any additional iteration step improves the accuracy significantly. In contrast, indirect methods on the average have more difficulty achieving highly accurate approximations of branch points. This does not mean, however, that the accuracy is required. The following table gives the values of λ_0 for four values of N (results partly from [2]):

N	λ_0
3	6.6905 . . .
7	6.7833 . . .
15	6.8080865 . . .
23	6.80811698 . . .

For instance, the effort of applying a direct method in case $N = 7$ is worthless because the discretization error is large. In fact, any continuation offers an approximation to λ_0 with an error smaller than this discretization error. The mere checking of a graphical output of a corresponding branching diagram produces enough accuracy. Even for $N = 15$ the discretization error is still worse than the accuracy that can be obtained by an indirect method as an easy by-product of a continuation (see Exercises 5.7–5.9, for instance). It is apparently futile in this particular example to apply a direct method such as that in Algorithm 5.1.

Exercise 5.13.
Derive the N^2 equations that are the discrete analog of the Dirichlet problem (5.25). Denote the approximation to

$$\Theta\left(\frac{i}{N+1}, \frac{j}{N+1}\right)$$

by u_{ij}, and approximate the associated Laplacian by

$$(N+1)^2(-4u_{ij} + u_{i-1,j} + u_{i+1,j} + u_{i,j-1} + u_{i,j+1})$$

(why?). Take $n = N^2$ and

$$y = (u_{11}, \ldots, u_{1N}, u_{21}, \ldots, u_{NN})$$

and constitute the n equations of $\mathbf{f}(\mathbf{y},\lambda) = \mathbf{0}$.

5.3.6 Direct Versus Indirect Methods

Direct methods for calculating branch points came up only recently and so far have scarcely been compared with indirect methods. Although research in this field is still going on, some tentative guidelines appear to be appropriate. The present section introduces direct methods that can be applied to general branch points. Because an introduction into variants and further developments is deferred to Section 5.5, we restrict our remarks to features inherent in all direct methods.

A judgment on what kind of method may be preferred is strongly affected by the demands for accuracy. In case one wants to calculate branch points with high accuracy, it may be worthwhile to derive and implement analytical expressions for the Jacobian in order to apply a direct method. A direct method does not replace part of a continuation, but it typically improves the guess produced during branch tracing by an indirect method. The direct method need not be applied when the medium accuracy achieved by the indirect method is sufficient.

There are two classes of problems for which it does not seem to make sense to request small error tolerances for $|\lambda_0 - \bar{\lambda}_0|$. First, equations resulting from discretization are subjected to discretization errors that typically exceed the error $|\lambda_0 - \bar{\lambda}_0|$ of a readily available indirect method. The second class of problems not requiring accurate approximations $\bar{\lambda}_0$ is the calculation of branch points that are unstable to perturbations. In this context we mentioned bifurcation points and hysteresis points; their calculation in one-parameter problems can be regarded as an ill-posed problem. Hence, for the approximation of bifurcation points during branch tracing and of branch points in equations resulting from discretization, indirect methods are sufficient. For all cases, indirect methods are well qualified and are to be recommended. In summary, we concede that a practical need for direct methods appears to be limited.

So far, we have made a sharp distinction between indirect and direct methods. Essentially our "direct" methods have met two criteria:

(a) The iteration is not confined to walk along the branch.
(b) Equations in standard form have been formulated, and the solution is delegated to standard software.

This understanding is by no means obvious. Relaxing criterion (b) qualifies a wide range of methods as direct methods or semidirect methods. In a way, then, even indirect methods might be formulated as "direct" methods, because (a) does not forbid walking along the branch. We therefore take the position that a direct method should meet both criteria. A free interpretation of what "delegate to standard software" means is a compromise between a meaningless and a too-restrictive definition of direct methods.

5.4 BRANCH SWITCHING

In this section we consider methods for calculating (at least) one solution on the emanating branch. We assume the following typical situation: some solutions $\mathbf{y}(\lambda)$ to $\mathbf{f}(\mathbf{y},\lambda) = \mathbf{0}$ have been calculated—call two of them $\mathbf{y}(\lambda_1)$ and $\mathbf{y}(\lambda_2)$. A bifurcation has been roughly located by an indirect method that travels along the branch. It makes sense to assume that at least one solution (say, $\mathbf{y}(\lambda_1)$) is situated somewhat close to the bifurcation $(\mathbf{y}_0,\lambda_0)$. We denote a solution on the emanating branch by $\mathbf{z}$—that is, for some λ the equation $\mathbf{f}(\mathbf{z},\lambda) = \mathbf{0}$ is satisfied (see Figure 5.14). As mentioned earlier, branch switching means to calculate one "emanating solution" $(\mathbf{z},\lambda)$. This "first solution" on the emanating branch then serves as the starting point for a subsequent tracing of the entire branch. As indicated in Figure 5.14, we assume that the calculated branch can be parameterized by λ. The other situation in which the calculated solutions are points on a branch that cannot be parameterized by λ are discussed at the end of this section.

All methods for switching branches can be regarded as consisting of two phases: First, an initial guess $(\bar{\mathbf{z}},\bar{\lambda})$ is to be constructed, then an

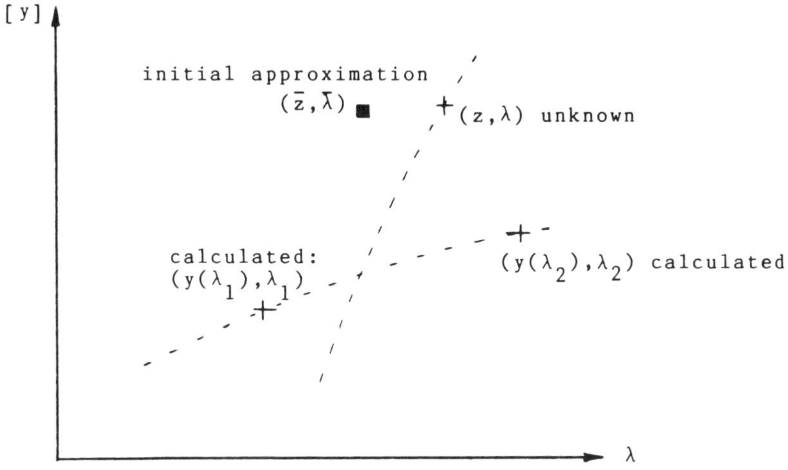

Figure 5.14

iteration that should converge to the new branch must be established. That is, branch switching can be seen as a matter of establishing suitable predictors and correctors.

5.4.1 Constructing a Predictor via the Tangent

A first idea is to calculate the tangent to the emanating branch in the bifurcation point. This requires a rather accurate calculation of the bifurcation point $(\mathbf{y}_0,\lambda_0)$ and of the derivatives of $\mathbf{f}(\mathbf{y},\lambda)$ in the bifurcation point. In what follows, these derivatives will be denoted by

$$\mathbf{f}_\mathbf{y}^0 = \mathbf{f}_\mathbf{y}(\mathbf{y}_0,\lambda_0), \quad \mathbf{f}_\lambda^0 = \mathbf{f}_\lambda(\mathbf{y}_0,\lambda_0).$$

The singularity of the Jacobian $\mathbf{f}_\mathbf{y}^0$ leads to the assumptions

$$\begin{aligned} &\text{null}(\mathbf{f}_\mathbf{y}^0) \text{ is spanned by } \mathbf{h}_0, \\ &\text{range}(\mathbf{f}_\mathbf{y}^0) = \text{null}(\Psi) \text{ with } \Psi^{tr}\mathbf{h}_0 = 1. \end{aligned} \tag{5.26}$$

Here $\text{null}(\Psi) = \{\mathbf{y} \mid \Psi^{tr}\mathbf{y} = 0\}$ denotes a null space, Ψ is a left eigenvector. Recall that a bifurcation point is characterized by

$$\mathbf{f}_\lambda^0 \in \text{range } \mathbf{f}_\mathbf{y}^0,$$

which can be written as

$$\Psi^{tr}\mathbf{f}_\lambda^0 = \mathbf{0}.$$

This condition implies the existence of a unique vector φ with

$$\mathbf{f}_\mathbf{y}^0\varphi + \mathbf{f}_\lambda^0 = \mathbf{0} \text{ with } \Psi^{tr}\varphi = 0. \tag{5.27}$$

Regard the equation $\Psi^{tr}\varphi = 0$ as a convenient way of saying that φ is in the range of $\mathbf{f}_\mathbf{y}^0$. The vector $\mathbf{h}_0$ in Eq. (5.26) is the eigenvector associated with the zero eigenvalue of $\mathbf{f}_\mathbf{y}^0$; it is part of the solution of the branching system (5.18). Let $(\mathbf{y}(s),\lambda(s))$ be a smooth branch passing through the bifurcation point, with $(\mathbf{y}(s_0),\lambda(s_0)) = (\mathbf{y}_0,\lambda_0)$. The parameter s stands for arclength or any other local parameter (see Section 4.2). The tangent defined by

$$\left(\frac{d\mathbf{y}}{ds}, \frac{d\lambda}{ds}\right)$$

satisfies in the bifurcation point the equation

$$\mathbf{f}_\mathbf{y}^0 \frac{d\mathbf{y}}{ds} + \mathbf{f}_\lambda^0 \frac{d\lambda}{ds} = \mathbf{0}. \tag{5.28}$$

Equations (5.28) and (5.27) and the property

$$\mathbf{f}_\mathbf{y}^0\mathbf{h}_0 = \mathbf{0}$$

imply

$$\frac{d\mathbf{y}}{ds} = \gamma_0\varphi + \gamma_1\mathbf{h}_0 \tag{5.29a}$$

with two constants γ_0,γ_1. The derivatives in Eqs. (5.28), (5.29a), and (5.29b) are evaluated at the bifurcation point. The constants γ_0 and γ_1 satisfy

$$\gamma_0 = \frac{d\lambda}{ds}, \gamma_1 = \Psi^{tr}\frac{d\mathbf{y}}{ds} \tag{5.29b}$$

(Exercise 5.15); they can be obtained from second-order derivatives.

The need for evaluating second-order derivatives comes from the singularity of the matrix

$$(\mathbf{f}_\mathbf{y}^0 \mid \mathbf{f}_\lambda^0)$$

in Eq. (5.28). Apart from bifurcation points, this matrix has full rank ($= n$) and the tangent can be calculated via Eq. (5.28); this involves only first-order derivatives. Recall that an equation corresponding to Eq. (5.28) was used in the context of continuation; compare Eq. (4.3a) and Eq. (4.4).

In order to derive equations for γ_0 and γ_1, we differentiate the identity

$$\mathbf{0} = \mathbf{f}_\mathbf{y}(\mathbf{y}(s),\lambda(s)) \frac{d\mathbf{y}(s)}{ds} + \mathbf{f}_\lambda(\mathbf{y}(s),\lambda(s)) \frac{d\lambda(s)}{ds}$$

with respect to s (Exercise 5.16). Evaluating the resulting expressions at the bifurcation point and indicating derivatives with respect to s by primes,

$$\mathbf{y}' = \frac{d\mathbf{y}}{ds}, \ \lambda' = \frac{d\lambda}{ds},$$

one has

$$\mathbf{f}^0_\mathbf{y}\mathbf{y}'' + \mathbf{f}^0_\lambda\lambda'' = -[\mathbf{f}^0_{\mathbf{yy}}\mathbf{y}'\mathbf{y}' + 2\mathbf{f}^0_{\mathbf{y}\lambda}\mathbf{y}'\lambda' + \mathbf{f}^0_{\lambda\lambda}\lambda'\lambda'].$$

Both terms on the left side are in the range of $\mathbf{f}^0_\mathbf{y}$. Hence the right-hand side must be in this range too,

$$\Psi^{tr}[\mathbf{f}^0_{\mathbf{yy}}\mathbf{y}'\mathbf{y}' + 2\mathbf{f}^0_{\mathbf{y}\lambda}\mathbf{y}'\lambda' + \mathbf{f}^0_{\lambda\lambda}\lambda'\lambda'] = 0.$$

Inserting Eq. (5.29a, b) and using the definitions

$$\begin{aligned} a &= \Psi^{tr}\mathbf{f}^0_{\mathbf{yy}}\mathbf{h}_0\mathbf{h}_0 \\ b &= \Psi^{tr}(\mathbf{f}^0_{\mathbf{yy}}\varphi + \mathbf{f}^0_{\mathbf{y}\lambda})\mathbf{h}_0 \\ c &= \Psi^{tr}(\mathbf{f}^0_{\mathbf{yy}}\varphi\varphi + 2\mathbf{f}^0_{\mathbf{y}\lambda}\varphi + \mathbf{f}^0_{\lambda\lambda}) \end{aligned} \tag{5.30}$$

yield the quadratic

$$a\gamma_1^2 + 2b\gamma_1\gamma_0 + c\gamma_0^2 = 0. \tag{5.31}$$

For a positive discriminant

$$b^2 - ac > 0 \tag{5.32}$$

there are two distinct pairs of solutions (γ_0,γ_1) to Eq. (5.31), referring to two distinct tangents in the bifurcation point. Condition (5.32) frequently serves as criterion for a simple bifurcation point, because it guarantees a transversal intersection of two branches in $(\mathbf{y}_0,\lambda_0)$; see, for instance, [44, 241]. That is, Eq. (5.32) can replace condition (4) of the definition of a simple bifurcation point (Definition 2.3). We note in passing that $a \neq 0$ is a criterion for transcritical bifurcation and that $a = 0$ is characteristic for pitchfork bifurcation [44, 366]. The above analysis leading to the quadratic Eq. (5.31) was carried out in a general setting by Keller [178].

Let us now return to the topic of branch switching. As we have seen, it is possible to calculate the tangents to branches in the bifurcation point. The vector **h** can be obtained either by the methods described in the preceding section or by inverse iteration [178, 293, 271, 309]. On how to calculate φ, refer to [178]. After the tangents are calculated, initial approximations $\bar{\mathbf{z}}$ to $\mathbf{z}$ are obtained the same way predictor steps are calculated in continuation. (Note that in this section $\bar{\mathbf{z}}$ is not a tangent but the result of the predictor.) The method of calculating the tangents to the branches in the bifurcation point is convincing in that it offers a systematic and reliable way of branch switching. Related methods were also applied in [241] and [200]. A severe disadvantage of calculating the tangents in the bifurcation point is high cost. The involved work is based on two expensive steps:

(a) The bifurcation point must be approximated quite accurately.

(b) The second-order derivatives must be evaluated or approximated.

In particular, the costs of calculating second-order derivatives do not appear to be generally tolerable, so we seek for cheaper alternatives.

Exercise 5.14.
Suppose that $(\mathbf{y}_0,\lambda_0)$ is a bifurcation point with Eq. (5.26). Show that there is a unique vector φ such that Eq. (5.27) holds.

Exercise 5.15.
Show that Eq. (5.29b) holds.

Exercise 5.16.
Derive the quadratic Eq. (5.31) in detail.

5.4.2 Predictors Based on Interpolation

Two branches intersecting transversely in a bifurcation point $(\mathbf{y}_0,\lambda_0)$ define a plane that is spanned by the two tangents to the branches in $(\mathbf{y}_0,\lambda_0)$. It would be nice to have a vector **d** in that plane, perpendicular to the λ-axis. If the curvature of the branches is small enough, this vector **d** can be regarded as an approximation to the difference between the branches, provided one knows an appropriate length of **d**. To say it in other words, define

$$\mathbf{d}(\bar{\lambda}) = \mathbf{z}(\bar{\lambda}) - \mathbf{y}(\bar{\lambda})$$

as this difference, where $\mathbf{f}(\mathbf{y},\bar{\lambda}) = \mathbf{f}(\mathbf{z},\bar{\lambda}) = \mathbf{0}$. Here, for ease of demonstration, both branches are taken as λ-parameterized. For $\bar{\lambda}$ close to λ_0, the vector $\mathbf{d}$ is almost parallel to the plane that is spanned by the tangents in $(\mathbf{y}_0,\lambda_0)$. If one knew this vector $\mathbf{d}$, one could easily switch directly from one branch to the other without taking the detour of calculating $(\mathbf{y}_0,\lambda_0)$. Such an approach, if possible, would be less affected by unfoldings of the bifurcation due to perturbations. The assumption that the distance $\mathbf{d} = \mathbf{z} - \mathbf{y}$ can be had without calculating the bifurcation point is out of the question, but a major part of the program, can be realized effectively. Related ideas are introduced in [312, 313, 314].

The problem of finding an approximation to $\mathbf{d}$ can be split into two subproblems (Figure 5.15):

(a) Find a direction $\bar{\mathbf{d}}$ pointing to the other branch, where $\bar{\mathbf{d}}$ is normalized by $\bar{\mathbf{d}}_k = 1$ (say).

(b) Find the distance δ between the branches, measured in the k-th component

$$\delta = z_k - y_k.$$

Such variables $\bar{\mathbf{d}}$ and δ establish a predictor by

$$\bar{\mathbf{z}} = \mathbf{y} + \delta\bar{\mathbf{d}}. \tag{5.33}$$

Notice that δ, $\bar{\mathbf{d}}$, $\mathbf{y}$, and $\bar{\mathbf{z}}$ depend on λ.

In Eq. (5.21) we introduced a variable $\bar{\mathbf{h}}$ as solution of a system of linear equations (Definition 5.1.). This vector $\bar{\mathbf{h}}$ serves as direction

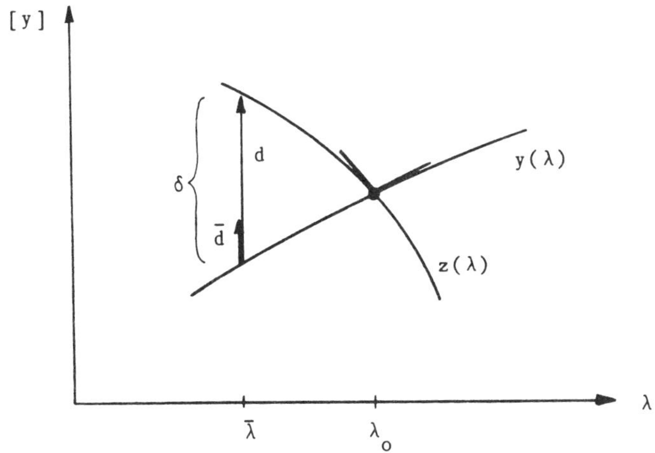

Figure 5.15

$\overline{\mathbf{d}}$. We know from Section 5.3.2 that $\overline{\mathbf{h}} \approx \mathbf{h}_0$ for $\overline{\lambda} \approx \lambda_0$. The vector $\mathbf{h}_0$ lies in the plane that is spanned by the two tangents in the bifurcation point, and $\mathbf{h}_0$ is perpendicular to the λ-axis. This holds for the majority of bifurcation problems. Hence, we can consider the easily available $\overline{\mathbf{h}}$ as an approximation to subproblem (a).

Finding the distance δ that fits the current value of the parameter $\lambda = \overline{\lambda}$ (Figure 5.15) appears to be impossible, at least for acceptable cost (no second-order derivatives, no accurate calculation of the bifurcation point). This difficulty can be shifted to the corrector, where it is much easier to cope with. This postponing and mitigation of difficulties can be accomplished by exchanging the roles of δ and λ. Instead of thinking about $\delta = \delta(\lambda)$, we prescribe δ and let the corrector take care of how to calculate $\lambda = \lambda(\delta)$. The value

$$|\,\delta\,| = \delta_0 \max\{1, |\,y_k(\overline{\lambda})\,|\} \tag{5.34}$$

with $\delta_0 = 0.02$, for example, has been used successfully in a wide range of practical problems. It remains to fix the sign of δ. We take a pragmatic point of view and take both signs. This means that two predictors $\overline{\mathbf{z}}$ are established and hence two solutions $\mathbf{z}$ of $\mathbf{f}(\mathbf{z},\lambda) = \mathbf{0}$ are to be calculated, one on either half-branch emanating from the bifurcation point. We summarize the calculation of the predictor:

Algorithm 5.2. Calculation of the Predictor.

Assume $(\overline{\mathbf{y}},\overline{\lambda})$ is calculated not far from $(\mathbf{y}_0,\lambda_0)$.
Calculate $\overline{\mathbf{h}}$ from Eq. (5.21).
Choose $|\,\delta\,|$ as in Eq. (5.34).
Two predictors are given by $\overline{\mathbf{z}} = \overline{\mathbf{y}} \pm |\,\delta\,|\,\overline{\mathbf{h}}$.

The success of the predictor in Algorithm 5.2, strongly depends on an appropriate corrector. In the case of pitchfork bifurcation, $\overline{\mathbf{z}}$ is close to the $\mathbf{z}$-branch, provided $\overline{\lambda}$ is close enough to λ_0. (This will become clear later.) For transcritical bifurcations it is not likely that $\overline{\mathbf{z}}$ is close to $\mathbf{z}$, and the corrector must compensate for a poor predictor. This situation calls for some care in designing a corrector (see next subsection).

Calculation of the predictor (see Algorithm 5.2) is based on results obtained at one solution $(\overline{\mathbf{y}},\overline{\lambda})$ only. In practice, at least two solutions $(\mathbf{y}^1,\lambda_1)$ and $(\mathbf{y}^2,\lambda_2)$ are available close to the bifurcation. Interpolating the data of two such solutions enables a calculation of better approximations to $\mathbf{h}_0$. The formulas are elementary; most of them were used in the context of indirect methods (Section 5.3.1). The interpolation requires the values of a branching test function τ. Linear interpolation yields the formulas

$$\xi = \tau(\mathbf{y}^1,\lambda_1)/[\tau(\mathbf{y}^1,\lambda_1) - \tau(\mathbf{y}^2,\lambda_2)]$$

$$\overline{\lambda}_0 = \lambda_1 + \xi(\lambda_2 - \lambda_1) \tag{5.35a}$$

$$\overline{\mathbf{y}}_0 = \mathbf{y}^1 + \xi(\mathbf{y}^2 - \mathbf{y}^1).$$

Let $\mathbf{h}^1$ and $\mathbf{h}^2$ be the vectors defined by Eq. (5.21), i.e.,

$$[(\mathbf{I} - \mathbf{e}_l\mathbf{e}_l^{tr})\mathbf{f}_\mathbf{y}(\mathbf{y}^i,\lambda_i) + \mathbf{e}_l\mathbf{e}_k^{tr}]\mathbf{h}^i = \mathbf{e}_l \qquad \text{for } i = 1,2,\ldots \tag{5.35b}$$

$$\overline{\mathbf{h}}_0 = \mathbf{h}^1 + \xi(\mathbf{h}^2 - \mathbf{h}^1)$$

$$\overline{\mathbf{z}} = \overline{\mathbf{y}}_0 \pm |\delta|\overline{\mathbf{h}}_0$$

with $|\delta|$ from Eq. (5.34). These formulas establish an approximation $(\overline{\mathbf{y}}_0,\overline{\lambda}_0)$ of the bifurcation $(\mathbf{y}_0,\lambda_0)$ and two predictors $\overline{\mathbf{z}}$ to emanating solutions. Equations (5.35) do not require that $(\mathbf{y}_0,\lambda_0)$ be straddled by $(\mathbf{y}^1,\lambda_1)$, $(\mathbf{y}^2,\lambda_2)$. The results of (5.35) are better approximations than those in Algorithm 5.2, provided the branching test function τ is smooth enough in the interval that includes λ_0, λ_1, λ_2. Hence, the continuation must provide $(\mathbf{y}^1,\lambda_1)$, $(\mathbf{y}^2,\lambda_2)$ close enough to $(\mathbf{y}_0,\lambda_0)$ such that linear interpolation makes sense. The indices l and k are the same indices used in the preceding section.

Let us stress the importance of the vectors $\mathbf{h}$ defined in Eq. (5.21) or (5.35), introduced in [306],

$$\mathbf{h} = \mathbf{h}(\mathbf{y},\lambda,l,k).$$

They provide a readily available basis for both a family of test functions and a guess at emanating solutions. The latter does not require calculation of the bifurcation point itself.

5.4.3 Correctors with Selective Properties

We now turn to the problem of how to proceed to a solution to $\mathbf{f}(\mathbf{y},\lambda) = \mathbf{0}$, starting from $(\overline{\mathbf{z}},\overline{\lambda}_0)$. An iteration toward the known branch (Figure 5.16a) must be avoided. This frequently happens when $\lambda = \overline{\lambda}_0$ is kept fixed and Newton iterations on $\mathbf{f}(\mathbf{y},\lambda) = \mathbf{0}$ are carried out. A convergence back to the known branch is considered a failure. In the case of pitchfork bifurcation, the probability that no emanating solution exists for $\lambda = \overline{\lambda}_0$ is 50 percent. Hence the seemingly obvious option of parameterizing the emanating branch by λ is ruled out. We require parameterizations or equations that have a greater affinity for the emanating branch than for the known branch. Such *selective properties* of

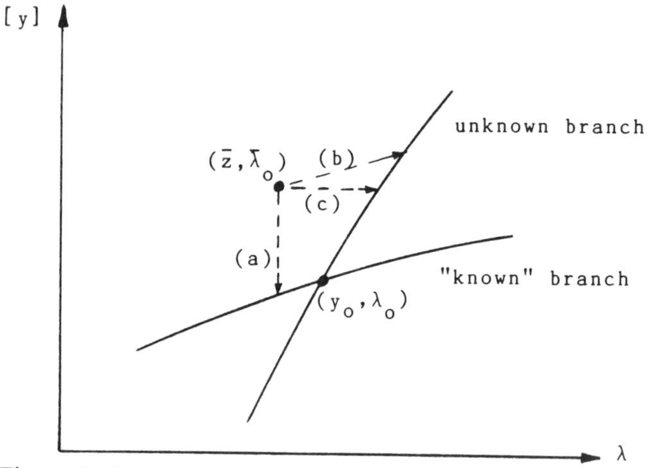

Figure 5.16

a corrector are badly needed because the predictor $(\bar{\mathbf{z}},\bar{\lambda}_0)$ might be a poor initial approximation only. In this subsection, we establish a corrector that proceeds parallel to the known branch ((b) in Figure 5.16) and a corrector that iterates parallel to the λ-axis ((c) in Figure 5.16). Here the term "parallel" is motivating rather than strict. In Section 5.4.6 we discuss correctors that exploit qualitative differences between the branches. All these correctors were introduced in [312, 313, 314].

The initial approximation $\bar{\mathbf{z}}$ is obtained by assuming a certain distance δ between the branches—that is, for example, that for the k-th component of the initial guess the relation

$$\bar{z}_k - \bar{y}_k = \delta$$

holds. Proceeding parallel to the known $\mathbf{y}$-branch means finding the value of λ that matches this prescribed distance δ,

$$\lambda = \lambda(\delta).$$

For that distance we want to find different solutions $\mathbf{y}$ and $\mathbf{z}$ to $\mathbf{f}(\mathbf{y},\lambda) = \mathbf{0}$ for the same parameter value λ,

$$z_k - y_k = \delta.$$

From what has already been said, this objective can be achieved by solving the equation

$$\begin{pmatrix} \mathbf{f}(\mathbf{y},\lambda) \\ \mathbf{f}(\mathbf{z},\lambda) \\ z_k - y_k - \delta \end{pmatrix} = \mathbf{0}. \tag{5.36}$$

The vector $\mathbf{Y} = (\mathbf{y},\lambda,\mathbf{z})$ with $2n + 1$ components enables writing Eq. (5.36) as the system of equations

$$\mathbf{F}(\mathbf{Y},\delta) = \mathbf{0}$$

in standard form, which can be solved by standard software ("CALL SOLVER"). For $\delta \neq 0$, each solution $\mathbf{Y}$ includes a solution on the emanating branch. The initial approximation required for solving Eq. (5.36) is

$$\mathbf{Y}^{(0)} = (\bar{\mathbf{y}},\bar{\lambda},\bar{\mathbf{z}}). \tag{5.37}$$

Solving Eq. (5.36) with standard software (that is, without exploiting the structure of (5.36)) must be considered as expensive because the number of equations is essentially doubled. This leads to storage requirements that larger values of n cannot tolerate. In addition, computing time increases with the number of equations.

As an alternative, we ask for the value of λ such that a solution $\mathbf{y}$ of $\mathbf{f}(\mathbf{y},\lambda) = \mathbf{0}$ has the same value as $\bar{\mathbf{z}}$ in the k-th component. This can be seen as proceeding parallel to the λ-axis when in Figure 5.16 the ordinate chosen is $[\mathbf{y}] = y_k$. Such a $\mathbf{y}$ is calculated by solving

$$\begin{pmatrix} \mathbf{f}(\mathbf{y},\lambda) \\ y_k - \bar{z}_k \end{pmatrix} = \mathbf{0}, \text{ initial guess } \begin{pmatrix} \mathbf{y} \\ \lambda \end{pmatrix} = \begin{pmatrix} \bar{\mathbf{z}} \\ \bar{\lambda}_0 \end{pmatrix}. \tag{5.38}$$

Because $\bar{\mathbf{z}}$ depends on δ, solutions of Eq. (5.38) can be seen as being parameterized by δ. Equation (5.38) is a special version of Eq. (4.10), which establishes local parameterization in the context of continuation. Again the equation is of standard form and amenable to SOLVER. Solving Eq. (5.38) is efficient because only $n + 1$ variables are involved. However, there is no guarantee that a solution of Eq. (5.38) lies on the emanating branch. Rather, there may be a solution on the known branch with the same value of $\bar{z}_k$. Hence, the selective properties of Eq. (5.38) are not fully satisfactory; much depends on the quality of the predictor. But because Eq. (5.38) is much more selective than the ordinary parameterization by λ, and because the predictor from Eq. (5.35) is close to an emanating solution at least for the pitchfork case, Eq. (5.38) is recommended. In a great majority of examples tested so far, the predictor/corrector combination, Eqs. (5.35) and (5.38), has been successful in switching branches. A branch switching approach with better selective properties will be described next.

5.4.4 Symmetry Breaking

Such functions as $\mathbf{f}(\mathbf{y},\lambda)$ often have specific properties. For example, $\mathbf{f}$ may be odd in $\mathbf{y}$,

$$\mathbf{f}(-\mathbf{y},\lambda) = -\mathbf{f}(\mathbf{y},\lambda). \tag{5.39}$$

Such a relation may hold for some components only. As an illustration, consider the example from Section 2.5:

$$f_1(y_1,y_2,\lambda) = 1 + \lambda(y_1^2 + y_2^2 - 1)$$

$$f_2(y_1,y_2,\lambda) = 10y_2 - \lambda y_2(1 + 2y_1^2 + y_2^2)$$

and the transformation $\mathbf{Sy}$ defined by

$$\mathbf{Sy} = \begin{pmatrix} 1 & 0 \\ 0 & -1 \end{pmatrix} \begin{pmatrix} y_1 \\ y_2 \end{pmatrix} = \begin{pmatrix} y_1 \\ -y_2 \end{pmatrix}.$$

For this transformation we find

$$\mathbf{f}(\mathbf{Sy},\lambda) = \mathbf{Sf}(\mathbf{y},\lambda). \tag{5.40}$$

In fact, replacing y_2 by $-y_2$ in f_1 and f_2 only changes the sign of f_2.

A relation like Eq. (5.40) with a matrix $\mathbf{S}$ obeying

$$\mathbf{S} \neq \mathbf{I},\ \mathbf{S}^2 = \mathbf{I}$$

is called a *symmetry condition.* A common type of transformation that satisfies Eq. (5.40) is a permutation; we shall handle a corresponding example in the following subsection. An n-dimensional vector $\mathbf{y}$ is called *symmetric* if $\mathbf{Sy} = \mathbf{y}$ holds, and it is called *antisymmetric* if $\mathbf{Sy} = -\mathbf{y}$. Symmetry conditions play a fundamental role in bifurcation [44, 210, 296, 366]. Among bifurcations, the following situation is widespread: the solutions of one of two intersecting branches are symmetric, while the solutions of the other branch are asymmetric. This common situation is referred to as *symmetry breaking.* The pitchfork type of bifurcation is intrinsic to this scenario.

To be more specific, assume that $(\mathbf{y}_0,\lambda_0)$ is a simple bifurcation point, where $\mathbf{y}_0$ is symmetric and $\mathbf{h}_0$ is antisymmetric,

$$\mathbf{Sy}_0 = \mathbf{y}_0,\ \mathbf{Sh}_0 = -\mathbf{h}_0. \tag{5.41}$$

Then the bifurcation point is of the pitchfork type and the emanating branch is asymmetric. The tangent to the emanating branch in $(\mathbf{y}_0,\lambda_0)$

is given by ($\mathbf{h}_0$,0). This explains why the the predictor, Algorithm 5.2 and Eq. (5.35), provides a close approximation to an emanating solution in the pitchfork situation. The symmetry breaking is summarized in Figure 5.17. As we shall see below, the frequent case of symmetry breaking has consequences both for branch switching and for the calculation of bifurcation points.

5.4.5 Coupled Cell Reaction

In [106] the trimolecular reaction

$$\begin{aligned} \mathrm{A} &\leftrightarrows X \\ 2X + Y &\leftrightarrows 3X \\ \mathrm{B} + X &\leftrightarrows Y + \mathrm{D} \\ X &\leftrightarrows \mathrm{E} \end{aligned}$$

was introduced. The concentrations of the chemicals A, B, D, E are assumed to remain at a constant level. Taking all direct kinetic constants equal to one, and neglecting the reverse processes, the reaction is described by the differential equations

$$\begin{aligned} \frac{\partial X}{\partial t} &= \mathrm{A} + X^2Y - \mathrm{B}X - X + \mathrm{D}_1 \frac{\partial^2 X}{\partial x^2} \\ \frac{\partial Y}{\partial t} &= \mathrm{B}X - X^2Y + \mathrm{D}_2 \frac{\partial^2 Y}{\partial x^2}. \end{aligned} \tag{5.42}$$

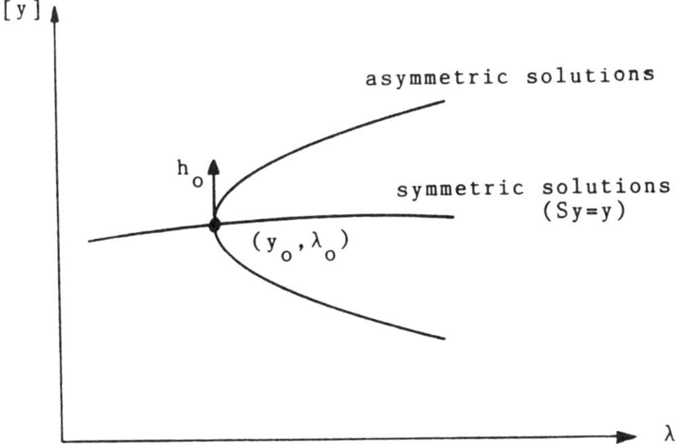

Figure 5.17

The variable x measures the length of the reactor. In what follows, we shall concentrate on the kinetic equations—that is, we neglect the diffusion terms ($D_1 = D_2 = 0$); for the impact of diffusion in this equation, see Section 6.1. Although Eq. (5.42) does not seem to have much chemical relevance [301], this model has become famous under the name "Brusselator," after the city in which the model was created. With various choices of boundary conditions, coefficients, and extensions, the Brusselator has been investigated often. A slightly more realistic modification of Eq. (5.42) allows diffusion of the initial components A and B [161]. For any nonlinear effect one may be interested in, there appears to be an appropriate variant of the Brusselator.

Here we consider the steady-state situation ($\dot{X} = \dot{Y} = 0$) in three connected reaction cells with coefficients A = 2, B = 6 [200]. The branching parameter λ is chosen as a coupling coefficient. The corresponding coupled cell reaction is described by six equations, $f_i(y_1, \ldots, y_6, \lambda) = 0,\ i = 1, \ldots, 6$,

$$\begin{aligned}
2 - 7y_1 + y_1^2 y_2 + \lambda(y_3 - y_1) &= 0\\
6y_1 - y_1^2 y_2 + 10\lambda(y_4 - y_2) &= 0\\
2 - 7y_3 + y_3^2 y_4 + \lambda(y_1 + y_5 - 2y_3) &= 0\\
6y_3 - y_3^2 y_4 + 10\lambda(y_2 + y_6 - 2y_4) &= 0\\
2 - 7y_5 + y_5^2 y_6 + \lambda(y_3 - y_5) &= 0\\
6y_5 - y_5^2 y_6 + 10\lambda(y_4 - y_6) &= 0.
\end{aligned} \tag{5.43}$$

Here y_1 and y_2 stand for X and Y in the first cell, and so on.

An inspection of Eq. (5.43) reveals that an exchange

$$y_1 \leftrightarrow y_5,\ y_2 \leftrightarrow y_6$$

yields the same equations. After a permutation of the order of equations, we arrive back at Eq. (5.43). This process of exchanging is described by the transformation

$$\mathbf{Sy} = \begin{pmatrix} 0 & 0 & 0 & 0 & 1 & 0\\ 0 & 0 & 0 & 0 & 0 & 1\\ 0 & 0 & 1 & 0 & 0 & 0\\ 0 & 0 & 0 & 1 & 0 & 0\\ 1 & 0 & 0 & 0 & 0 & 0\\ 0 & 1 & 0 & 0 & 0 & 0 \end{pmatrix} \begin{pmatrix} y_1\\ y_2\\ y_3\\ y_4\\ y_5\\ y_6 \end{pmatrix}. \tag{5.44}$$

We realize that Eq. (5.44) and $\mathbf{f}(\mathbf{y},\lambda)$ from Eq. (5.43) satisfy the symmetry condition

$$\mathbf{f}(\mathbf{Sy},\lambda) = \mathbf{Sf}(\mathbf{y},\lambda).$$

Hence, in this example, we expect pitchfork bifurcations with symmetry breaking.

As inspection of Eqs. (5.42) and (5.43) shows, there is a stationary solution for all λ,

$$X = \mathrm{A},\ Y = \frac{\mathrm{B}}{\mathrm{A}}.$$

In Eq. (5.43) this solution is given by

$$\mathbf{y} = (2, 3, 2, 3, 2, 3). \tag{5.45}$$

We carry out a numerical analysis of Eq. (5.43) for the parameter range $0 < \lambda < 5$, using branch switching and continuation methods of [314, 315]. The resulting branching diagram is depicted in Figure 5.18, and a detail for small values of λ is shown in Figure 5.19. As these figures show, there are a great number of turning points, pitchfork bifurcations, and transcritical bifurcations. The branch points are listed in Table 5.1 (rounded to six digits). These points were calculated by the Algorithm 5.1.

TABLE 5.1. BRANCH POINTS OF EQ. (5.43)

No.	λ_0	$y_1(\lambda_0)$	Type of branch point
1	0.0203067	2.64153	turning point (TP)
2	0.0295552	2.00000	transcritical bif. pt.
3	0.0335279	0.838479	TP
4	0.0341459	1.265004	TP
5	0.0872084	0.764918	TP
6	0.0872084	3.98597	TP
7	0.0886656	2.00000	pitchfork bif. pt.
8	0.0893470	1.35524	TP
9	0.0893470	2.87172	TP
10	0.0976019	4.80000	TP
11	0.0976019	0.653982	TP
12	0.155218	2.75475	pitchfork bif. pt.
13	1.50378	2.00000	transcritical bif. pt.
14	1.53561	1.57162	pitchfork bif. pt.
15	1.60171	1.76993	TP
16	4.51133	2.00000	pitchfork bif. pt.

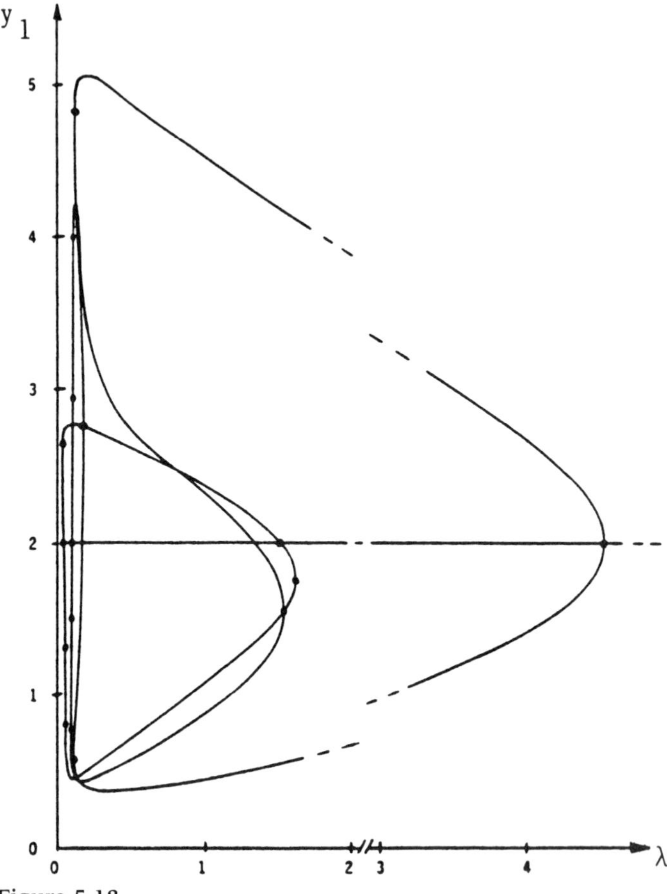

Figure 5.18

The nontrivial branches are connected with the trivial branch, Eq. (5.45), via bifurcation points 2, 7, 13, 16. Because only two are pitchfork bifurcations, the way symmetries are broken or preserved is especially interesting. The trivial branch reflects a "higher" regularity than the symmetry condition with Eq. (5.44) because

$$y_1 = y_3 = y_5, \; y_2 = y_4 = y_6 \tag{5.46}$$

holds. Figure 5.20 condenses the bifurcation behavior of Figures 5.18 and 5.19 in a qualitative way. Symmetric solutions (with respect to Eq. (5.44)) are indicated by solid lines, and branches of asymmetric solutions are dotted. The symmetry of Eq. (5.44) is preserved at the transcritical bifurcation points 2 and 13, but at these points the regularity

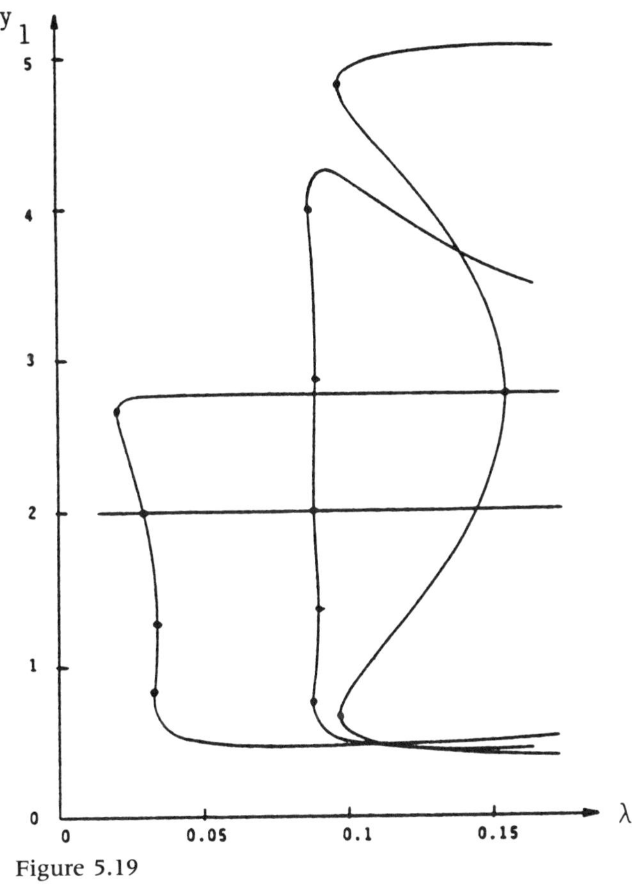

Figure 5.19

responsible for Eq. (5.46) is broken. The symmetry breaking takes place at the pitchfork bifurcations 7, 12, 14, 16. The difference between the transcritical bifurcations and the pitchfork bifurcations is reflected in the $\mathbf{h}_0$ vectors: At the pitchfork points these vectors are antisymmetric,

$$h_1 = -h_5, h_2 = -h_6,$$

while at the transcritical bifurcation points 2 and 13 the $\mathbf{h}_0$ vectors remain symmetric,

$$h_1 = h_5, h_2 = h_6.$$

This different bifurcation behavior can be discovered during the tracing of the symmetric branches by inspecting the approximations $\overline{\mathbf{h}}_0$.

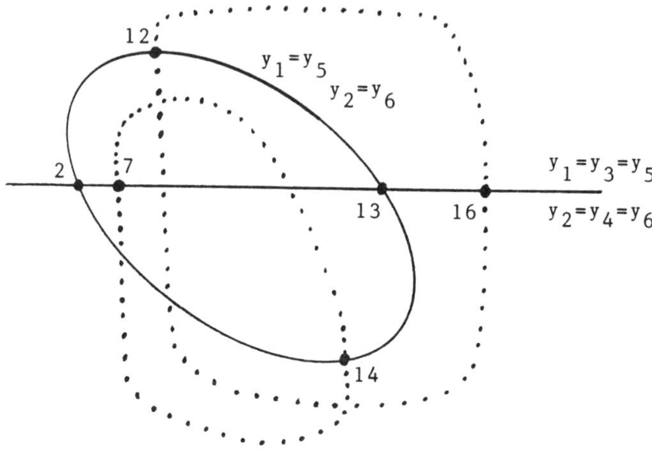

Figure 5.20

5.4.6 Parameterization by Irregularity

The common scenario of symmetry breaking can be exploited for numerical purposes [314]. Instead of trying to find numerically a symmetry relation $\mathbf{S}$ with $\mathbf{f}(\mathbf{Sy},\lambda) = \mathbf{Sf}(\mathbf{y},\lambda)$, the numerical approach concentrates on readily available *symptoms* of a symmetry condition. That is, the components of $\mathbf{y}$ are investigated with regard to regularities. For example, a related algorithm easily detects whether two components of a current solution $\mathbf{y}$ are identical,

$$y_i = y_m.$$

In case two such indices i and m are found, the next step is to check the vector $\overline{\mathbf{h}}_0$ to determine whether its corresponding components are distinct. In this way a symmetry breaking is detected reliably.

The details are as follows. After having located a symmetry breaking (or *regularity breaking*) with the two indices i and m, the equation

$$\begin{pmatrix} \mathbf{f}(\mathbf{y},\lambda) \\ y_i - y_m - \gamma \end{pmatrix} = \mathbf{0} \tag{5.47a}$$

with

$$\gamma = \delta(\overline{h}_{0i} - \overline{h}_{0m}) \tag{5.47b}$$

is solved, starting from the initial approximation

$$\mathbf{Y}^{(0)} = (\overline{\mathbf{z}},\overline{\lambda}_0). \tag{5.47c}$$

By construction (5.35), this initial approximation obeys the attached equation $y_i - y_m - \gamma = 0$. Equation (5.47a) forms a system $\mathbf{F}(\mathbf{Y},\gamma) = \mathbf{0}$ or $\mathbf{F}(\mathbf{Y},\delta) = \mathbf{0}$ consisting of $n + 1$ components. The parameter γ measures the asymmetry. Equation (5.47a) enables parameterizing the emanating branch by asymmetry; for $\gamma = 0$ one obtains a symmetric solution.

Comparing the corrector established by (5.47a–c) to the corrector (5.38) shows that both are equally efficient. The additional checking for asymmetry required for Eq. (5.47) can be automated at negligible cost. Concerning selectivity, Eq. (5.47) is preferable by far because it excludes solutions that are regular in a sense such as $y_i = y_m$. The branch switching approach, Eqs. (5.35) and (5.47), is reliable, fast, and easy to implement. The process of checking solutions for a symmetry of the type $y_i = y_m$ can be extended to other kinds of regularities (such as $y_i = 0$ for some i). Equation (5.47) must then be adapted accordingly. The method advocated here can be regarded as *parameterization by irregularity*.

5.4.7 Other Methods

Apart from the methods based on tangents (Section 5.4.1) and the methods based on interpolation combined with selective correctors (Sections 5.4.2, 5.4.3, and 5.4.6), other methods have been proposed—for example, the methods of Keller [178], Langford [208], Rheinboldt [271], and Scheurle [299]. Other approaches were based on simplicial principles (see [8, 253] and the references therein). Instead of describing these approaches, we turn our attention to a technique that can be practical.

The idea is to perturb the system $\mathbf{f}(\mathbf{y},\lambda)$ in a way that destroys the underlying regularity that is intrinsic to the current bifurcation. In literature this approach is referred to as *unfolding*. The problem is to find a parameter value γ_0 in $\mathbf{f}$ that is critical for the underlying regularity. This value is then varied to $\gamma \neq \gamma_0$ so that the regularity is destroyed and the bifurcation disappears. The parameter may occur naturally in $\mathbf{f}$, or it can be added artificially. Frequently, just adding γ to the first component (say) helps,

$$f_1(y,\lambda) + \gamma;$$

in this situation $\gamma_0 = 0$. By choosing either $\gamma > \gamma_0$ or $\gamma < \gamma_0$, branches are connected in such a way that branch switching is carried out by standard branch tracing (cf. Figure 2.39). Branch switching by unfolding is not always easy to apply. Not knowing in advance where to

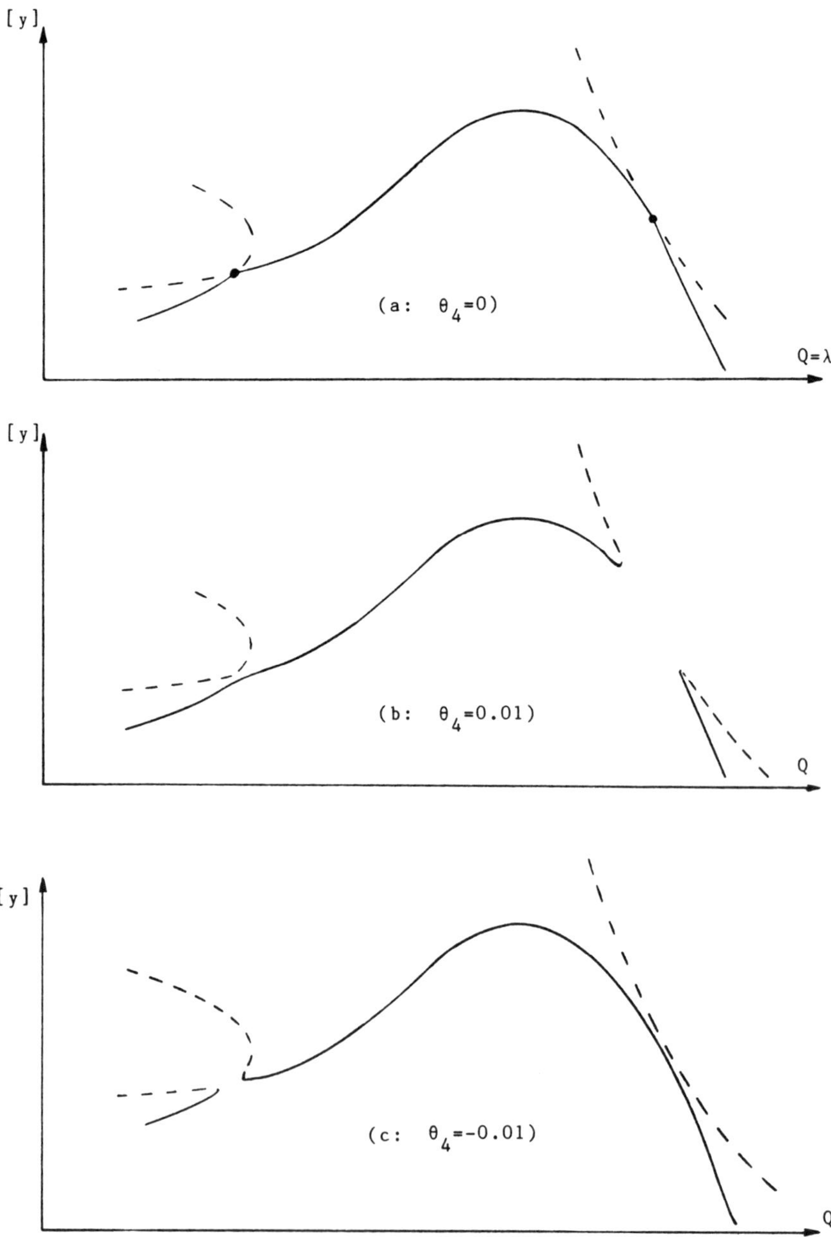

Figure 5.21

impose the perturbation and what values of γ to choose makes it difficult.

We illustrate branch switching by unfolding in Figure 5.21. In (a) this figure recalls Figure 3.12 of Section 3.4. The unfolding parameter is $\gamma = \Theta_4$. We show two unfoldings, for $\Theta_4 = +0.01$ (in Figure 5.21b) and for $\Theta_4 = -0.01$ (Figure 5.21c). Stable and unstable solutions are marked following the usual conventions. For small $|\Theta_4|$ the continuation step size must be reduced significantly to avoid skipping the gaps.

Using a sophisticated branch tracing algorithm, it can be a fun game to trace all branches in ON-LINE mode by switching between the two values $\Theta_4 = \pm 0.01$. The reader may try to figure out where to switch Θ_4 in order to follow all branches, beginning at small values of Q and eventually come back. Care must be taken not to trace the branches too far, lest a bifurcation into spurious solutions prevent a comeback.

So far, we have discussed methods for branch switching under the assumption that some solutions were calculated on a branch that can be parameterized by λ. Now we comment briefly on the other situation, in which branch switching starts from a branch that cannot be parameterized by λ close to a bifurcation. We visualize this scenario as a pitchfork bifurcation and assume that solutions on the "vertical" branch are available (Figure 5.22). For switching branches in this situation, no particular method is required. One starts from an approximation $(\bar{\mathbf{y}}_0, \bar{\lambda}_0)$ to the bifurcation point, which can be obtained, for instance, by an

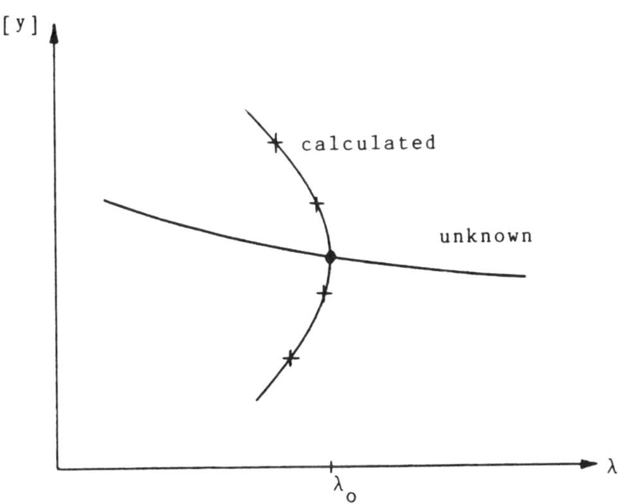

Figure 5.22

indirect method. Then solve $\mathbf{f}(\mathbf{y},\lambda) = \mathbf{0}$ for

$$\lambda = \bar{\lambda}_0 \pm |\delta| \tag{5.48}$$

for a small value of $|\delta|$ chosen as in Eq. (5.34), say. Because on one side of λ_0 locally only the unknown branch exists, this approach is selective for one sign in Eq. (5.48). In general, the appropriate sign becomes apparent when the shape of the "vertical" branch is inspected. For example, in a situation like that depicted in Figure 5.22, the positive sign must be chosen in Eq. (5.48). This branch switching can be incorporated into Eq. (5.38), choosing $k = n + 1$.

Exercise 5.17.
Assume that solutions $\mathbf{y}$ to $\mathbf{f}(\mathbf{y},\lambda) = \mathbf{0}$ have been approximated with an accuracy such that the absolute error in each component of $\mathbf{y}$ is bounded by ϵ_1 ($\epsilon_1 = 10^{-4}$, for instance). Assume further that the error in $\bar{\mathbf{h}}_0$ (from Eq. (5.35)) is bounded by ϵ_2 (for instance, $\epsilon_2 = 50\epsilon_1$). Design an algorithm that checks $\mathbf{y}$ and $\bar{\mathbf{h}}_0$ for symmetry/regularity breaking. Based on the result of this investigation, an appropriate corrector equation must be chosen. To this end, unify Eq. (5.38) and Eq. (5.47a) by introducing the general $(n + 1)$-st component

$$y_i - \zeta y_m - \gamma = 0.$$

Fix i, m, ζ, and γ, depending on the outcome of the regularity check, in such a way that Eq. (5.38) is carried out in case no regularity breaking is detected.

5.5 METHODS FOR CALCULATING SPECIFIC BRANCH POINTS

In Section 5.3 we discussed indirect and direct methods for calculating branch points of a general type. To simplify, one could say that any of these methods can calculate without change both bifurcation and turning points. In particular, these methods are recommended for their favorable relation overhead (required software) versus applicability.

Allowing specific methods for specific purposes (relaxing criteria C3 and C5 from the list in Section 5.3.2), methods that are more efficient can be constructed (improvements in C2 and C4). In recent years many such methods have been proposed, most of them direct methods. Several of the approaches are modifications of the methods described earlier. In particular, the branching system Eqs. (5.15) or (5.18) introduced in [306] has been shown to be a starting point for further developments.

Systems of equations that are solved to obtain specific branch points have been called *determining equations.* This reflects interest in finding different equations for different types of branch points. Such a desire was indicated in [118], for example, which announced a system for Hopf points that does not admit turning points as solutions. From such a remark and from the expression "determining equation," one might envision being able to have as many different systems as types of branch points, each equation being selective in that other branch points are not admitted as solutions. Although this would be fascinating from a theoretical point of view, it does not appear to be practical, because too much overhead is required. In addition, the current type of branch point is usually *not known in advance,* so there is some uncertainty in applying a specific apparatus. It is desirable that a method perform reasonably well in a variety of phenomena.

On the other hand, improvements in the efficiency of a specialized method can be so significant that it is worthwhile to discuss related ideas. This is the object of the present section. Concerning gains in efficiency, we distinguish between speeding up computing time and reducing storage. While implications of the former appear to be marginal in one-parameter problems (because the handling of branch points is only a small fraction of the overall branch tracing efforts), a reduction of storage is more rewarding.

5.5.1 A Special Implementation for the Branching System

Moore and Spence [242] propose a significant storage reduction that takes into account the structure of the branching system Eq. (5.18),

$$\mathbf{0} = \begin{pmatrix} \mathbf{f}(\mathbf{y},\lambda) \\ \mathbf{f_y}(\mathbf{y},\lambda)\mathbf{h} \\ h_k - 1 \end{pmatrix} = \mathbf{F}\begin{pmatrix} \mathbf{y} \\ \lambda \\ \mathbf{h} \end{pmatrix} = \mathbf{F}(\mathbf{Y}).$$

The reduction is based on the assumption that the classical Newton method is used for solving $\mathbf{F}(\mathbf{Y}) = \mathbf{0}$. We report on this implementation in detail because similar approaches can also be used in other instances. Each iteration step of the Newton method must solve a linear system of the form

$$\left(\begin{array}{c|c|c} \mathbf{f_y} & \mathbf{f}_\lambda & \mathbf{0} \\ \hline \mathbf{f_{yy}h} & \mathbf{f}_{\mathbf{y}\lambda}\mathbf{h} & \mathbf{f_y} \\ \hline 0 \ldots 0 & 0 & 1\ 0 \ldots 0 \end{array}\right) \begin{pmatrix} \Delta\mathbf{y} \\ \Delta\lambda \\ \Delta\mathbf{h} \end{pmatrix} = -\mathbf{F}\begin{pmatrix} \mathbf{y} \\ \lambda \\ \mathbf{h} \end{pmatrix}. \tag{5.49}$$

Here $\Delta\mathbf{y}$, $\Delta\lambda$, $\Delta\mathbf{h}$ are the current unknown correction vectors (the symbol Δ is no factor). The index k is chosen as $k = 1$. The matrix in block structure is the Jacobian of $\mathbf{F}$; the arguments $(\mathbf{y},\lambda)$ are suppressed. With the nomenclature

$$\mathbf{J} = \mathbf{f_y}(\mathbf{y},\lambda), \quad \mathbf{B} = \mathbf{f_{yy}}(\mathbf{y},\lambda)\mathbf{h}, \quad \mathbf{c} = \mathbf{f}_{\mathbf{y}\lambda}(\mathbf{y},\lambda)\mathbf{h},$$

Eq. (5.49) can be written as two subsystems:

$$\mathbf{J}\Delta\mathbf{y} + \mathbf{f}_\lambda\Delta\lambda = -\mathbf{f} \tag{5.50a}$$

and

$$\mathbf{J}\Delta\mathbf{h} + \mathbf{c}\Delta\lambda + \mathbf{B}\Delta\mathbf{y} = -\mathbf{Jh}. \tag{5.50b}$$

The first component of $\Delta\mathbf{h}$ is known, $\Delta h_1 = 0$. Following [242], we first solve subsystem Eq. (5.50a) and then use the results in solving the second system Eq. (5.50b). Because Eq. (5.50a) has one unknown more than scalar equations, one of the $n + 1$ unknowns

$$\Delta\lambda, \Delta y_1, \ldots, \Delta y_n$$

is temporarily regarded as being known. Choosing Δy_1 as such a parameter takes advantage of $h_1 = 1$ and $\Delta h_1 = 0$ and allows us to replace the first column of $\mathbf{J}$ in the left-hand sides of both subsystems by a suitable vector. This can be seen as follows: Rewriting (5.50) yields

$$\mathbf{J}[\Delta\mathbf{y} - \Delta y_1\mathbf{h}] + \mathbf{f}_\lambda\Delta\lambda = -\mathbf{f} - \Delta y_1\mathbf{Jh} \tag{5.51a}$$

and

$$\mathbf{J}\Delta\mathbf{h} + \mathbf{f}_\lambda\Delta\lambda = -\mathbf{Jh} - \mathbf{c}\Delta\lambda - \mathbf{B}\Delta\mathbf{y} + \mathbf{f}_\lambda\Delta\lambda. \tag{5.51b}$$

Equation (5.51a) is obtained from Eq. (5.50a) by subtracting $\Delta y_1\mathbf{Jh}$. This produces a zero in the first component of the bracketed vector. Now in both the Eqs. (5.51a) and (5.51b), $\Delta\lambda$ can take the position of the first component of the unknown vectors on the left side—that is, the vector $\mathbf{f}_\lambda$ replaces the first column of $\mathbf{J}$. Let us introduce the unknown vectors $\Delta\mathbf{a}$, $\Delta\mathbf{b}$ by

$$\Delta\mathbf{a} = \begin{pmatrix} \Delta\lambda \\ \Delta y_2 - \Delta y_1 h_2 \\ \vdots \\ \Delta y_n - \Delta y_1 h_n \end{pmatrix}, \quad \Delta\mathbf{b} = \begin{pmatrix} \Delta\lambda \\ \Delta h_2 \\ \vdots \\ \Delta h_n \end{pmatrix} \tag{5.52}$$

and the matrix $\mathbf{A}$ by

$$\mathbf{A} = \begin{pmatrix} \mathbf{f}_\lambda & (\mathbf{J}) \end{pmatrix}.$$

With this notation, the left-hand sides of Eqs. (5.51a) and (5.51b) can be written as $\mathbf{A}\Delta\mathbf{a}$ and $\mathbf{A}\Delta\mathbf{b}$ with the same matrix $\mathbf{A}$.

The solution of Eq. (5.51) is obtained by splitting (5.51a) into two systems of equations for two unknown vectors $\mathbf{c}^{(1)}$, $\mathbf{c}^{(2)}$,

$$\mathbf{A}\mathbf{c}^{(1)} = -\mathbf{f}$$

$$\mathbf{A}\mathbf{c}^{(2)} = -\mathbf{J}\mathbf{h}.$$

The solution $\Delta\mathbf{a}$ is given by linear combination,

$$\Delta\mathbf{a} = \mathbf{c}^{(1)} + \Delta y_1 \mathbf{c}^{(2)}.$$

This constitutes $\Delta\lambda, \Delta y_2, \ldots, \Delta y_n$ in dependence of Δy_1,

$$\Delta\lambda = c_1^{(1)} + \Delta y_1 c_1^{(2)} \tag{5.53}$$

and

$$\Delta y_i = c_i^{(1)} + \Delta y_1 (c_i^{(2)} + h_i) \qquad \text{for } i = 2, \ldots, n.$$

These expressions are substituted into Eq. (5.51b), leading to the equation

$$\mathbf{A}\Delta\mathbf{b} = \mathbf{r}^{(1)} + \Delta y_1 \mathbf{r}^{(2)}$$

with two vectors, $\mathbf{r}^{(1)}$ and $\mathbf{r}^{(2)}$. We take $\Delta\mathbf{b}$ in the form

$$\Delta\mathbf{b} = \mathbf{c}^{(3)} + \Delta y_1 \mathbf{c}^{(4)}$$

and solve two further systems of linear equations (Exercise 5.18), ending up with the relation

$$\Delta\lambda = c_1^{(3)} + \Delta y_1 c_1^{(4)} \tag{5.54}$$

together with relations for Δh_i ($i = 2, \ldots, n$). The two equations, Eq. (5.53) and Eq. (5.54), give the final values of $\Delta\lambda$, Δy_1, which establishes Δy_i, Δh_i for $i > 1$. As proved in [242], the final system of two equations for $\Delta\lambda$, Δy_1 is nonsingular at turning points. Apart from calculating second-order derivatives, only one **LU** decomposition per step is required, because the four linear systems involve the same matrix **A**.

To summarize, this special implementation of the Newton method requires less storage than solving the branching system Eq. (5.18) by standard software. The method is designed for turning points. No quasi-Newton method is possible in this particular implementation. The size of n should be such that **LU** decompositions are convenient.

In light of the limited value of direct methods in case the equations result from discretizations (see Section 5.3.6), the latter requirement does not impose a further restriction. Note the dichotomy that the savings of this implementation are most pronounced for large n resulting from discretizations, the case in which a direct method is most questionable. In [242], 10 or 11 digits are calculated, although because of inherent discretization errors, only four digits are accurate.

5.5.2 Regular Systems for Bifurcation Points

As mentioned in Section 5.3.2, the Jacobian of the branching system Eq. (5.18) is singular at a bifurcation point. Following a proposal of Abbott [1, 2], regularity can be achieved by replacing suitable components of **f** by symmetry relations. Werner and Spence [366] pointed out that the branching system is regular when **y** and **h** are confined to subdomains of symmetric and antisymmetric vectors (see Sections 5.4.4 and 5.3.2). This restriction can be enforced by reducing Eq. (5.18) appropriately [366]. This reduced branching system consists of only $n + 1$ scalar equations.

Following a suggestion of Moore [241, 119], the singularity can be removed by adding a suitable vector to **f**. We define

$$\tilde{\mathbf{f}}(\mathbf{y},\lambda,\gamma) = \mathbf{f}(\mathbf{y},\lambda) + \gamma\mathbf{r} \tag{5.55}$$

for a parameter γ and a vector **r** that satisfies

$$\mathbf{r} \notin \text{range}\,(\mathbf{f}_{\mathbf{y}}(\mathbf{y}_0,\lambda_0) \mid \mathbf{f}_\lambda(\mathbf{y}_0,\lambda_0)).$$

One considers γ as new branching parameter and sets $\tilde{\mathbf{y}} = (\mathbf{y},\lambda)$. For the modified function $\tilde{\mathbf{f}}$,

$$\partial\tilde{\mathbf{f}}/\partial\gamma = \mathbf{r} \notin \text{range}(\partial\tilde{\mathbf{f}}/\partial\tilde{\mathbf{y}})$$

holds. Hence the bifurcation point of $\mathbf{f}(\mathbf{y},\lambda) = \mathbf{0}$ has become a turning point of $\tilde{\mathbf{f}}(\tilde{\mathbf{y}},\gamma) = \mathbf{0}$. Because this modified equation possesses a rectangular Jacobian, its singularity is characterized by the existence of a *left* null vector $\mathbf{g}$,

$$\mathbf{0} = \mathbf{g}^{tr}(\partial\tilde{\mathbf{f}}/\partial\tilde{\mathbf{y}}),$$

rather than by the right null vector $\mathbf{h}$, as in the branching system Eq. (5.18). This is equivalent to the two equations

$$\mathbf{g}^{tr}\mathbf{f}_{\mathbf{y}} = \mathbf{0}, \quad \mathbf{g}^{tr}\mathbf{f}_{\lambda} = \mathbf{0}.$$

Collecting these equations together with $\tilde{\mathbf{f}}(\tilde{\mathbf{y}},\gamma) = \mathbf{0}$ and a scaling condition $\mathbf{g}^{tr}\mathbf{r} = 1$ leads to the system

$$\begin{pmatrix} \mathbf{f}(\mathbf{y},\lambda) + \gamma\mathbf{r} \\ (\mathbf{f}_{\mathbf{y}}(\mathbf{y},\lambda))^{tr}\mathbf{g} \\ \mathbf{g}^{tr}\mathbf{f}_{\lambda}(\mathbf{y},\lambda) \\ \mathbf{g}^{tr}\mathbf{r} - 1 \end{pmatrix} = \mathbf{0} \tag{5.56a}$$

for the $2n + 2$ unknowns $\mathbf{y}$, λ, γ, and $\mathbf{g}$. The ambiguity of the choice of $\mathbf{r}$ can be removed by identifying $\mathbf{r}$ with $\mathbf{g}$:

$$\mathbf{r} = \mathbf{g}. \tag{5.56b}$$

Solutions of Eq. (5.56) are called "imperfect bifurcations" [241]. The Jacobian of Eq. (5.56) is regular in a (simple) bifurcation point. Solving (5.56) yields for the artificial perturbation parameter the value $\gamma = 0$.

5.5.3 Methods for Turning Points

The literature on the calculation of turning points is rich; for references and first comparisons, see [233]. The methods of Abbott have already been mentioned. Pönisch and Schwetlick proposed two algorithms [260, 261]. An approach of Rheinboldt is described in [273]. Griewank and Reddien [119] generalized the principles underlying various determining equations and sketched a direct method for the turning point case. This method requires only one additional scalar equation; hence, a specific $(n + 1)$-system is to be solved. In this point the approach in [119] resembles suggestions of [1, 2] and [260, 261] (see also [259]). As [171] shows, differences in numerical performance of various direct methods do not appear to be significant.

5.5.4 Methods for Hopf Bifurcation Points

Hopf points are characterized by a simple pair of purely imaginary eigenvalues $\pm i\beta$ of the Jacobian $\mathbf{f}_{\mathbf{y}}(\mathbf{y}_0,\lambda_0)$ (see Section 2.7). Hence the equation

$$\mathbf{f_y}(\mathbf{y}_0,\lambda_0)\mathbf{w} = i\beta\mathbf{w} \tag{5.57}$$

holds for a nonzero complex vector $\mathbf{w} = \mathbf{h} + i\mathbf{g}$. This complex system is equivalent to two real systems for the vectors of the real part $\mathbf{h}$ and the imaginary part $\mathbf{g}$,

$$\mathbf{f_y h} = -\beta\mathbf{g}, \quad \mathbf{f_y g} = \beta\mathbf{h}. \tag{5.58}$$

In order to normalize $\mathbf{w}$, impose, for instance,

$$w_k = 1,$$

which is equivalent to two real normalizing conditions. Collecting these equations yields the determining system

$$\begin{pmatrix} \mathbf{f}(\mathbf{y},\lambda) \\ \mathbf{f_y}(\mathbf{y},\lambda)\mathbf{h} + \beta\mathbf{g} \\ \mathbf{f_y}(\mathbf{y},\lambda)\mathbf{g} - \beta\mathbf{h} \\ h_k - 1 \\ g_k \end{pmatrix} = \mathbf{0}. \tag{5.59}$$

With $1 \le k \le n$, k is an index. System (5.59) can be solved to calculate a Hopf bifurcation point [154, 163]. In case of stationary branching, $\beta = 0$ and one of the two linear subsystems is redundant. This shows that Eq. (5.59) extends the branching system (5.18) into the complex case. Solving equation (5.59) by calling standard software may be advisable for small or moderate values of n only, since Eq. (5.59) consists of $(3n + 2)$ scalar equations. An efficient implementation of Eq. (5.59) is proposed in [118]. A variant of Eq. (5.59) requires fewer equations but must work with the squared matrix $(\mathbf{f_y}(\mathbf{y},\lambda))^2$ (see [154, 284] and Exercise 5.19).

These methods for Hopf points are direct methods. The branching system in the corresponding ODE-form can also be applied (solving (7.24), see Section 7.6). The above direct methods have analogs in the ODE/PDE situation (see Section 6.7). Branch switching from stationary branch to periodic branch is relegated to Section 7.6 because we shall require elements of boundary-value problems that will be introduced in the following chapter.

We emphasize the effective indirect methods described earlier (Section 5.3.1) that are readily applied to Hopf bifurcations. A sophisticated algorithm has been developed by Hassard [135] (see also [136]).

5.5.5 Other Methods

In this subsection we give some hints on further methods. There are approaches that do not work with any derivatives [101, 175]. The repeatedly occurring extended systems can be solved by specific block elimination methods [57]. Deflation techniques as suggested in [45] can be helpful for branch switching purposes [1]. Relations between different approaches for calculating turning points were investigated in [303]. In [36, 234] the calculation of multiple bifurcation points is treated. Related methods in a general setting were proposed in [7, 369]; these papers include multiparameter problems.

Exercise 5.18.
Carry out in detail the special implementation of the Newton method as indicated by Eqs. (5.49) through (5.54). In particular, calculate the vectors $\mathbf{r}^{(1)}$ and $\mathbf{r}^{(2)}$ and give a solution $\Delta\lambda$, Δy_1 to Eqs. (5.53) and (5.54).

Exercise 5.19.
Find another determining equation for Hopf points. (Hint: Combine the two systems of Eq. (5.58) in a suitable way.)

5.6 CONCLUDING REMARKS

The reader may now have the (correct) impression that there are many methods for a numerical bifurcation analysis. Instead of attempting to report on all details, we have restricted ourselves to basic principles. Nevertheless, the preceding sections may have been confusing for those who are just looking for help with practical stability and bifurcation analysis. Hence we offer some preliminary advice.

For the location of turning points, no particular method will be needed. A simple device like that proposed by Exercises 5.7 through 5.9 can be readily implemented in any continuation algorithm. In this way a turning point is approximated with reasonable accuracy. The situation is not complicated because it is difficult to miss a turning point during branch tracing.

A location of bifurcation points is much more complex. Branching test functions have to be involved, and they are expensive when based on eigenvalues. In the context of branch tracing, one must be aware that the approximation of stationary bifurcation points in high accuracy is an ill-posed problem, because the bifurcation is destroyed by a perturbation of the underlying regularity. Therefore, indirect methods as described in Section 5.3.1 are appropriate. Occasions in which a specific direct method is preferable seem to be rare; in some cases it may be advisable to apply some package (see Appendix 6). As far as branch switching is concerned, the methods of Sections 5.4.2, 5.4.3, and 5.4.6 are extremely easy to implement. At the present stage, the question of what method is "best" cannot be answered, but such questions are not too important because many aspects of a bifurcation treatment just do not occur in a two-parameter setting.

5.7 TWO-PARAMETER PROBLEMS

So far we have studied one-parameter families of equation $\mathbf{f}(\mathbf{y},\lambda) = \mathbf{0}$. Here no method for calculating hysteresis points, isola centers, multiple bifurcatoin points, or Hopf centers is required (see Section 2.9). This situation might change when a two-parameter family of equations is studied,

$$\mathbf{f}(\mathbf{y},\lambda,\gamma) = \mathbf{0}.$$

Recall that there may be organizing centers, such as those illustrated in Figures 2.43 and 2.44. For example, consider the hysteresis center $(\lambda_0^{HY},\gamma_0^{HY})$, in which curves of turning points coalesce. Figure 5.23 shows the part of a (λ,γ) parameter chart that contains two critical boundaries, (1) and (2), of parameter combinations for which turning points exist. These curves characterize a hysteresis phenomenon (see

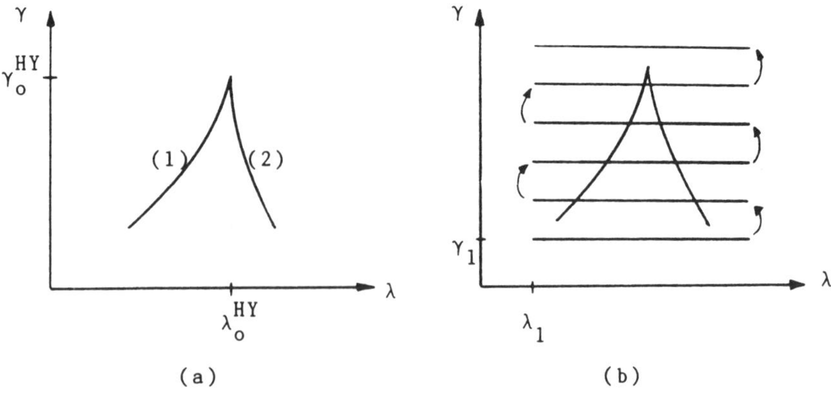

Figure 5.23

Figures 2.37 and 2.38 and the discussion of Eq. (2.21)). Figure 5.23 illustrates three principles for calculating an organizing center such as a hysteresis point:

(i) Calculate the center *point* $(\lambda_0^{\text{HY}}, \gamma_0^{\text{HY}})$ directly.

(ii) Calculate the *curves* (1) and (2) and trace them until they coalesce.

(iii) Calculate the solution *manifold*, for instance, as follows: Choose (λ_1, γ_1) and start a standard continuation varying one parameter (say, λ) and keeping the other parameter (γ) fixed in a systematic fashion, as illustrated by Figure 5.23b. The critical boundaries (1) and (2) can be generated by interpolation.

Before mentioning specific methods, let us discuss differences between the above three approaches. Approach (i) is fully direct and produces only the one particular solution of the organizing center, here for parameter value $(\lambda_0^{\text{HY}}, \gamma_0^{\text{HY}})$. On its own, this approach does not provide sufficient information. Approach (ii) may be called semidirect, because each turning point on (1) or (2) can be calculated by a direct method. The results of approach (ii) are highly informative because it produces an approximation to the hysteresis point and shows where the range of hysteresis extends. Finally, approach (iii) is fully indirect and produces the greatest amount of information on the solution structure. With (iii), an approximation of the hysteresis point is obtained as by-product, because the main objective is to approximate the manifold (surface) defined by $\mathbf{f}(\mathbf{y}, \lambda, \gamma) = \mathbf{0}$. In case turning points are calculated by indirect methods, the illustration of a class-ii approach resembles Figure 5.24. Both (ii) and (iii) can apply one-parameter strategies. Note that the amount of work in applying (ii) is proportional to the number of curves of critical boundaries. For example, in the heart model (2.23) in Section 2.9, we encounter both turning-point curves and Hopf-point curves (Figure 2.44). Applying a class-iii-type method in this example produces both kinds of curves without extra effort (interpolation is not counted). A fully direct method (class i) cannot be seen as alternative to a class-ii or class-iii approach because the latter are required anyway in order to calculate global information. This situation resembles the discussion "direct versus indirect methods" in Section 5.3.6.

Next we briefly mention specific methods falling into the above classes. Direct procedures for calculating hysteresis and Hopf points have been proposed by Roose in [205, 287]. Equations defining organizing centers have been discussed by Beyn [36]. Methods for tracing critical boundaries are given by suitable direct methods. We illustrate this by means of the method described in Section 5.3.2. The enlarged

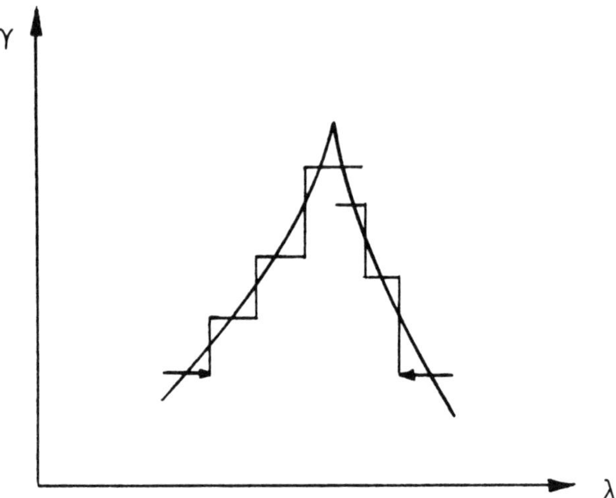

Figure 5.24

branching system Eq. (5.18) depends on the second parameter,

$$\mathbf{F}(\mathbf{Y},\gamma) = \begin{pmatrix} \mathbf{f}(\mathbf{y},\lambda,\gamma) \\ \mathbf{f_y}(\mathbf{y},\lambda,\gamma)\mathbf{h} \\ h_k - 1 \end{pmatrix} = \mathbf{0}. \tag{5.60}$$

A standard continuation with respect to γ traces critical boundaries. As an example for results produced by this approach, we consider a specific

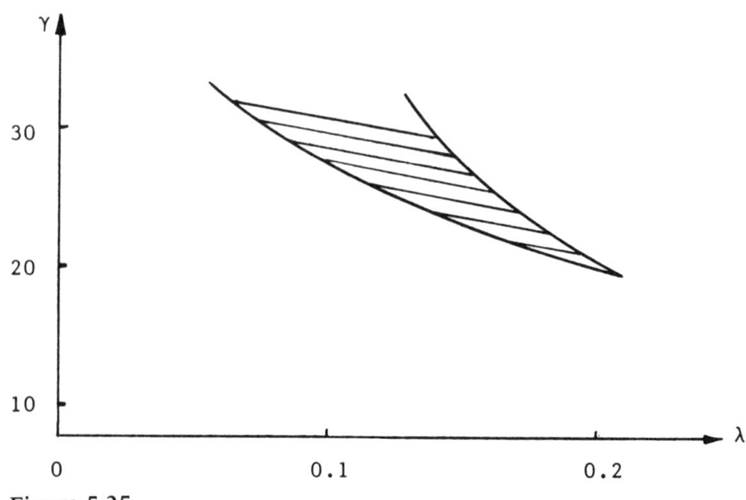

Figure 5.25

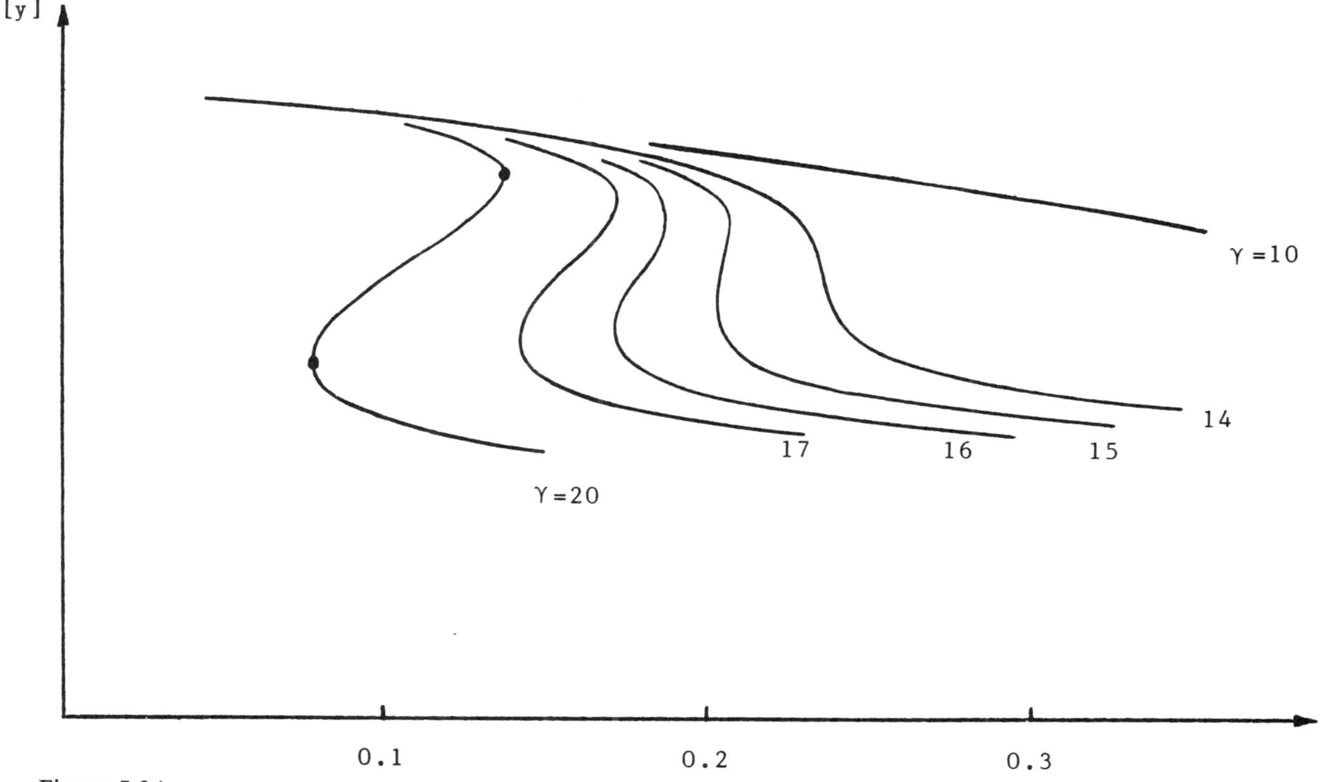

Figure 5.26

catalytic reaction with two parameters λ and γ. The equations will be given in Eq. (6.15) in Section 6.2.2; here we focus on the two-parameter aspect of solutions. Figure 5.25 shows two critical boundaries that were produced by the semidirect method (from [307]). Other methods for calculating turning points can be similarly applied, including indirect methods (Figure 5.24). Further methods of class (ii) are proposed in [273, 330, 331]. In general, such methods are not restricted to the handling of hysteresis points, but apply also to the calculation of bifurcations and isola centers. This situation emphasizes the importance of being able to calculate turning points. The calculation of bifurcation points (gap centers), isola centers, and hysteresis centers can be reduced to repeated calculations of turning points.

Finally, we come to approach (iii), which approximates a part of the manifold defined by $\mathbf{f}(\mathbf{y},\lambda,\gamma) = \mathbf{0}$. Considering a scalar measure $[\mathbf{y}]$ of solutions to $\mathbf{f}(\mathbf{y},\lambda,\gamma) = \mathbf{0}$, the results form a surface in the three-dimensional $([\mathbf{y}],\lambda,\gamma)$ space. Figure 5.23b suggests how to apply standard one-parameter continuation. In this way, cuts through the surface are obtained. Accuracy can be improved by varying the spacing. Figure 5.26 illustrates results as they are obtained by approximating the surface in the way indicated by Figure 5.23b. Figure 5.26 again refers to the mentioned calalytic reaction and thus matches Figure 5.25. In the approach indicated by Figure 5.23b or Figure 5.26, it is helpful to take advantage of calculated results on the previous parallel line in the parameter chart. Alternatives to the simple parallel slicing construct a triangulation on the surface. Related efficient approaches have been developed by Allgower and Schmidt [10] and Rheinboldt (in [205]).

6 Calculating Branching Behavior of ODE Boundary-Value Problems

The topic of this chapter is the calculation of branching behavior of the one-parameter family of two-point boundary-value problems

$$\mathbf{y}' = \mathbf{f}(t,\mathbf{y},\lambda), \quad \mathbf{r}(\mathbf{y}(a),\mathbf{y}(b)) = \mathbf{0}. \tag{6.1}$$

The variable $\mathbf{y}(t)$ consists as usual of n scalar functions $y_1(t), \ldots, y_n(t)$. The right-hand side $\mathbf{f}(t,\mathbf{y},\lambda)$ is a vector function; the boundary conditions (the second Eq. (6.1)) consist of n scalar equations,

$$\begin{aligned} r_1(y_1(a), \ldots, y_n(a), y_1(b), \ldots, y_n(b)) &= 0 \\ \vdots \qquad\qquad\qquad & \quad \vdots \\ r_n(y_1(a), \ldots, y_n(a), y_1(b), \ldots, y_n(b)) &= 0. \end{aligned}$$

The independent variable $t (a \leq t \leq b)$ need not be time; accordingly, the derivative with respect to t is denoted by a prime $'$ rather than a dot: $\mathbf{y}' = d\mathbf{y}/dt$. The branching parameter λ can occur in the boundary conditions:

$$\mathbf{r}(\mathbf{y}(a),\mathbf{y}(b),\lambda) = \mathbf{0}.$$

However, because the methods discussed in this chapter are not affected by the dependence of $\mathbf{r}$ on λ, the notation $\mathbf{r}(\mathbf{y}(a),\mathbf{y}(b))$ of Eq. (6.1) will be retained.

The methods for boundary-value problems resemble methods explained in the previous chapter. Following [306], our approach stays in the "infinite-dimensional space" of ODEs and lets standard software take care of the transition to the finite-dimensional world of numerical approximation. The methods discussed in this chapter are not dependent on any particular solution procedure or software. As an alternative to the methods set out in this chapter, the methods of the preceding chapter can be applied, provided Eq. (6.1) is discretized first. As an example of such a discretization, see the PDE model in Section 5.3.6, which was transformed into a system of algebraic equations.

6.1 ENLARGED BOUNDARY-VALUE PROBLEMS

Specific tasks (such as continuation or branch switching) can be reformulated into boundary-value problems of the general type

$$\mathbf{Y}' = \mathbf{F}(t,\mathbf{Y}), \quad \mathbf{R}(\mathbf{Y}(a),\mathbf{Y}(b)) = \mathbf{0}. \tag{6.2}$$

Typically, Eq. (6.2) includes Eq. (6.1) as subsystem. Because solvers for two-point boundary-value problems, Eq. (6.2), are well established, we consider that the specific task is solved whenever we succeed in

(1) formulating Eq. (6.2) equivalent to the specific task and

(2) constructing a reasonable initial guess to the solution.

In this section we use two examples to explain elementary ways of constructing boundary-value problem (6.2).

The most widely applicable enlarged boundary-value problem is

$$\begin{pmatrix} \mathbf{y} \\ \lambda \end{pmatrix}' = \begin{pmatrix} \mathbf{f}(t,\mathbf{y},\lambda) \\ 0 \end{pmatrix}, \quad \begin{pmatrix} \mathbf{r}(\mathbf{y}(a),\mathbf{y}(b)) \\ y_k(a) - \eta \end{pmatrix} = \mathbf{0}. \tag{6.3}$$

The system (6.3) of dimension $n + 1$ is of problem type (6.2), with vector function $\mathbf{Y} = (\mathbf{y},\lambda)$. System (6.3) is applied the same way the related Eq. (4.10) is applied in continuation of systems of equations. In particular, system (6.3) is appropriate for passing turning points. Prescribing a suitable index k and a value of η enables the calculation of λ as dependent variable. The "trivial" differential equation $\lambda' = 0$ characterizes λ as constant. Solving Eq. (6.3) automatically yields a value of λ that matches the underlying k and η. Concerning the choice of k and η, the strategy outlined in Section 4.2.3 is applied. If one of the n boundary conditions (r_j, say) is of the form

$$r_j = y_\nu(a) - \text{constant} = 0,$$

the index k must be different from ν in order to avoid a contradiction. A strategy such as Eq. (4.12), evaluated for $t = a$, prevents k from conflicting with an initial condition. Choosing $k = n + 1$ in Eq. (6.3), the parameter λ becomes the control parameter.

Example 6.1. Brusselator with Diffusion.
We reconsider the Brusselator model (see Section 5.4.5) now without neglecting the diffusion. Concentrating on the steady-state situation $\dot{X} = \dot{Y} = 0$ yields two differential equations of the second order, namely,

$$\begin{aligned} 0 &= \mathrm{A} + X^2 Y - \mathrm{B}X - X + \mathrm{D}_1 \frac{\partial^2 X}{\partial x^2} \\ 0 &= \mathrm{B}X - X^2 Y + \mathrm{D}_2 \frac{\partial^2 Y}{\partial x^2} . \end{aligned} \tag{6.4}$$

These ordinary differential equations describe the spatial dependence of the two chemicals X and Y along a reactor with length L, $0 \leq x \leq \mathrm{L}$. We impose fixed boundary conditions,

$$X = \mathrm{A} \qquad \text{for } x = 0,\ x = \mathrm{L}$$

$$Y = \mathrm{B/A} \qquad \text{for } x = 0,\ x = \mathrm{L}.$$

Scaling the independent variable according to $t = x/\mathrm{L}$ and writing $\lambda = \mathrm{L}^2$ leads to the system of four ODEs of the first order

$$\begin{aligned} y_1' &= y_2 \\ y_2' &= -\lambda[\mathrm{A} + y_1^2 y_3 - (\mathrm{B} + 1)y_1]/\mathrm{D}_1 \\ y_3' &= y_4 \\ y_4' &= -\lambda[\mathrm{B}y_1 - y_1^2 y_3]/\mathrm{D}_2 \end{aligned} \tag{6.5a}$$

with boundary conditions

$$y_1(0) = y_1(1) = \mathrm{A}, \quad y_3(0) = y_3(1) = \mathrm{B/A} \tag{6.5b}$$

(Exercise 6.1). We adopt the constants

$$\mathrm{D}_1 = 0.0016, \quad \mathrm{D}_2 = 0.008, \quad \mathrm{A} = 2, \quad \mathrm{B} = 4.6$$

from [202]. As Eq. (6.5a) shows, the independent variable t in Eq. (6.1) need not occur explicitly in the right-hand side; the right-hand side of Eq. (6.5a) is of the form $\mathbf{f}(\mathbf{y},\lambda)$.

Because the boundary conditions in Eq. (6.5b) include two initial conditions (imposed on y_1 and y_3), the index k in Eq. (6.3) can take only the values $k = 2$ or $k = 4$. In order to extend Eq. (6.5) to the system (6.3), define $y_5 = \lambda$ and attach the differential equation and boundary condition

$$y_5' = 0, \quad y_k(0) = \eta. \tag{6.5c}$$

Writing the boundary conditions in the form used in Eqs. (6.1), (6.2), or (6.3) yields

$$\begin{aligned} y_1(0) - A &= 0 \\ y_1(1) - A &= 0 \\ y_3(0) - B/A &= 0 \\ y_3(1) - B/A &= 0 \\ y_k(0) - \eta &= 0. \end{aligned} \tag{6.5d}$$

This concludes the formal preparations that transform the original problem, Eq. (6.4), into the standard forms of Eq. (6.1) or Eq. (6.2).

After having implemented the right-hand side of Eq. (6.5a) together with $y_5' = 0$ in a routine for **F**, and of Eq. (6.5d) in a routine for **R**, the next step is to call SOLVER in order to calculate solutions and to trace branches. The specific Brusselator model Eq. (6.5) has a great number of solutions [202]; some are represented in the branching diagram of Figure 6.1. One nontrivial branch point is marked. The closed branches have been calculated by using the $k - \eta$ strategy in Eqs. (4.12), (4.13), or (4.16).

The above Brusselator has served as a first example illustrating the transformation from a particular model into the two-point boundary-value problem, Eqs. (6.1) or (6.3), in standard form. The second example represents a different class of solutions—namely, the time-dependent periodic solutions.

Example 6.2. Forced Duffing Equation.
We reconsider the Duffing equation introduced in Section 2.3,

$$\ddot{u} + 0.04\dot{u} - 0.2u + 8u^3/15 = 0.4 \cos \omega t.$$

The harmonic forcing term on the right-hand side provokes a response of the system (beam, electric current). We confine ourselves to *harmonic oscillations* $u(t)$, which have the same period T as the excitation,

$$T = \frac{2\pi}{\omega}. \tag{6.6}$$

Hence harmonic oscillations obey the boundary conditions

$$u(0) = u(T), \quad \dot{u}(0) = \dot{u}(T).$$

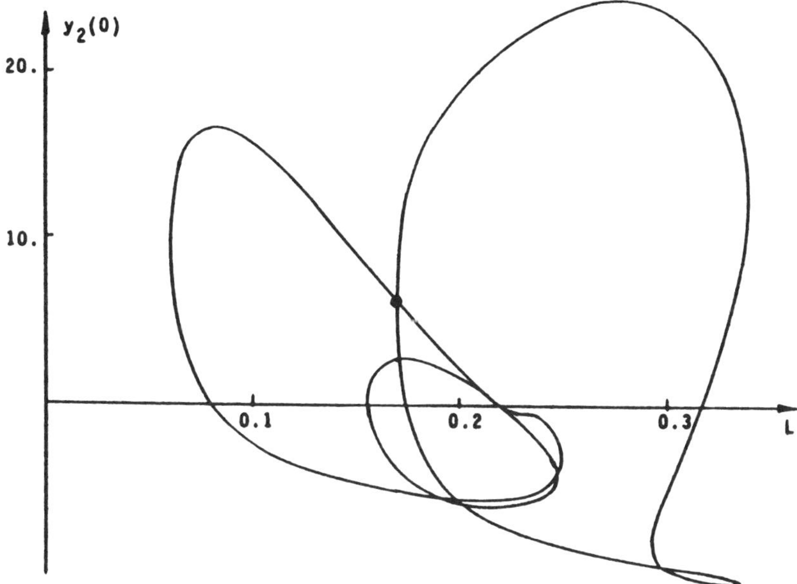

Figure 6.1

Because the period T varies with ω, the integration interval $0 \leq t \leq T$ must be adapted whenever ω is changed. This is inconvenient when nodes of a numerical solution procedure must be adapted too, so we transform the integration interval to unit length, thereby shifting the dependence on ω to the right-hand side of the differential equation. The normalized time $\tilde{t}$ satisfies

$$T\tilde{t} = t, \quad 0 \leq \tilde{t} \leq 1. \tag{6.7}$$

The transformation

$$y_1(\tilde{t}) = u(t), \quad y_2(\tilde{t}) = \dot{u}(t) \tag{6.8}$$

leads to the first-order system

$$\begin{aligned} y_1' &= Ty_2 \\ y_2' &= T(-0.04y_2 + 0.2y_1 - 8y_1^3/15 + 0.4 \cos 2\pi\tilde{t}) \end{aligned} \tag{6.9}$$

with boundary conditions

$$y_1(0) - y_1(1) = 0$$

$$y_2(0) - y_2(1) = 0.$$

Redefining the normalized time by t, we obtain a boundary-value problem of the standard form in Eq. (6.1). With $y_3 = \omega$, the corresponding extended system in Eqs. (6.2) and (6.3) is

$$y_1' = 2\pi y_2/y_3$$

$$y_2' = 2\pi(-0.04y_2 + 0.2y_1 - 8y_1^3/15 + 0.4\cos 2\pi t)/y_3$$

$$y_3' = 0$$

$$\begin{aligned} y_1(0) - y_1(1) &= 0 \\ y_2(0) - y_2(1) &= 0 \\ y_k(0) - \eta &= 0. \end{aligned} \tag{6.10}$$

Harmonic solutions can be calculated by solving Eq. (6.10). With appropriate choices for k and η, the branching diagram in Figure 2.7 results. The branching behavior is rich for small values of ω. As Figure 6.2 shows, there are many branches, turning points, and bifurcation points for values of the parameter $\omega < 0.5$. The solid curve in this branching diagram represents oscillations, with phase diagrams being symmetric with respect to the origin (Figure 6.3). For decreasing ω, each loop of this "main" branch attaches a further wiggle to the os-

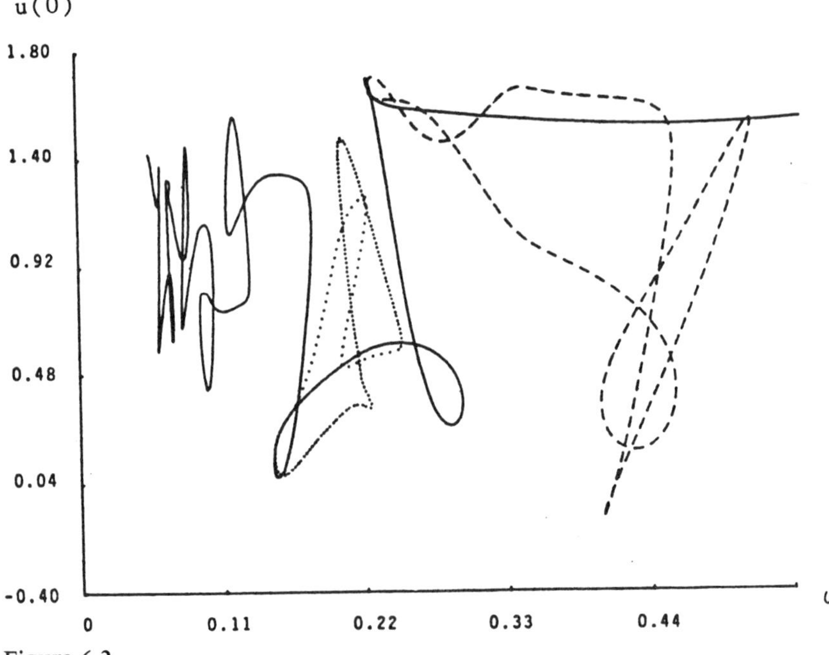

Figure 6.2

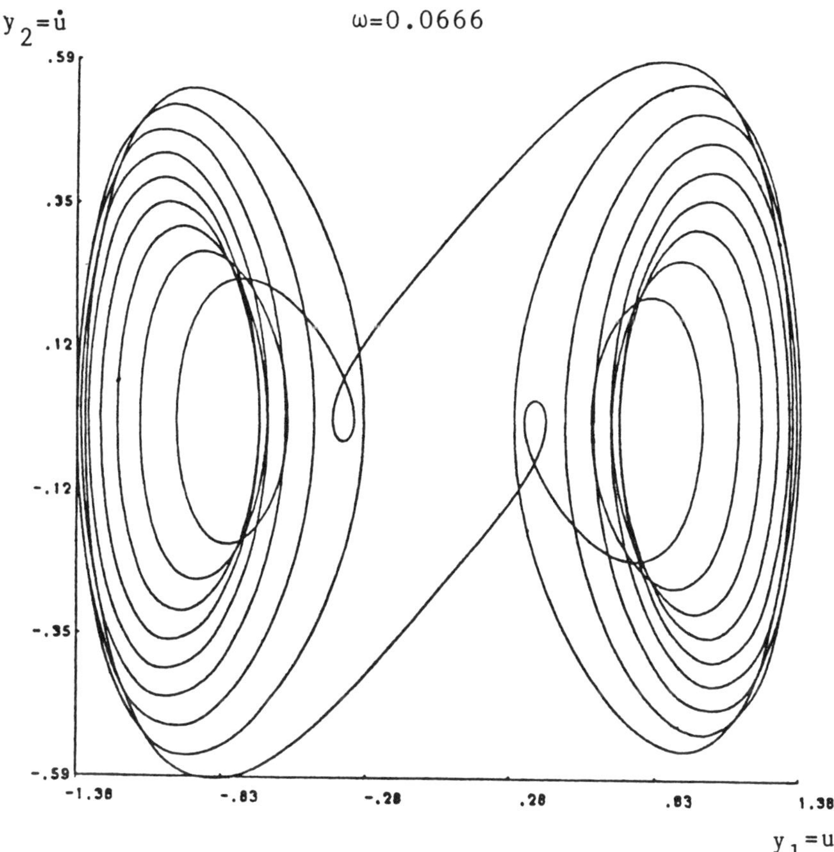

Figure 6.3

cillation. The wiggles indicate that the small-amplitude oscillation around any of the stable equilibria (Section 2.3) takes much time to collect enough energy before a transition to the other attracting basin is possible; compare the phase diagram in Figure 6.3 and the time dependence in Figure 6.4 ($\omega = 0.0666$).

Each of the closed branches in Figure 6.2 (dashed and dotted curves) is attached to the main branch via two pitchfork bifurcations. Accordingly, the phase diagrams of these "secondary" branches are asymmetric with respect to the origin (Figure 6.5). Further secondary branches were calculated but not included in Figure 6.2. The closed branch drawn in a dashed line appears to have sharp corners, and one might expect difficulties in tracing this particular branch. The plotted behavior refers only to the dependence $y_1(0)$ versus λ; the correspond-

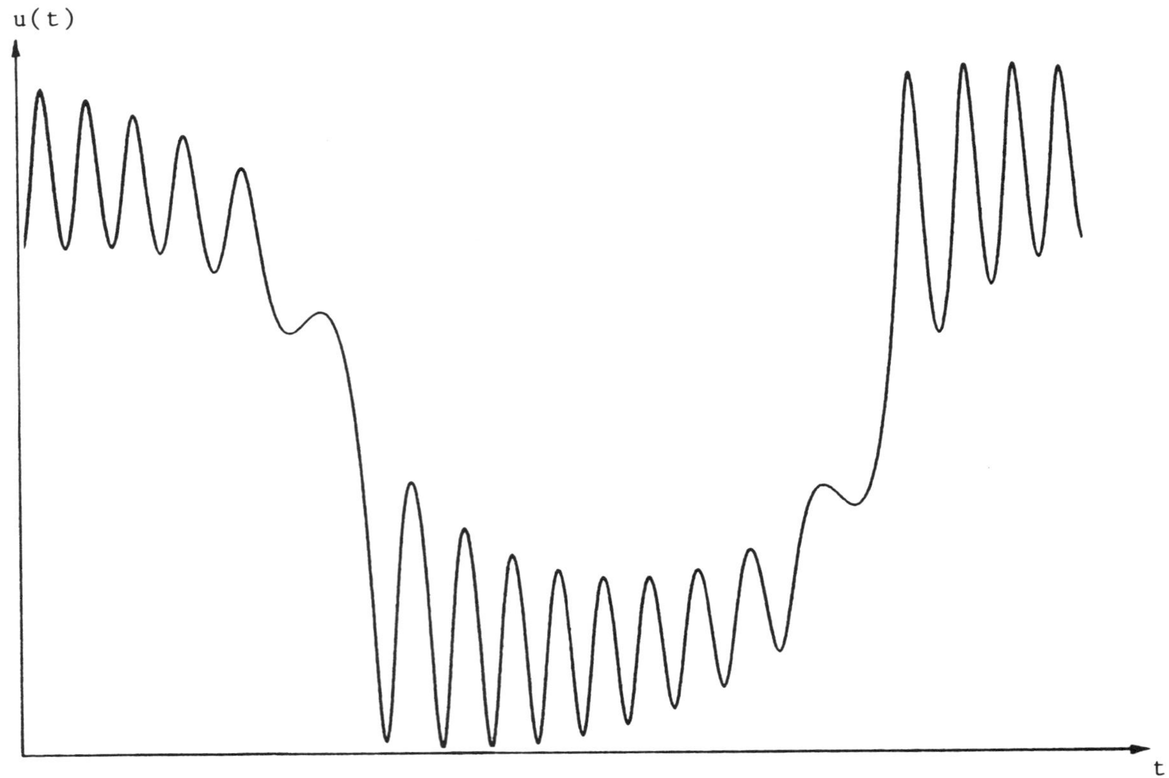

Figure 6.4

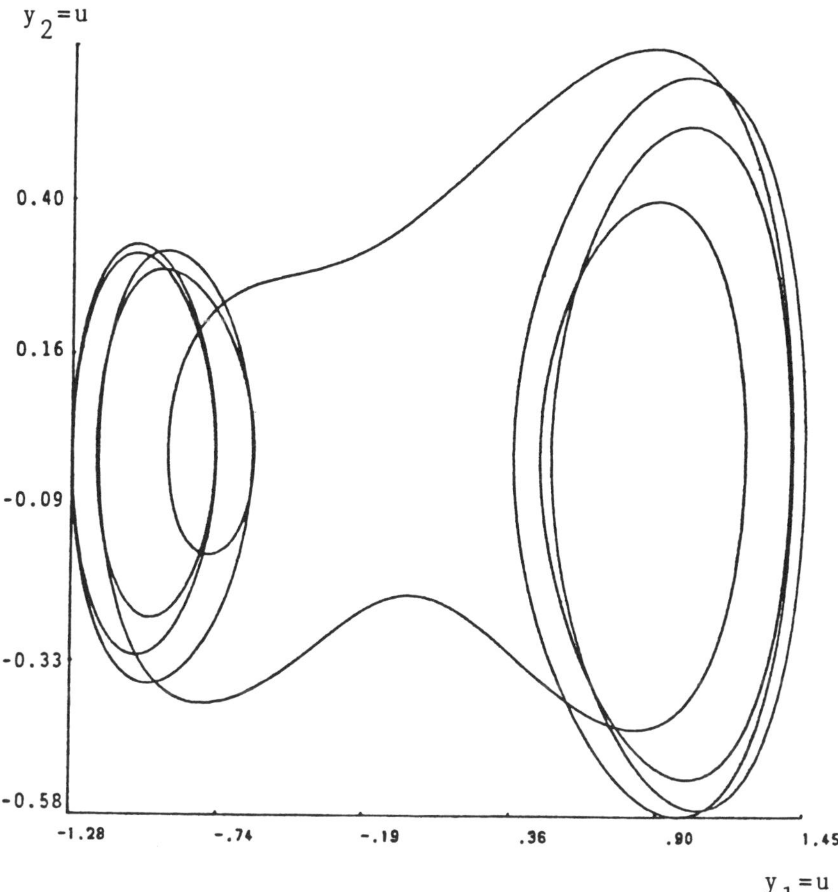

Figure 6.5

ing graph of $y_2(0)$ behaves more smoothly (see Figure 6.6). Accordingly, a local parameterization does not encounter difficulty in finding an appropriate index k. The closed branch depicted in Figure 6.6 illustrates how difficult it is to decide when the tracing of the branch is completed. As outlined in Section 3.3, it appears to be barely possible to implement a criterion that recognizes safely that a part of a branch is already known.

For periodic oscillations it is interesting to know the amplitude. The amplitude can often be measured sufficiently accurately from related plots (as in Figures 6.3, 6.4, and 6.5). If more accuracy is required, one must resort

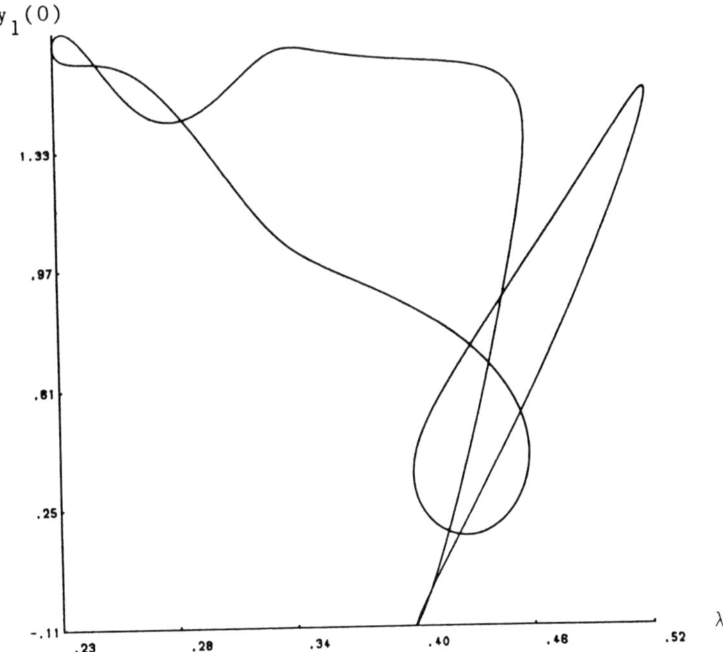
$y_1(0)$
1.33
.97
.61
.25
-.11
.23
.28
.34
.40
.46
.52
λ

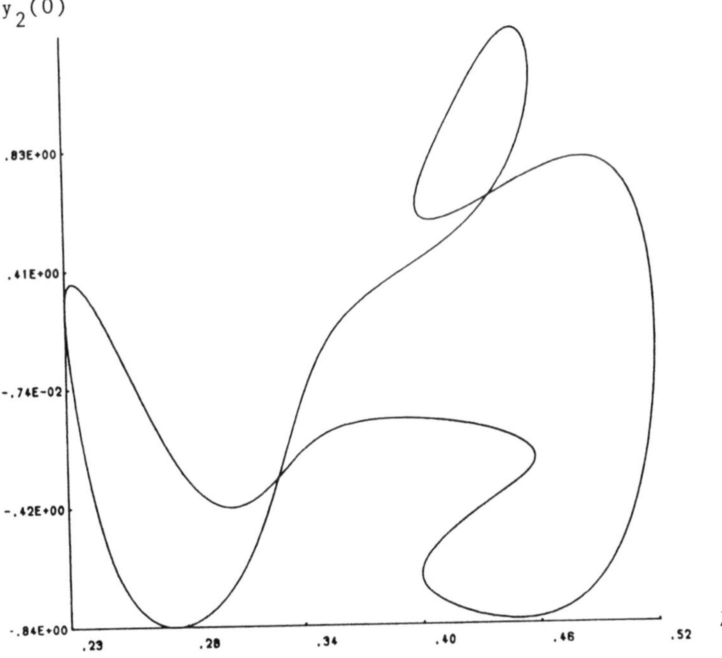
$y_2(0)$
.83E+00
.41E+00
-.74E-02
-.42E+00
-.84E+00
.23
.28
.34
.40
.46
.52
λ

Figure 6.6

to numerical approximations. The amplitude of an oscillation can be calculated with high accuracy as part of the solution procedure [28]. As an alternative, the amplitude can be approximated after the solution is calculated. To this end, a spline that approximates the current solution can be constructed. Evaluating the zeros of the derivative of the spline yields the values of the amplitude (Exercise 6.3). A table of 20 branch points of the above Duffing equation with accurate values of the amplitude can be found in [28].

Exercise 6.1.

(a) Derive Eqs. (6.5a) and (6.5b) from Eq. (6.4).

(b) What are the boundary conditions that replace those in (6.5b) when *zero flux* is considered,

$$\frac{\partial X}{\partial x} = \frac{\partial Y}{\partial x} = 0 \qquad \text{for } x = 0,\ x = \mathrm{L}?$$

(c) Write down explicitly the functions f_i and r_i $(i = 1, \ldots, 4)$ from Eq. (6.1) that define Eqs. (6.5a) and (6.5b).

(d) Calculate all first-order partial derivatives of f_i and r_i with respect to y_j $(i,j = 1, \ldots, 4)$ and arrange them in matrices. Notice that the derivatives of $\mathbf{r}$ include $2n^2$ entries.

(e) Which constant solution does satisfy Eqs. (6.5a) and (6.5b)? Where is this solution found in Figure 6.1?

Exercise 6.2.

Derive in detail boundary-value problem (6.10). Start from the original Duffing equation and use Eqs. (6.7) and (6.8).

Exercise 6.3 (project).

Assume that a periodic oscillation has been calculated with vector of initial values $\mathbf{y}(\mathbf{0})$ and period T. The right-hand side of the differential equation is $\mathbf{f}$. Suppose further that N values $y(t_j)$ at equidistant nodes t_j are required (for plotting purposes, say), together with an approximation of the amplitude. Design and implement an algorithm that calculates the required approximations. To this end, modify an integration routine with step control (as, for instance, an RKF method; see Appendix 4) as follows: Each function evaluation of $\mathbf{f}$ is to be used to calculate locally cubic splines [72, 334]. Establish formulas that calculate the extrema of the splines. Use the splines also for calculating the desired approximations of $y(t_j)$; in this way the step control need not be manipulated.

6.2 CALCULATION OF BRANCH POINTS

The calculation of branch points in boundary-value problems follows along the lines outlined in the previous chapter. We establish a general direct method and deduce a branching test function on the ODE level.

6.2.1 Branching System

The linearization of Eq. (6.1) with respect to $\mathbf{y}$ is the boundary-value problem

$$\begin{aligned} \mathbf{h}' &= \mathbf{f}_{\mathbf{y}}(t,\mathbf{y},\lambda)\mathbf{h} \\ \mathbf{A}\mathbf{h}(a) + \mathbf{B}\mathbf{h}(b) &= \mathbf{0}. \end{aligned} \tag{6.11}$$

The vector-valued function $\mathbf{h}(t)$ corresponds to the $\mathbf{h}$ of Chapter 5. $\mathbf{A}$ and $\mathbf{B}$ are the n^2 matrices of the linearization of the boundary conditions,

$$\mathbf{A} = \frac{\partial \mathbf{r}(\mathbf{y}(a),\mathbf{y}(b))}{\partial \mathbf{y}(a)}, \quad \mathbf{B} = \frac{\partial \mathbf{r}(\mathbf{y}(a),\mathbf{y}(b))}{\partial \mathbf{y}(b)}. \tag{6.12}$$

Note that $\mathbf{A}$ and $\mathbf{B}$ in general vary with $\mathbf{y}$. But since $\mathbf{A}$ and $\mathbf{B}$ are frequently constant matrices, we suppress the dependence on $\mathbf{y}$ for the sake of convenience. Evaluated at a branch point $(\mathbf{y}_0,\lambda_0)$, the linearized problem in Eq. (6.11) has a nontrivial solution $\mathbf{h} \neq 0$. For $(\mathbf{y},\lambda) \neq (\mathbf{y}_0,\lambda_0)$, the only solution to Eq. (6.11) is $\mathbf{h} = 0$. Along the lines set out in the previous chapter, we impose the equation

$$h_k(a) = 1$$

on Eq. (6.11); this initial condition enforces $\mathbf{h} \neq 0$. Hence, a branch point is characterized by

$$\zeta = h_k(a) - 1 = 0$$

for a solution $\mathbf{h}$ to Eq. (6.11). This characteristic feature formally replaces the $(n + 1)$st boundary condition of Eq. (6.3). In this way the value λ_0 of a branch point is fixed as a solution of

$$\begin{pmatrix} \mathbf{y} \\ \lambda \end{pmatrix}' = \begin{pmatrix} \mathbf{f}(t,\mathbf{y},\lambda) \\ 0 \end{pmatrix}, \quad \begin{pmatrix} \mathbf{r}(\mathbf{y}(a),\mathbf{y}(b)) \\ \zeta(\mathbf{y},\lambda) \end{pmatrix} = \mathbf{0}. \tag{6.13}$$

One way of implementing this equation is to attach the linearization Eq. (6.11), which is implicitly included in $\zeta = 0$. This leads to the branching system of ODEs [306, 307],

$$\begin{pmatrix} \mathbf{y} \\ \lambda \\ \mathbf{h} \end{pmatrix}' = \begin{pmatrix} \mathbf{f}(t,\mathbf{y},\lambda) \\ 0 \\ \mathbf{f}_\mathbf{y}(t,\mathbf{y},\lambda)\mathbf{h} \end{pmatrix}, \quad \begin{pmatrix} \mathbf{r}(\mathbf{y}(a),\mathbf{y}(b)) \\ h_k(a) - 1 \\ \mathbf{Ah}(a) + \mathbf{Bh}(b) \end{pmatrix} = \mathbf{0}. \tag{6.14}$$

The boundary-value problem Eq. (6.14) is of the standard form in Eq. (6.2) and is thus amenable to SOLVER. Because the branching system involves $2n + 1$ components, storage requirements grow rapidly with n. This usually restricts Eq. (6.14) to small or moderate values of n. The index k is the same as the k of the previous section and the previous chapter. In particular, indices that refer to an initial condition in the boundary conditions of Eq. (6.1) are not feasible (Exercise 6.5). Other implementations of Eq. (6.13) lead to an indirect method (to be described next) or to a semidirect method (Exercise 6.7). The branching system Eq. (6.14) applies both to bifurcation and to turning points, as is pointed out in the next chapter, it also applies to Hopf bifurcation and other bifurcation points. Because Eq. (6.14) includes Eq. (5.18) as a special case, we can regard (6.14) as a most general extended system devoted to the calculation of branch points. Before we discuss such topics as the calculation of an initial guess, we study an example.

6.2.2 Catalytic Reaction

Heat and mass transfer within a porous catalyst particle can be described by the boundary-value problem

$$\frac{d^2y}{dx^2} + \frac{a}{x}\frac{dy}{dx} = \vartheta^2 y^n \exp\left[\frac{\gamma\beta(1 - y)}{1 + \beta(1 - y)}\right]. \tag{6.15}$$

The meaning of the variables (see [149]) is:

y: dimensionless concentration

x: dimensionless coordinate

a: coefficient defining shape of particle ($a = 0$: flat plate; $a = 1$: cylindrical; $a = 2$: spherical)

ϑ: Thiele modulus

γ: dimensionless energy of activation

β: dimensionless parameter of heat evolution.

The boundary conditions are

$$\frac{dy(0)}{dx} = 0, \quad y(1) = 1.$$

We choose the simple case of a flat particle ($a = 0$) with first-order reaction ($n = 1$) and take as a branching parameter the modified Thiele modulus $\lambda = \vartheta^2$.

The resulting first-order system is

$$\begin{aligned} y_1' &= y_2 \\ y_2' &= \lambda y_1 e(y_1), \\ y_1(1) = 1,& \quad y_2(0) = 0. \end{aligned} \tag{6.16}$$

Here $e(y)$ serves as an abbreviation for the exponential term,

$$e(y_1) = \exp[\gamma\beta(1 - y_1)/(1 + \beta(1 - y_1))].$$

The Jacobian matrix $\mathbf{f_y}$ associated with Eq. (6.16) is

$$\mathbf{f_y}(t,\mathbf{y},\lambda) = \begin{pmatrix} 0 & 1 \\ \lambda\left[e(y_1) + \dfrac{de}{dy_1} y_1\right] & 0 \end{pmatrix}.$$

We write $y_3 = \lambda$, $y_4 = h_1$, $y_5 = h_2$ and obtain for the linearized system

$$h_1' = y_4' = h_2 = y_5$$

$$h_2' = y_5' = y_3 y_4 \left[e(y_1) + \frac{de}{dy_1} y_1\right].$$

Because the boundary conditions are linear (Exercise 6.6), the boundary conditions of the linearization can be written down immediately as

$$h_1(1) = 0, \quad h_2(0) = 0.$$

Collecting all parts, one obtains the branching system

$$\begin{aligned} y_1' &= y_2 \\ y_2' &= y_1 y_3 e(y_1) \\ y_3' &= 0 \\ y_4' &= y_5 \\ y_5' &= y_3 y_4 e(y_1)[1 - y_1\gamma\beta(1 + \beta(1 - y_1))^{-2}] \\ y_1(1) = 1,& \quad y_2(0) = 0, \\ y_4(1) = 0,& \quad y_5(0) = 0, \quad y_4(0) = 1 \qquad (k = 1). \end{aligned} \tag{6.17}$$

For $\gamma = 20$, $\beta = 0.4$, and positive values of λ, there is a branch that exhibits hysteresis behavior; see the branching diagram in Figure 6.7. During the course of the continuation, the two turning points have been calculated by solving the branching system Eq. (6.17). The resulting values are

$$\lambda_0 = 0.07793, \quad y_0(0) = 0.2273$$

$$\lambda_0 = 0.13756, \quad y_0(0) = 0.7928.$$

For values of λ between these two critical values, there are three different concentrations (see Figure 6.8). Only the upper and lower profiles are stable.

Section 5.7 showed how the hysteresis behavior varies with γ (see Figures 5.25 and 5.26). The critical boundaries of Figure 5.25 can be obtained by continuation of the branching system Eq. (6.17) with respect to γ or by indirect methods (see Section 5.7). For completeness we remark that for $\lambda < 0$ another branch can be found. These solutions are not physical because concentrations become negative or larger than one. From a mathematical point of view, however, the new branch is interesting, because it branches off to infinity for $\lambda \to 0$.

6.2.3 Initial Approximation and Branching Test Function

Consider the situation that $(\bar{\mathbf{y}},\bar{\lambda})$ is a solution to Eq. (6.1), calculated not far from a branch point $(\mathbf{y}_0,\lambda_0)$. The linearized boundary-value

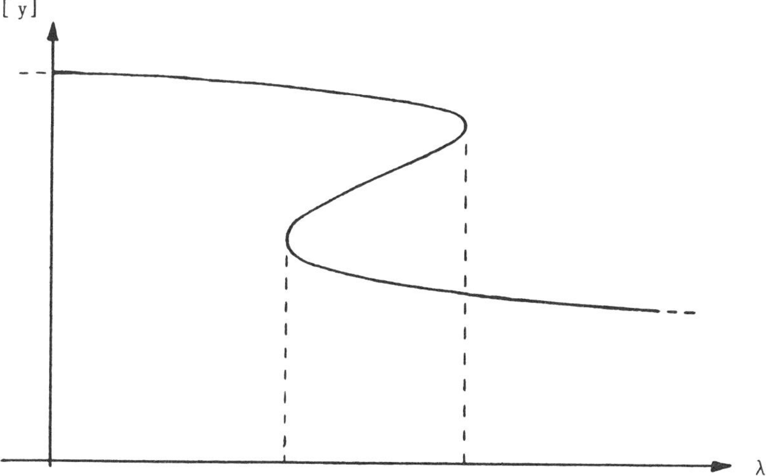

Figure 6.7

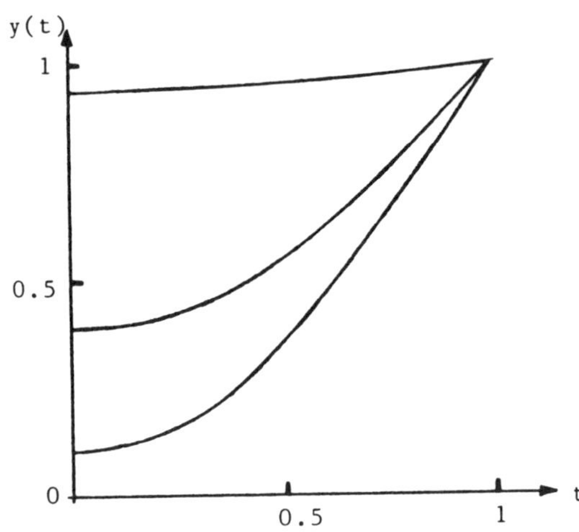

Figure 6.8

problem Eq. (6.11) has only the trivial solution if evaluated about $(\bar{\mathbf{y}},\bar{\lambda}) \neq (\mathbf{y}_0,\lambda_0)$. As in the methods described in Section 5.3.2, the l-th boundary condition (for an index l, $1 \leq l \leq n$) is removed to obtain a solvable system,

$$\begin{gathered} \mathbf{h}' = \mathbf{f_y}(t,\bar{\mathbf{y}},\bar{\lambda})\mathbf{h} \\ (\mathbf{I} - \mathbf{e}_l\mathbf{e}_l^{tr})(\mathbf{Ah}(a) + \mathbf{Bh}(b)) = \mathbf{0}, \quad h_k(a) = 1. \end{gathered} \tag{6.18}$$

We denote the solution to Eq. (6.18) by $\bar{\mathbf{h}}$. This vector function serves as the concluding part of the initial approximation $(\bar{\mathbf{y}},\bar{\lambda},\bar{\mathbf{h}})$ of the solution $(\mathbf{y}_0,\lambda_0,\mathbf{h}_0)$ to the branching system Eq. (6.14). One way of calculating $\bar{\mathbf{h}}$ is to solve the boundary-value problem

$$\begin{pmatrix} \mathbf{y} \\ \lambda \\ \mathbf{h} \end{pmatrix}' = \begin{pmatrix} \mathbf{f}(t,\mathbf{y},\lambda) \\ 0 \\ \mathbf{f_y}(t,\mathbf{y},\lambda)\mathbf{h} \end{pmatrix}, \quad \begin{pmatrix} \mathbf{y}(a) - \bar{\mathbf{y}}(a) \\ \lambda - \bar{\lambda} \\ (\mathbf{I} - \mathbf{e}_l\mathbf{e}_l^{tr})(\mathbf{Ah}(a) + \mathbf{Bh}(b)) \\ h_k(a) - 1 \end{pmatrix} = \mathbf{0} \tag{6.19}$$

which differs from Eq. (6.14) only in the boundary conditions. Solving Eq. (6.19) is not problematic because it is a linear problem; the nonlinear part is not effective because its solution $\bar{\mathbf{y}}$ is already calculated. In the following section, we see how $\bar{\mathbf{h}}$ can be approximated more efficiently.

By construction, the value of the removed boundary condition is a branching test function τ,

$$\tau = \mathbf{e}_l^{\mathrm{tr}}(\mathbf{A}\mathbf{h}(a) + \mathbf{B}\mathbf{h}(b)). \tag{6.20}$$

One possibility for calculating τ is to use $\mathbf{h}(a)$, $\mathbf{h}(b)$ from Eq. (6.19); an economical way to calculate τ is presented in the next section. As in Chapter 5, the test function depends on the choice of the indices l and k. As in the method in Exercise 5.11, the branching system Eq. (6.14) can be embedded in a set of boundary-value problems parameterized by τ from Eq. (6.20).

On the accuracy that can be obtained efficiently by using Eq. (6.14), see Section 5.3.2. The same observations are valid in the ODE situation. The singularity in the case of bifurcation points can be overcome by following a proposal in [241] (see Section 5.5.2). Alternatively, underlying symmetries can be exploited to remove the singularity.

Exercise 6.4.
Establish the linearized boundary-value problems in Eq. (6.11) for the Brusselator in Eq. (6.5) and the Duffing equation in Eq. (6.9).

Exercise 6.5.
Show that a boundary condition $r_j = y_\nu(a) - \text{constant} = 0$ implies $h_\nu(a) = 0$.

Exercise 6.6.
Boundary conditions are frequently linear, $\mathbf{r} = \tilde{\mathbf{A}}\mathbf{y}(a) + \tilde{\mathbf{B}}\mathbf{y}(b) - \mathbf{c}$, with an n-vector $\mathbf{c}$ and matrices $\tilde{\mathbf{A}}$ and $\tilde{\mathbf{B}}$.

(a) What are the boundary conditions of the linearization?
(b) What are $\tilde{\mathbf{A}}$, $\tilde{\mathbf{B}}$, and $\mathbf{c}$ in the case of periodic boundary conditions?

6.3 STEPPING DOWN FOR AN IMPLEMENTATION

Although from a theoretical point of view $\mathbf{h}$ and τ can be satisfactorily defined and calculated via ODE boundary-value problems, this approach is not practical. Notice that Eq. (6.19) represents a boundary-value problem of double the size. In order to construct an efficient implementation, we temporarily step down from the ODE level to a finite-dimensional approximation. The vector function $\mathbf{h}(t)$ can be approximated for discrete values of t_j, which will suffice for both branch switching and calculating τ.

Information required for evaluating $\mathbf{h}$ and τ is hidden in the data calculated during the approximation of a solution $\mathbf{y}$ to Eq. (6.1) or Eq. (6.3). It is possible to reveal this information, thereby benefiting from data that are available without extra cost. Different methods incorporated in SOLVER produce data in different ways. Accordingly, the analysis of revealing $\mathbf{h}$ and τ depends on the type of method applied to solve Eq. (6.1) or Eq. (6.3). A corresponding analysis has been carried out for multiple shooting [307]; the results are as follows.

Assume that $\mathbf{y}$ was calculated by multiple shooting. In a shooting implementation in *condensed form,* an n^2 iteration matrix, which is the core of the internal Newton iteration, is established [334]. This iteration matrix is

$$\mathbf{E} = \mathbf{A} + \mathbf{B}\mathbf{G}_{m-1}\cdot\ldots\cdot\mathbf{G}_1. \tag{6.21}$$

The matrices $\mathbf{A}$ and $\mathbf{B}$ are those of the linearized boundary conditions in Eq. (6.12). The matrices $\mathbf{G}_j$ are defined as

$$\mathbf{G}_j = \mathbf{G}(t_{j+1}), \tag{6.22}$$

where $\mathbf{G}(t)$ solves the matrix initial-value problem

$$\mathbf{G}' = \mathbf{f}_{\mathbf{y}}(t,\mathbf{y},\lambda)\mathbf{G}, \quad \mathbf{G}(t_j) = \mathbf{I}.$$

Here $a = t_1 < t_2 < \ldots < t_m = b$ are the nodes of multiple shooting ($m = 2$ for simple shooting). When SOLVER terminates, the internal iteration matrix $\mathbf{E}$ and the matrices $\mathbf{G}_j$ can be made available without extra cost. As in the method in Section 5.3, we replace the l-th row of $\mathbf{E}$ by the k-th unit vector to obtain the matrix $\mathbf{E}_{lk}$. This can be written formally as

$$\mathbf{E}_{lk} = (\mathbf{I} - \mathbf{e}_l\mathbf{e}_l^{tr})\mathbf{E} + \mathbf{e}_l\mathbf{e}_k^{tr}, \tag{6.23}$$

compare Eq. (5.20). The matrix $\mathbf{E}_{lk}$ depends on the solution $(\mathbf{y},\lambda)$ or $(\overline{\mathbf{y}},\overline{\lambda})$, at which $\mathbf{E}$ was evaluated. If the indices l and k are chosen in a way such that $\mathbf{E}_{lk}$ is nonsingular (not problematic), then Eq. (6.18) possesses a unique solution $\mathbf{h}(t)$. The initial vector $\mathbf{h}(a)$ is the solution of the linear equation

$$\mathbf{E}_{lk}\mathbf{h}(a) = \mathbf{e}_l. \tag{6.24}$$

Test functions $\tau = \tau_{lk}$ are given by the scalar product of $\mathbf{h}(a)$ with the l-th row of $\mathbf{E}$,

$$\tau_{lk} = \mathbf{e}_l^{tr}\mathbf{E}\mathbf{h}(a). \tag{6.25}$$

Because SOLVER outputs $\mathbf{y}$ at the discrete shooting nodes, $\mathbf{y}(t_j)$, it is sufficient to calculate $\mathbf{h}$ at the same points t_j. This is furnished by

$$\mathbf{h}(t_j) = \mathbf{G}_{j-1}\mathbf{h}(t_{j-1}) \qquad \text{for } j = 2, \ldots, m. \tag{6.26}$$

Summarizing, the branching test function τ can be calculated by solving one linear system, Eq. (6.24), and evaluating one scalar product, Eq. (6.25). This amount of work compares favorably to solving boundary-value problem Eq. (6.19). Because the matrices $\mathbf{A}$, $\mathbf{B}$, and $\mathbf{G}_j$ are approximated by numerical differentiation, the iteration matrix $\mathbf{E}$ is subjected to discretization errors. This affects the accuracy of $\mathbf{h}$ and τ. The remarks in Section 5.1 are valid here too—that is, the accuracy of τ depends on the strategy of rank-one approximations in SOLVER and on the history of previous continuation steps.

Exercise 6.7 (laborious project).
Construct a direct method for calculating branch points as follows: Implement a special shooting code that solves Eq. (6.13), exploiting Eq. (6.24) and $\mathbf{E}\cdot\mathbf{h}(a) = 0$ internally.

6.4 BRANCH SWITCHING AND SYMMETRY

We end the excursion to a finite-dimensional approximation and come back to the ODE level. Concerning branch switching, approaches similar to those described earlier are valid. We confine ourselves to a brief list of suitable boundary-value problems; for motivations and illustrating figures, see Section 5.4.

Predictors $\bar{\mathbf{z}}$ can be based on interpolation,

$$\bar{\mathbf{z}}(t) = \bar{\mathbf{y}}_0(t) \pm |\delta| \bar{\mathbf{h}}_0(t). \tag{6.27}$$

Here $\bar{\mathbf{y}}_0(t)$ and $\bar{\mathbf{h}}_0(t)$ are linear interpolations based on the values of $\mathbf{y}$ and $\mathbf{h}$ obtained at two solutions not far from a bifurcation point; Eq. (5.35) holds true for boundary-value problems when the Jacobian $\mathbf{f_y}$ is replaced by the matrix $\mathbf{E}$. The distance δ is given by Eq. (5.34); in this formula, y_k is evaluated at $t = a$. If $\mathbf{h}$ is approximated as described in the previous section, the predictor $\bar{\mathbf{z}}$ is given only at the discrete nodes t_j. This is no disadvantage, because SOLVER does not require more.

The boundary-value problem that establishes a parallel computation of two distinct solutions is written in standard form, Eq. (6.2), with $\mathbf{Y} = (\mathbf{y},\lambda,\mathbf{z})$,

$$\mathbf{Y}' = \mathbf{F}(t,\mathbf{Y}) = \begin{pmatrix} \mathbf{f}(t,\mathbf{y},\lambda) \\ 0 \\ \mathbf{f}(t,\mathbf{z},\lambda) \end{pmatrix}$$
$$\mathbf{R}(\mathbf{Y}(a),\mathbf{Y}(b),\delta) = \begin{pmatrix} \mathbf{r}(\mathbf{y}(a),\mathbf{y}(b)) \\ z_k(a) - y_k(a) - \delta \\ \mathbf{r}(\mathbf{z}(a),\mathbf{z}(b)) \end{pmatrix} = \mathbf{0}. \tag{6.28}$$

The initial approximation is $(\bar{\mathbf{y}}_0,\bar{\lambda}_0,\bar{\mathbf{z}})$. We shall prefer equations of the simple dimension $n + 1$ to the dimension $2n + 1$ of Eq. (6.28). A first possibility is given by Eq. (6.3); initial approximation and η are

$$(\mathbf{y},\lambda) = (\bar{\mathbf{z}},\bar{\lambda}_0), \quad \eta = \bar{z}_k(a). \tag{6.29}$$

Better selective properties are obtained by exploiting symmetry breaking. To classify and investigate symmetries, consider a *reflection* $\tilde{y}_i(t)$ of $y_i(t)$. For example,

$$\tilde{y}_i(t) = y_i(a + b - t)$$

defines the reflection with respect to the line $t = \frac{1}{2}(a + b)$. We define the component y_i to be symmetric if

$$\tilde{y}_i(t) = y_i(t)$$

holds for all t in the interval $a \le t \le b$. We shall say that a boundary-value problem *supports* symmetry if it is solved by $\tilde{\mathbf{y}}$ whenever it is solved by $\mathbf{y}$. In particular, two kinds of symmetries occur frequently in bifurcation problems: even functions (6.30a) and odd functions (6.30b):

$$y_i(t) = \tilde{y}_i(t) \qquad \text{for } \tilde{y}_i(t) = y_i(a + b - t) \tag{6.30a}$$

$$y_i(t) = \tilde{y}_i(t) \qquad \text{for } \tilde{y}_i(t) = -y_i(a + b - t). \tag{6.30b}$$

In Eq. (6.30b) the reflection is with respect to the point $(y,t) = (0,\frac{1}{2}(a + b))$. Note that a component y_i can be odd while another component of the same solution is even. Two other types of symmetry are characteristic for bifurcations of periodic solutions with period $b - a$:

$$y_i(t) = \tilde{y}_i(t) \qquad \text{for } \tilde{y}_i(t) = -y_i\left(t + \frac{a + b}{2}\right) \tag{6.30c}$$

$$y_i(t) = \tilde{y}_i(t) \qquad \text{for } \tilde{y}_i(t) = y_i\left(t + \frac{a + b}{2}\right). \tag{6.30d}$$

A periodic solution of type (6.30d) actually has a period $\frac{1}{2}(a + b)$.

For a practical evaluation of symmetry, we first check a current solution to see whether its components are symmetric. Because it is not practical to check the above criteria for all t in the interval $a \leq t \leq b$, we confine ourselves to $t = a$. This establishes necessary criteria for the onset of symmetries. For a small error tolerance ϵ_1 (not smaller than the error tolerance of SOLVER) we check the criteria

$$|\, y_i(b) - y_i(a) \,| < \epsilon_1 \tag{6.31a}$$

$$|\, y_i(b) + y_i(a) \,| < \epsilon_1 \tag{6.31b}$$

$$\left| y_i(a) + y_i\left(\frac{a+b}{2}\right) \right| < \epsilon_1 \tag{6.31c}$$

$$\left| y_i(a) - y_i\left(\frac{a+b}{2}\right) \right| < \epsilon_1 . \tag{6.31d}$$

If for one i (say, k) one of the inequalities holds, we assume a symmetry of the corresponding type, Eq. (6.30). The next question is whether this symmetry is broken in the component z_k of the emanating solution. Because the emanating solution $\mathbf{z}$ is yet unknown, we resort to the predictor $\bar{\mathbf{z}}$ from Eq. (6.27). For a symmetric $\bar{\mathbf{z}}$, the following value of γ is zero:

$$\begin{aligned}
\gamma &= \bar{z}_k(b) - \bar{z}_k(a) && \text{in case (6.31a)} \\
\gamma &= \bar{z}_k(b) + \bar{z}_k(a) && \text{in case (6.31b)} \\
\gamma &= \bar{z}_k(a) + \bar{z}_k\left(\frac{a+b}{2}\right) && \text{in case (6.31c)} \\
\gamma &= \bar{z}_k(a) - \bar{z}_k\left(\frac{a+b}{2}\right) && \text{in case (6.31d).}
\end{aligned} \tag{6.32}$$

Hence, asymmetry of $\bar{\mathbf{z}}$ manifests itself by

$$|\, \gamma \,| > \epsilon_2 ; \tag{6.33}$$

here we choose an error tolerance ϵ_2 slightly larger than ϵ_1. Numerical experience indicates that the test in Eqs. (6.31) through (6.33) safely detects a symmetry breaking. This success is to be expected, because $\bar{\mathbf{z}}$ is close to an emanating solution in case of pitchfork bifurcations—that is, symmetry breaking can be detected in advance during branch tracing not only when τ is monitored but also when $\mathbf{h}$ is. The above test for symmetry breaking is easily automated. Note that the test is based on solutions instead of on equations.

Assume now that for $i = k$ one of the criteria in Eq. (6.31) holds and that Eq. (6.33) is satisfied. For each kind of symmetry breaking there is a corresponding boundary-value problem with an appropriate selective boundary condition:

$$\begin{pmatrix}\mathbf{y}\\ \lambda\end{pmatrix}' = \begin{pmatrix}\mathbf{f}(t,\mathbf{y},\lambda)\\ 0\end{pmatrix}, \quad \begin{pmatrix}\mathbf{r}(\mathbf{y}(a),\mathbf{y}(b))\\ r_{n+1}\end{pmatrix} = \mathbf{0} \tag{6.34}$$

with γ from Eq. (6.32) and the boundary condition

$$r_{n+1} = y_k(b) - y_k(a) - \gamma \qquad \text{in case (6.31a)}$$

$$r_{n+1} = y_k(b) + y_k(a) - \gamma \qquad \text{in case (6.31b)}$$

$$r_{n+1} = y_k(a) + y_k\left(\frac{a+b}{2}\right) - \gamma \qquad \text{in case (6.31c)}$$

$$r_{n+1} = y_k(a) - y_k\left(\frac{a+b}{2}\right) - \gamma \qquad \text{in case (6.31d).}$$

The initial approximation is $(\bar{\mathbf{z}},\bar{\lambda}_0)$. In cases (6.31c and d) the boundary-value problem is a three-point boundary-value problem; a transformation to a two-point boundary-value problem is possible [313]. Case (6.30d) refers to period doubling; this can be seen by setting $T = \frac{1}{2}(b - a)$. The corresponding bifurcation means there is a transition from period T to period $2T$. This kind of bifurcation will be discussed in Chapter 7.

The above numerical test for symmetry breaking dispenses with an analytical investigation of Eq. (6.1), which can be cumbersome (Exercise 6.10). There is no guarantee that the criteria in Eq. (6.31) detect all the relevant regularities. For example, Eq. (6.31b) fails if Eq. (6.30b) is generalized to

$$y_i(a + b - t) + \vartheta = \vartheta - y_i(t)$$

for a nonzero value of ϑ. Fortunately, if y_i is a solution of a second-order differential equation, the derivative or antiderivative is even and Eq. (6.31a) works.

Example 6.3. Superconductivity Within a Slab.
In [247] a model of a superconducting slab in a parallel magnetic field is discussed. The Ginzburg-Landau equations are equivalent to the boundary-value problem of two second-order ODEs

$$\begin{aligned}\Theta'' &= \Theta(\Theta^2 - 1 + \lambda\Psi^2)\kappa^2 \\ \Psi'' &= \Theta^2\Psi\end{aligned} \tag{6.35}$$

for $0 \le t \le d$,

$$\Theta'(0) = \Theta'(d) = 0, \quad \Psi'(0) = \Psi'(d) = 1.$$

The meaning of the variables is

λ: square of external field
d: thickness of the slab ($d = 5$)
κ: Ginzburg-Landau parameter ($\kappa = 1$)
Θ: an order parameter that characterizes different superconducting states
Ψ: potential of the magnetic field.

Investigation of the above boundary-value problem reveals that it supports symmetry. To see this, define

$$\tilde{\Theta}(t) = \Theta(d - t), \quad \tilde{\Psi}(t) = -\Psi(d - t)$$

and realize that the signs are the same after differentiating twice with respect to t. Hence $\tilde{\Theta}$ and $\tilde{\Psi}$ satisfy the differential equations of Eq. (6.35) if Θ and Ψ do. Analyzing the boundary conditions, as, for example,

$$\tilde{\Theta}'(0) = -\Theta'(d) = 0,$$

shows that $\tilde{\Theta}$ and $\tilde{\Psi}$ satisfy the boundary conditions. Hence Eq. (6.35) supports even Θ and odd Ψ. This does *not* imply that solutions must be even or odd. In fact, Eq. (6.35) has asymmetric solutions. See the solution in Figure 6.9, which depicts both the solutions Θ,Ψ and $\tilde{\Theta},\tilde{\Psi}$ for the parameter value $\lambda = 0.6629$.

In order to calculate solutions, we rewrite the boundary-value problem as a first-order system. Denoting $y_1 = \Theta$, $y_2 = \Theta'$, $y_3 = \Psi$, and $y_4 = \Psi'$ yields

$$\begin{aligned} y_1' &= y_2 \\ y_2' &= y_1(y_1^2 - 1 + \lambda y_3^2) \\ y_3' &= y_4 \\ y_4' &= y_1^2 y_3 \end{aligned} \tag{6.36}$$

$$y_2(0) = y_2(5) = 0, \quad y_4(0) = y_4(5) = 1.$$

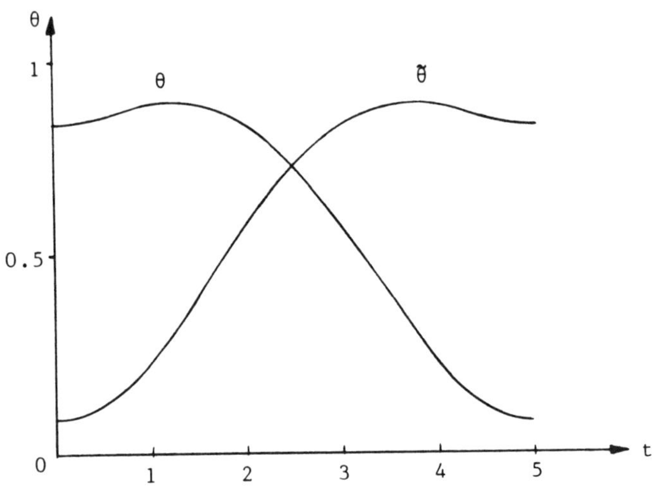

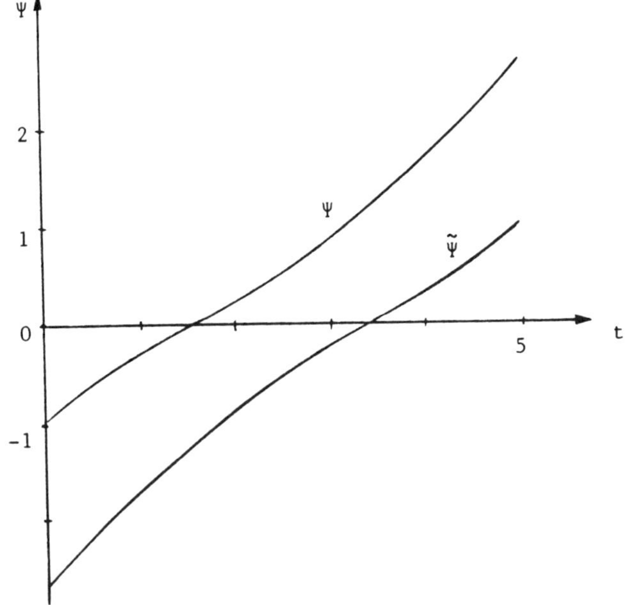

Figure 6.9

Solutions are summarized in the branching diagram in Figure 6.10. A branch of asymmetric solutions branches off a branch of symmetric solutions at the pitchfork bifurcation point. The data of the bifurcation point and of four turning points (TP) are listed in Table 6.1. The values in that table have been calculated by solving the branching system. The solution profiles of the bifurcation point are shown in Figure 6.11; notice the symmetries. This bifurcation point is straddled between the two turning points with asymmetric solutions, which are depicted in Figure 6.9. Compare Figure 6.9 and Figure 6.11 to see how this straddling becomes evident in the solution profiles.

TABLE 6.1

λ_0	$\Theta_0(0)$	$\Psi_0(0)$	Type
0.66294	0.08996	−2.687	TP, asymmetric sol.
0.66294	0.8461	−1.016	TP, asymmetric sol.
0.91196	0.5853	−1.407	pitchfork bif. pt.
0.92070	0.5214	−1.496	TP, symmetric sol.
0.84769	0.1923	−2.095	TP, symmetric sol.

Boundary-value problem Eq. (6.34) is designed to calculate asymmetric solutions. For the sake of completeness, we mention that boundary-value problems can be reformulated in a way that admits only symmetric solutions. We illustrate this possibility by means of the above example. Symmetric solutions to Eq. (6.35) satisfy

$$\Theta'\left(\frac{d}{2}\right) = 0, \quad \Psi\left(\frac{d}{2}\right) = 0$$

(see Figure 6.11). This allows us to reduce the integration interval from $0 \leq t \leq d$ to

$$0 \leq t \leq \frac{d}{2}, \text{ or, generally, } a \leq t \leq \frac{a+b}{2}.$$

Replacing the previous boundary conditions at $t = d$ by the midpoint conditions leads to the new boundary conditions

$$\Theta'(0) = 0, \quad \Psi'(0) = 1, \quad \Theta'\left(\frac{d}{2}\right) = 0, \quad \Psi\left(\frac{d}{2}\right) = 0.$$

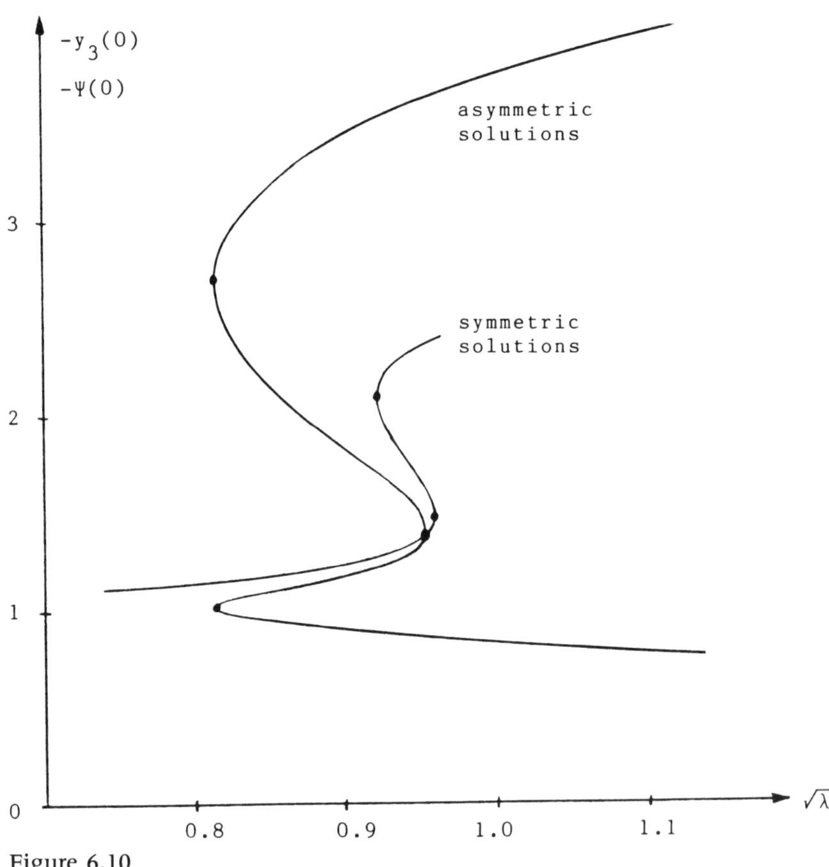

Figure 6.10

Solving a corresponding boundary-value problem yields the symmetric solutions supported by this example. Note that for this modified boundary-value problem the symmetry-breaking bifurcation *does not exist.*

Let us discuss the matter in more general terms. The "space" of symmetric functions is a (small) subspace of the space of all functions. The smaller space reflects certain assumptions that are imposed on the underlying model (here it is the assumption of symmetry). Relaxing a restricting assumption means enlarging the space of admissible functions. As a result, the bifurcation behavior may become richer. Another example of this phenomenon is in Section 7.4.1.

Exercise 6.8.
For each of the cases (6.30a) through (6.30d), sketch an example.

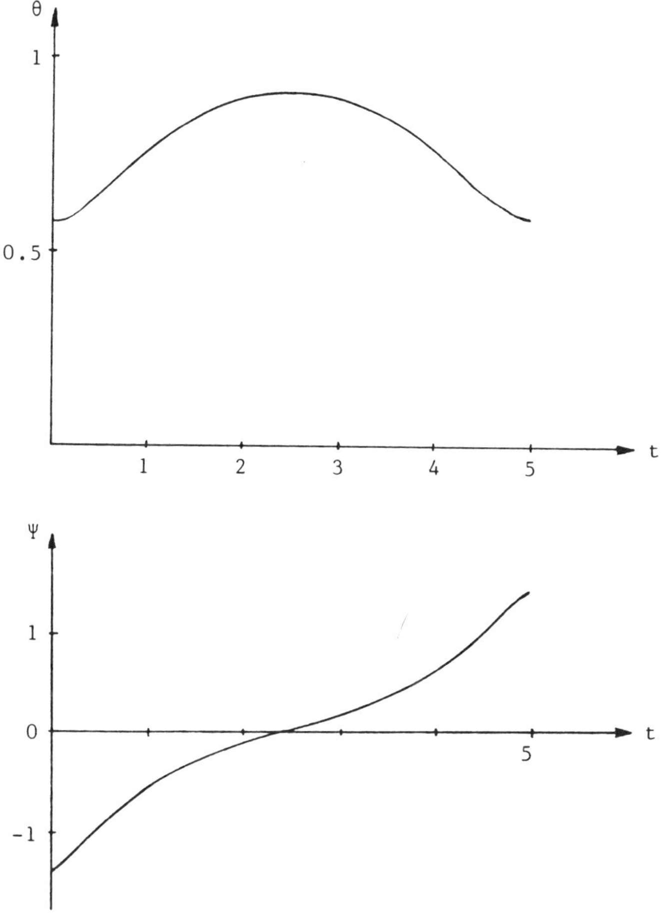

Figure 6.11

Exercise 6.9.
Assume that $u(t)$ and $v(t)$ are two scalar variables defined for $-1 \leq t \leq 1$. $u(t)$ is supposed to be odd, and $v(t)$ is even. What can be said about u', v', u'', v'', u^2, uv?

Exercise 6.10.
Show that the Brusselator model Eq. (6.4) supports even solutions. (Hint: Substitute $X^* = X - A$, $Y^* = Y - B/A$, and investigate the resulting system in X^*, Y^*.)

Exercise 6.11.
Show that the Duffing equation in Eq. (6.9) supports a symmetry of the type in case (6.30c).

Exercise 6.12.
Consider a periodic solution of a second-order oscillator (such as the Duffing equation in Eq. (6.9)) and its graphical representation in a phase plane. What can be said about the graphs of $u(t)$ and $\tilde{u}(t)$ when there is a symmetry of the type in case (6.30c)?

Exercise 6.13.
Find the trivial solution(s) of the superconductivity example in Eq. (6.35).

6.5 TRIVIAL BIFURCATION

Before modern computers and powerful numerical methods became generally available, a substantial part of the bifurcation research concentrated on problems where one *basic solution* y^B is known analytically for all λ. This drastic restriction is theoretically equivalent to the assumption

$$\mathbf{f}(t,\mathbf{0},\lambda) = \mathbf{0}, \quad \mathbf{r}(\mathbf{0},\mathbf{0}) = \mathbf{0}. \tag{6.37}$$

This can be seen by applying the transformation

$$\mathbf{y}^* = \mathbf{y} - \mathbf{y}^B$$

to the original problem (Exercise 6.10). We refer to bifurcations from the trivial solution $\mathbf{y} = \mathbf{0}$ as *trivial bifurcations* or *primary bifurcations.*

In principle, all the methods described so far apply to trivial bifurcation; some of the relevant methods simplify significantly. The branching system Eq. (6.14) reduces to

$$\begin{pmatrix} \mathbf{h} \\ \lambda \end{pmatrix}' = \begin{pmatrix} \mathbf{f_y}(t,\mathbf{0},\lambda)\mathbf{h} \\ 0 \end{pmatrix}, \quad \begin{pmatrix} \mathbf{Ah}(a) + \mathbf{Bh}(b) \\ h_k(a) - 1 \end{pmatrix} = \mathbf{0}. \tag{6.38}$$

In contrast to Eq. (6.14), this boundary-value problem of the simple dimension must be considered an efficient tool for calculating bifurcation points, at least as far as storage requirements are concerned. The Jacobian $\mathbf{f_y}$, however, is still required analytically. Indirect methods simplify drastically in the case of trivial bifurcations [312]; they are a recommended alternative to solving Eq. (6.38). Note that the techniques reported here and in Section 6.2.3 are readily applied to linear *eigenvalue problems*—namely, to the calculation of eigenvalues and eigenvectors.

The assumption in Eq. (6.37) often enables a simple analytical evaluation of bifurcation points. Although we generally put emphasis on numerical methods, let us take a break and carry out such a hand calculation.

Example 6.4.
Consider the simple second-order system

$$u'' + u^3 = \lambda u, \quad u(0) = u(\pi) = 0. \tag{6.39}$$

Clearly, Eq. (6.37) is satisfied. The linearization is

$$v'' = \lambda v, \quad v(0) = v(\pi) = 0.$$

Because both $\cos(\sqrt{\lambda} t)$ and $\sin(\sqrt{\lambda} t)$ solve the linearized differential equation, the general solution of the equation must be of the form

$$v(t) = c_1 \cos\sqrt{\lambda} t + c_2 \sin\sqrt{\lambda} t.$$

The constants and eigenvalues are determined by the boundary conditions. We obtain $c_1 = 0$ and $\sin(\sqrt{\lambda} t) = 0$, which fixes the eigenvalues

$$\lambda = j^2, \qquad j = 1, 2, 3, \ldots .$$

In this example, all the eigenvalues are bifurcation points. Simple eigenvalues of the linearization are always bifurcation points of the nonlinear problem. We remark in passing that the emanating solutions of Eq. (6.39) can be calculated analytically in terms of elliptic functions [305].

We close this section by investigating a more demanding example:

Example 6.5. Buckling of a Rod.
Consider a rod of length L that is subject to end loading (load $P > 0$). The shape of the rod is described by the tangent angle $u(x)$ as a function of the material points x on the rod in its originally straight position [248]. The variation of the angle with x is zero at the ends of the rod,

$$u'(0) = u'(L) = 0. \tag{6.40a}$$

A deformation of the rod depends on the pressure P, on the stiffness K of the rod material, and on compressibility effects. Neglecting the latter effects, the tangent angle satisfies the differential equation

$$u'' + \frac{P}{K}\sin(u) = 0. \tag{6.40b}$$

This is the famous Euler buckling problem [86]; a finite-element analog is discussed in Exercise 2.12. Euler showed that the undeformed state $u = 0$ is the only soution for P/K less than the first eigenvalue of the linearization,

$$\frac{P}{K} < \left(\frac{\pi}{L}\right)^2 .$$

A bent state is stable for values of P/K exceeding this critical value. Solutions in closed form can be found in [333].

A rod with nonlinear compressibility properties is discussed in [248]; the differential equation is

$$u'' + \frac{P}{K}\exp[C(P\cos u(x)) - C(0)]\sin u = 0 \tag{6.41}$$

with $C(\zeta)$ chosen to be

$$C(\zeta) = \tfrac{1}{5}\tanh[5(3 - 5\zeta + 2\zeta^2)].$$

In order to make it fit into the analytical framework of [248], this problem was simplified by fixing a unit load $P = 1$. This means varying the stiffness K or the parameter

$$\lambda = \frac{\exp(-C(0))}{K} = \frac{0.81873\ldots}{K}$$

in the boundary-value problem

$$u'' + \lambda\exp[C(\cos(u))]\sin(u) = 0, \quad u'(0) = u'(L) = 0. \tag{6.42}$$

We follow this simplification in order to obtain comparable results; numerical methods can handle the original problem, Eq. (6.41), taking $\lambda = P$ as parameter.

The boundary-value problem Eq. (6.42) has been solved for $L = 2$; Figure 6.12 shows a branching diagram. The eigenvalues of the linearization (Exercise 6.15) can be calculated numerically by solving Eq. (6.38). Alternatively, approximations $\bar{\lambda}_0, \bar{\mathbf{h}}_0$ are obtained effectively by the method of Section 6.3, which simplifies

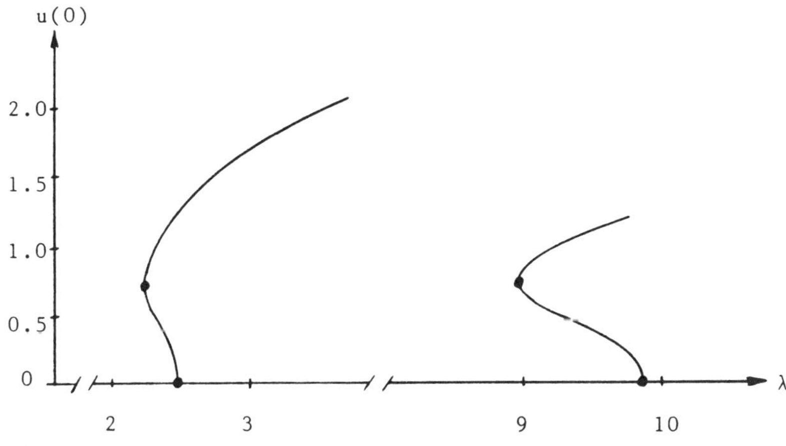

Figure 6.12

significantly in case of trivial bifurcation. Starting from these approximations, the emanating branch is traced by means of the boundary-value problem in Eq. (6.3). This reveals that the nontrivial branches locally branch off to the left, but globally the branches "soon" turn back to the right. The related turning points are easily obtained as by-products of the continuation; the methods indicated in Exercises 5.7 through 5.9 apply without change to boundary-value problems. The turning point with the smallest value of λ is

$$\lambda_0 = 2.242, \quad u(0) = 0.7293.$$

Exercise 6.14.
For the boundary-value problem

$$-u'' + u^3 = \lambda u, \quad u(0) = u(\pi), \quad u'(0) = u'(\pi)$$

calculate the eigenvalues of the linearization about $u = 0$.

Exercise 6.15.
Consider the buckling rod problem in Eqs. (6.40a) and (6.40b). Calculate the eigenvalues of the linearization about $u = 0$. Show that the linearization of Eq. (6.42) has the same eigenvalues.

6.6 TESTING STABILITY

This section looks at the stability of solutions to boundary-value problems. Recall the situation encountered with nonlinear equations

$$\mathbf{0} = \mathbf{f}(\mathbf{y},\lambda).$$

If this equation represents equilibria of a system of ODEs, the dynamic behavior is governed by the *evolution equation*

$$\dot{\mathbf{y}} = \frac{d\mathbf{y}}{dt} = \mathbf{f}(\mathbf{y},\lambda).$$

The situation of second-order ODE boundary-value problems

$$\mathbf{0} = \mathbf{y}'' - \mathbf{f}(x,\mathbf{y},\mathbf{y}',\lambda), \quad \mathbf{r}(\mathbf{y}(a),\mathbf{y}(b),\mathbf{y}'(a),\mathbf{y}'(b)) = \mathbf{0} \tag{6.43}$$

is analogous. Assume that Eq. (6.43) represents a stationary solution of a partial differential equation with the independent space variable x varying in the interval $a \le x \le b$. An evolution equation corresponding to second-order differential equations with linear boundary conditions is the system of PDEs

$$\frac{\partial \mathbf{y}}{\partial t} = \mathbf{D}\frac{\partial^2 \mathbf{y}}{\partial x^2} + \mathbf{f}\left(x,\mathbf{y},\frac{\partial \mathbf{y}}{\partial x},\lambda\right),$$

$$\mathbf{A}_0\mathbf{y}(a,t) + \mathbf{A}_1\frac{\partial \mathbf{y}(a,t)}{\partial x} = \mathbf{c}_a \tag{6.44}$$

$$\mathbf{B}_0\mathbf{y}(b,t) + \mathbf{B}_1\frac{\partial \mathbf{y}(b,t)}{\partial x} = \mathbf{c}_b.$$

Solutions to Eq. (6.44) depend on both time t and space variable x, $\mathbf{y}(x,t)$ for $a \le x \le b$, $t \ge 0$. $\mathbf{D}$, $\mathbf{A}_0$, $\mathbf{A}_1$, $\mathbf{B}_0$, and $\mathbf{B}_1$ are n^2 matrices, and $\mathbf{c}_a$ and $\mathbf{c}_b$ are n-vectors, fixing linear boundary conditions for $x = a$ and $x = b$. Often, the matrices are diagonal. With

$$\mathbf{y}' = \frac{\partial \mathbf{y}}{\partial x}, \quad \mathbf{y}'' = \frac{\partial^2 \mathbf{y}}{\partial x^2}$$

we write the ODE problem that represents the steady states of Eq. (6.44) in a form similar to Eq. (6.43),

$$\mathbf{0} = \mathbf{D}\mathbf{y}'' + \mathbf{f}(x,\mathbf{y},\mathbf{y}',\lambda)$$

$$\mathbf{A}_0\mathbf{y}(a) + \mathbf{A}_1\mathbf{y}'(a) = \mathbf{c}_a \tag{6.45}$$

$$\mathbf{B}_0\mathbf{y}(b) + \mathbf{B}_1\mathbf{y}'(b) = \mathbf{c}_b.$$

The Brusselator model is an example for the pair of equations (6.44) and (6.45). The full PDE problem of the Brusselator is listed in Eq. (5.42); the steady-state problem is in Eq. (6.4).

It makes sense to investigate the stability of a solution $\mathbf{y}_s(x)$ to the steady-state problem in Eq. (6.45) that has the evolution equation in Eq. (6.44) as background. Clearly, $\mathbf{y}(x,t) = \mathbf{y}_s(x)$ is a solution to the full system in Eq. (6.44) for all t. The question is whether nearby solutions approach the steady state or leave its neighborhood. Testing the stability of a steady state is cumbersome [141, 162, 198, 255, 285, 317]. In what follows we describe some basic strategies.

One method for testing a steady state for stability simulates the dynamic behavior. To this end, $\mathbf{y}_s(x)$ is taken as an initial profile for $t = 0$. Then Eq. (6.44) is solved numerically as an initial-value problem starting from

$$\mathbf{y}(x,0) = \mathbf{y}_s(x)$$

or from an artificially perturbed $\tilde{\mathbf{y}}_s$. Initial deviations eventually die out or grow. Stability is indicated when $\mathbf{y}(x,t)$, for increasing t, converges to $\mathbf{y}_s(x)$ or remains close. On the other hand, instability is signaled when $\mathbf{y}(x,t)$ takes values that are significantly distinct from the initial profile. The difficulty with simulation lies in the uncertainty about how to perturb $\tilde{\mathbf{y}}_s$ and until which value of t to carry out the integration.

A second method is based on a semidiscretization in the space variable—namely, the *method of lines*. For example, perform a simple discretization using the standard difference quotients

$$\begin{aligned} \frac{\partial^2 \mathbf{y}(x)}{\partial x^2} &\approx \frac{\mathbf{y}(x+\Delta) - 2\mathbf{y}(x) + \mathbf{y}(x-\Delta)}{\Delta^2} \\ \frac{\partial \mathbf{y}(x)}{\partial x} &\approx \frac{\mathbf{y}(x+\Delta) - \mathbf{y}(x-\Delta)}{2\Delta} \end{aligned} \tag{6.46}$$

for a small increment Δ. This is easily organized by imposing an equidistant grid on the interval $a \leq x \leq b$, using N interior grid points,

$$\Delta = \frac{b-a}{N+1}, \quad x_j = a + j\Delta \qquad \text{for } j = 0, 1, \ldots, n+1. \tag{6.47}$$

The components $y_i(x,t)$ of $\mathbf{y}(x,t)$ are approximated along the discrete "lines" $x = x_j$. This gives rise to nN scalar functions

$$y_i(x_j,t),$$

which depend on time only. Introducing an nN-vector $\mathbf{Y}$ by

$$Y_\nu(t) = y_i(x_j,t) \text{ with } \nu = (j-1)n + i, \qquad i = 1, \ldots, n, j = 1, \ldots, N \quad (6.48)$$

and inserting Eq. (6.46) into Eq. (6.44) yields an ordinary differential equation $\dot{\mathbf{Y}} = \mathbf{F}(\mathbf{Y},\lambda)$. The generation of $\mathbf{F}(\mathbf{Y},\lambda)$ is best done by computer; some care in implementing the boundary conditions is required (Exercise 6.16). After the ODE system in standard form is constituted, standard methods for checking the stability apply.

This second approach may be attractive because it is more deterministic than simulation is, but there are many difficulties to overcome. An equidistant grid with a constant spacing Δ is not adequate for many applications. Steep gradients in the $\mathbf{y}_s(x)$ profile call for a nonequidistant grid. Also, to keep the dimension small, difference quotients of an order higher than the second-order discretizations in Eq. (6.46) are preferred. Such improvements lead to highly complex algorithms. In any case, the dimension nN will be large, and the evaluation of the eigenvalues of the Jacobian of F is expensive. One must frequently resort to a small value of N, hoping that the stability of the stationary PDE solution is reflected even by a coarse ODE approximation.

Summarizing, we point out that it requires much effort to test the stability of a solution that was the straightforward result of some ODE solver. Consequently, during continuation stability is tested only occasionally, in order to save on cost.

Exercise 6.16.

Assume that the scalar dependent variable $u(x,t)$ satisfies the boundary conditions of Eq. (6.44) (consider $\mathbf{A}_0$, $\mathbf{A}_1$, $\mathbf{B}_0$, $\mathbf{B}_1$ to be scalars). Suppose further that the standard difference quotients in Eq. (6.46) with equidistant spacing are applied with $N > 1$.

(a) Show that

$$\frac{\partial u(a,t)}{\partial x} = \frac{1}{2\Delta}[-3u(a,t) + 4u(x_1,t) - u(x_2,t)] + O(\Delta^2).$$

(b) Use (a) to show

$$u(a,t) \approx [2\Delta \mathbf{c}_a - \mathbf{A}_1(4u(x_1,t) - u(x_2,t))]/(2\Delta \mathbf{A}_0 - 3\mathbf{A}_1).$$

(c) Show the analog

$$u(b,t) \approx [2\Delta \mathbf{c}_b + \mathbf{B}_1(4u(x_N,t) - u(x_{N-1},t))]/(2\Delta \mathbf{B}_0 + 3\mathbf{B}_1).$$

(d) How can these expressions be used to generate the differential equation $\dot{\mathbf{Y}} = \mathbf{F}(\mathbf{Y},\lambda)$ that belongs to Eq. (6.48)?

6.7 HOPF BIFURCATION IN PDEs

The stability of solutions to ODE boundary-value problems can be lost in similar ways, as is common with solutions to equations $\mathbf{f}(\mathbf{y},\lambda) = \mathbf{0}$. The way a critical eigenvalue crosses the imaginary axis is crucial [66, 167].

Consider the PDE in Eq. (6.44) and a solution $\mathbf{y}_s(x)$ to the steady-state equations in Eq. (6.45). Let us denote the difference between $\mathbf{y}_s(x)$ and a solution $\mathbf{y}(x,t)$ to Eq. (6.44) by $\mathbf{d}$,

$$\mathbf{y}(x,t) = \mathbf{y}_s(x) + \mathbf{d}(x,t).$$

Substituting $\mathbf{y}$ into the PDE in Eq. (6.44) leads to

$$\frac{\partial \mathbf{d}}{\partial t} = \mathbf{D}\mathbf{y}_s'' + \mathbf{D}\mathbf{d}'' + \mathbf{f}(x,\mathbf{y}_s + \mathbf{d},\mathbf{y}_s' + \mathbf{d}',\lambda).$$

Linearizing $\mathbf{f}$ about $\mathbf{y}_s,\mathbf{y}_s'$ yields the linearized PDE

$$\frac{\partial \mathbf{d}}{\partial t} = \mathbf{D}\mathbf{d}'' + \mathbf{f}_{\mathbf{y}}(x, \mathbf{y}_s,\mathbf{y}_s',\lambda)\mathbf{d} + \mathbf{f}_{\mathbf{y}'}(x,\mathbf{y}_s,\mathbf{y}_s',\lambda)\mathbf{d}'$$

$$\mathbf{A}_0\mathbf{d}(a,t) + \mathbf{A}_1\mathbf{d}'(a,t) = \mathbf{0} \tag{6.49}$$

$$\mathbf{B}_0\mathbf{d}(b,t) + \mathbf{B}_1\mathbf{d}'(b,t) = \mathbf{0}.$$

For small $|\mathbf{d}|$, the linear PDE in Eq. (6.49) reflects the dynamics of the solutions to Eq. (6.44) that are close to $\mathbf{y}_s(x)$. This analysis parallels the ODE stability analysis outlined in Section 1.2 and in Section 5.5.4. The ansatz

$$\mathbf{d}(x,t) = e^{\mu t}\mathbf{w}(x),$$

substituted into Eq. (6.49), yields the ODE eigenvalue problem

$$\mu\mathbf{w} = \mathbf{D}\mathbf{w}'' + \mathbf{f}_{\mathbf{y}}(x,\mathbf{y}_s,\mathbf{y}_s',\lambda)\mathbf{w} + \mathbf{f}_{\mathbf{y}'}(x,\mathbf{y}_s,\mathbf{y}_s',\lambda)\mathbf{w}'$$

$$\mathbf{A}_0\mathbf{w}(a) + \mathbf{A}_1\mathbf{w}'(a) = \mathbf{0} \tag{6.50}$$

$$\mathbf{B}_0\mathbf{w}(b) + \mathbf{B}_1\mathbf{w}'(b) = \mathbf{0}.$$

For complex μ and $\mathbf{w}$, this boundary-value problem is equivalent to two boundary-value problems for real variables. The two real systems are obtained by substituting

$$\mu = \alpha + i\beta, \quad \mathbf{w}(x) = \mathbf{h}(x) + i\mathbf{g}(x)$$

into Eq. (6.50). After collecting real and imaginary parts, we have

$$\begin{aligned}
\mathbf{D}\mathbf{h}'' + \mathbf{f}_{\mathbf{y}}(x,\mathbf{y}_s,\mathbf{y}_s',\lambda)\mathbf{h} + \mathbf{f}_{\mathbf{y}'}(x,\mathbf{y}_s,\mathbf{y}_s',\lambda)\mathbf{h}' &= \alpha\mathbf{h} - \beta\mathbf{g} \\
\mathbf{D}\mathbf{g}'' + \mathbf{f}_{\mathbf{y}}(x,\mathbf{y}_s,\mathbf{y}_s',\lambda)\mathbf{g} + \mathbf{f}_{\mathbf{y}'}(x,\mathbf{y}_s,\mathbf{y}_s',\lambda)\mathbf{g}' &= \beta\mathbf{h} + \alpha\mathbf{g} \\
\mathbf{A}_0\mathbf{h}(a) + \mathbf{A}_1\mathbf{h}'(a) &= \mathbf{0} \\
\mathbf{B}_0\mathbf{h}(b) + \mathbf{B}_1\mathbf{h}'(b) &= \mathbf{0} \\
\mathbf{A}_0\mathbf{g}(a) + \mathbf{A}_1\mathbf{g}'(a) &= \mathbf{0} \\
\mathbf{B}_0\mathbf{g}(b) + \mathbf{B}_1\mathbf{g}'(b) &= \mathbf{0}.
\end{aligned} \tag{6.51}$$

This coupled system consists of $2n$ second-order ODEs with boundary conditions. The difficulty with the eigenvalue problem in Eq. (6.51) is that it has an infinite number of eigenvalues. Hence, it can hardly be calculated whether, for instance, all eigenvalues have negative real parts. Branch points are characterized in the usual way: The α and β depend on λ, and for one of the α we have $\alpha(\lambda_0) = 0$ at the parameter value λ_0 of a branch point. In the real case, $\beta(\lambda_0) = 0$, we encounter a turning point or a stationary bifurcation point. In the imaginary case, $\beta(\lambda_0) \neq 0$, we have a Hopf bifurcation (see Section 2.6).

This situation enables us to set up direct methods for calculating branch points. As in the branching system in Eq. (6.14), we keep the solution $\mathbf{y}_s(x)$ available by attaching Eq. (6.45) to Eq. (6.51). The α terms in Eq. (6.51) can be dropped because $\alpha(\lambda_0) = 0$. For the stationary bifurcation, the β terms too are dropped. In the latter case, the two systems in Eq. (6.51) are identical, and one can be removed (the $\mathbf{g}$ system, say). Imposing a normalizing equation such as $h_k(a) = 1$, and attaching the trivial differential equation $\lambda' = 0$, reproduces the branching system in Eq. (6.14), here written for second-order ODEs. This shows that the branch points calculated by Eq. (6.14) are candidates for gain or loss of stability. We conclude that it is sufficient to test stability only once on either side of a branch point.

The direct method for calculating Hopf points is now easily set up. Because $\beta(\lambda_0) \neq 0$ is an unknown constant, we set up the trivial differential equation $\beta' = 0$. Imposing a normalizing condition on $\mathbf{g}(a)$

and collecting all equations, one has

$$
\begin{aligned}
\mathbf{D}\mathbf{y}'' + \mathbf{f}(x,\mathbf{y},\mathbf{y}',\lambda) &= \mathbf{0} \\
\mathbf{D}\mathbf{h}'' + \mathbf{f}_{\mathbf{y}}(x,\mathbf{y},\mathbf{y}',\lambda)\mathbf{h} + \mathbf{f}_{\mathbf{y}'}(x,\mathbf{y},\mathbf{y}',\lambda)\mathbf{h}' &= -\beta\mathbf{g} \\
\mathbf{D}\mathbf{g}'' + \mathbf{f}_{\mathbf{y}}(x,\mathbf{y},\mathbf{y}',\lambda)\mathbf{g} + \mathbf{f}_{\mathbf{y}'}(x,\mathbf{y},\mathbf{y}',\lambda)\mathbf{g}' &= \beta\mathbf{h} \\
\lambda' = 0, \quad \beta' &= 0 \\
\mathbf{A}_0\mathbf{y}(a) + \mathbf{A}_1\mathbf{y}'(a) = \mathbf{c}_a, \quad \mathbf{B}_0\mathbf{y}(b) + \mathbf{B}_1\mathbf{y}'(b) &= \mathbf{c}_b \qquad (6.52) \\
\mathbf{A}_0\mathbf{h}(a) + \mathbf{A}_1\mathbf{h}'(a) = \mathbf{0}, \quad \mathbf{B}_0\mathbf{h}(b) + \mathbf{B}_1\mathbf{h}'(b) &= \mathbf{0} \\
\mathbf{A}_0\mathbf{g}(a) + \mathbf{A}_1\mathbf{g}'(a) = \mathbf{0}, \quad \mathbf{B}_0\mathbf{g}(b) + \mathbf{B}_1\mathbf{g}'(b) &= \mathbf{0} \\
h_k(a) = 1, \; g_k(a) &= 1.
\end{aligned}
$$

In order to use standard software for ODEs, each of the second-order systems is transformed into a system of $2n$ first-order ODEs. Hence, altogether, Eq. (6.52) represents a first-order boundary-value problem consisting of $6n + 2$ differential equations. Numerical results with $n = 2$ are reported in [198]. Severe problems in finding reasonable initial approximations were indicated. The calculation of Hopf points in PDEs has been the object of recent research [104, 285, 317].

The above approach can be generalized to PDEs with more than one space variable. For ease of notation, we suppress boundary conditions and write down the equations for the simple PDE

$$\frac{\partial \mathbf{y}}{\partial t} = \nabla^2\mathbf{y} + \mathbf{f}(\mathbf{x},\mathbf{y},\lambda).$$

Here the independent space variable $\mathbf{x}$ is two-dimensional or three-dimensional. For Hopf bifurcation, the system that corresponds to Eq. (6.52) is based on

$$
\begin{aligned}
\nabla^2\mathbf{y} + \mathbf{f}(\mathbf{x},\mathbf{y},\lambda) &= \mathbf{0} \\
\nabla^2\mathbf{h} + \mathbf{f}_{\mathbf{y}}(\mathbf{x},\mathbf{y},\lambda)\mathbf{h} &= -\beta\mathbf{g} \\
\nabla^2\mathbf{g} + \mathbf{f}_{\mathbf{y}}(\mathbf{x},\mathbf{y},\lambda)\mathbf{g} &= \beta\mathbf{h}.
\end{aligned}
$$

The system reduces in the case of stationary bifurcation ($\beta = 0$).

Example 6.6. Nonstationary Heat and Mass Transfer.
A nonstationary heat and mass transfer inside a porous catalyst particle can be described by the two parabolic PDEs [198, 285]:

$$Lw\,\frac{\partial c}{\partial t} = \frac{\partial^2 c}{\partial x^2} - \frac{\lambda c}{\gamma\beta}\exp\left(\frac{\Theta}{1 + \Theta/\gamma}\right)$$
$$\frac{\partial \Theta}{\partial t} = \frac{\partial^2 \Theta}{\partial x^2} + \lambda c\cdot\exp\left(\frac{\Theta}{1 + \Theta/\gamma}\right) \tag{6.53a}$$

with boundary conditions

$$\frac{\partial c(0,t)}{\partial x} = \frac{\partial \Theta(0,t)}{\partial x} = 0,$$
$$c(1,t) = 1, \quad \Theta(1,t) = 0. \tag{6.53b}$$

The meaning of the variables is

$c(x,t)$: concentration
$\Theta(x,t)$: temperature
λ: Frank-Kamenetskii parameter
γ: activation energy ($= 20$)
β: heat evolution parameter ($= 0.2$)
Lw: Lewis number ($= 10$).

The steady state satisfies the boundary-value problem

$$c'' = \frac{\lambda c}{\gamma\beta}\exp\left(\frac{\Theta}{1 + \Theta/\gamma}\right)$$
$$\Theta'' = -\lambda c\cdot\exp\left(\frac{\Theta}{1 + \Theta/\gamma}\right)$$
$$c'(0) = 0,\ c(1) = 1, \quad \Theta'(0) = 0, \quad \Theta(1) = 0.$$

After evaluating the Jacobian $\mathbf{f_y}$, the remaining parts of Eq. (6.52) can be easily written down. Denoting the exponential terms

$$E_0 = \exp\left(\frac{\Theta}{1 + \Theta/\gamma}\right)$$
$$E_1 = \frac{c}{(1 + \Theta/\gamma)^2}\,E_0,$$

one obtains

$$h_1'' - \frac{\lambda}{\gamma\beta}E_0 h_1 - \frac{\lambda}{\gamma\beta}E_1 h_2 = -\beta\cdot Lw\cdot g_1$$
$$h_2'' + \lambda E_0 h_1 + \lambda E_1 h_2 = -\beta g_2$$

$$g_1'' - \frac{\lambda}{\gamma\beta} E_0 g_1 - \frac{\lambda}{\gamma\beta} E_1 g_2 = \beta h_1$$

$$g_2'' + \lambda E_0 g_1 + \lambda E_1 g_2 = \beta h_2$$

$$h_1'(0) = h_1(1) = h_2'(0) = h_2(1) = 0$$

$$g_1'(0) = g_1(1) = g_2'(0) = g_2(1) = 0.$$

The coupling with the steady state is hidden in the exponential terms. Writing down the 14 first-order differential equations explicitly is beyond the scope of this book. Such systems are seldom written down explicitly because they can be conveniently set up in a computer working with n^2 matrices and subvectors. In order to give an impression of how the system in Eq. (6.52) can be organized, we write down the components of the resulting extended vector **Y**:

$$Y_1 = c, \quad Y_2 = c', \quad Y_3 = \Theta, \quad Y_4 = \Theta',$$

$$Y_5 = h_1, \quad Y_6 = h_1', \quad Y_7 = h_2, \quad Y_8 = h_2',$$

$$Y_9 = g_1, \quad Y_{10} = g_1', \quad Y_{11} = g_2, \quad Y_{12} = g_2',$$

$$Y_{13} = \lambda, \quad Y_{14} = \beta.$$

7 Stability of Periodic Solutions

Periodic solutions of an equation are time-dependent or space-dependent orbits, cycles, or oscillations. In this chapter a major concern is time-dependent periodicity ("time periodicity" for short) of solutions of autonomous systems of ODEs

$$\dot{\mathbf{y}} = \mathbf{f}(\mathbf{y},\lambda). \tag{7.1}$$

(Recall that "autonomous" means that $\mathbf{f}$ does not depend on t.) For time-periodic solutions, there is a time interval T (the "period") after which the system returns to its original state:

$$\mathbf{y}(t + T) = \mathbf{y}(t)$$

for all t.

Before discussing time periodicity, let us comment on space-dependent periodicity. Space periodic phenomena are abundant in nature, ranging from the stripes on a zebra, to various rock formations, to sand dunes on a beach. PDEs are appropriate for describing spatial *patterns*. To give a simple example, consider the PDE

$$\nabla \times \mathbf{y} + \mathbf{y} = \mathbf{0}. \tag{7.2}$$

Here $\mathbf{y}$ consists of three components

$$y_i(x_1,x_2,x_3), \qquad i = 1, 2, 3$$

that depend on three spatial variables. Equation (7.2) is solved by

$$\mathbf{y} = \begin{pmatrix} c_2 \sin(x_2) + c_3 \cos(x_3) \\ c_3 \sin(x_3) + c_1 \cos(x_1) \\ c_1 \sin(x_1) + c_2 \cos(x_2) \end{pmatrix},$$

the constants c_1, c_2, c_3 are arbitrary (Exercise 7.1).

Systems arise in fluid dynamics, chemical reactions, biology, and many other areas that exhibit both spatial and temporal oscillations—

for example, the Belousov-Zhabotinskii reaction [91, 116, 191]. In this model there is a pattern of spirals and rings that reflects spatial differences of concentrations; temporal oscillations are indicated by changes in color. Models in morphogenesis that exhibit regular patterns are discussed in [172]. Wave-type patterns were mentioned in Section 1.3.2, and references to waves in excitable media will be given at the end of Section 7.4. Regular patterns frequently arise through bifurcation. For example, the waves discussed in the context of nerve models [278], and pulsating fronts in combustion problems [229], are related to Hopf bifurcation. For further hints and examples, refer to [116, 129, 130, 212, 338].

Exercise 7.1.
Consider the PDE in Eq. (7.2). Imagine the solution to be a velocity field; draw streamlines in the (x_1,x_2)-plane for

$$0 \leq x_1 \leq 3\pi, \quad 0 \leq x_2 \leq 2\pi$$

and $c_1 = c_2 = 1$, $c_3 = 0$. In which parts of the plane is the flow upward? In which parts is it downward?

7.1 PERIODIC SOLUTIONS OF AUTONOMOUS SYSTEMS

Because the right-hand side $\mathbf{f}(\mathbf{y},\lambda)$ of Eq. (7.1) is autonomous, $\mathbf{y}(t + \zeta)$ is a solution of Eq. (7.1) whenever $\mathbf{y}(t)$ is a solution. This holds for all constants ζ. Therefore one is free to start measuring the period T of a periodic orbit at any point $\mathbf{y}(t_0)$ along the profile (see Figure 7.1). We can fix an "initial" moment $t_0 = 0$ wherever we like. Such a normalization is set by a *phase condition.*

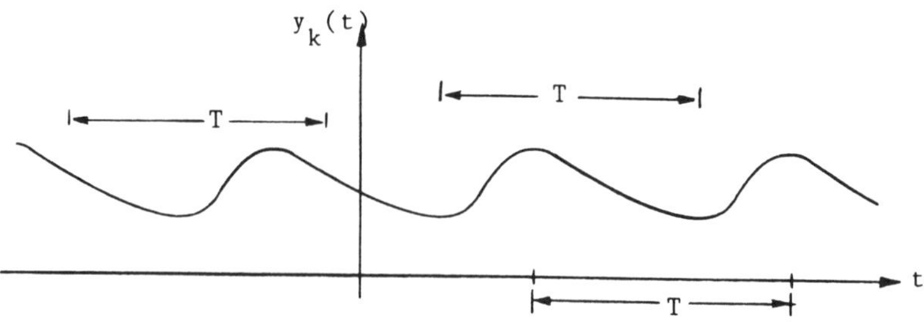

Figure 7.1

The relation

$$p(\mathbf{y}(0),\lambda) = y_k(0) - \eta = 0 \tag{7.3}$$

is a simple phase condition [11, 155]. Here, $\eta = \eta(k,\lambda)$ is a prescribed value in the range of $y_k(t)$. Choosing k and η requires an extra device to determine the range of $y_k(t)$. Once an index k has been chosen, it is advisable to stick to that value of k as long as possible in order to facilitate the adjusting of η. The phase condition in Eq. (7.3) is related to the local parameterization equation in Eq. (4.10) (cf. Eq. (6.3)).

Another phase condition is given by the relation

$$p(\mathbf{y}(0),\lambda) = f_k(\mathbf{y}(0),\lambda) = 0, \tag{7.4}$$

which requires that $t_0 = 0$ be a critical point of y_k: $\dot{y}_k(0) = 0$. This normalization, proposed in [311], does not require any adjustment of k or other quantities. As pointed out in Section 2.8, the corresponding initial value $y_k(0)$ is well suited to be the ordinate of the branching diagrams. The index k is arbitrary $(1 \leq k \leq n)$; choose $k = 1$ or pick the component of f that has the simplest structure.

The phase condition in Eqs. (7.3) and (7.4) are not the only possibilities; an integral condition was proposed in [81].

The period T of a particular periodic solution $\mathbf{y}(t)$ to Eq. (7.1) is usually not known beforehand. It must be calculated together with $\mathbf{y}$. Once the phase is set by a phase condition such as Eq. (7.3) or Eq. (7.4), it makes sense to impose the periodicity condition

$$\mathbf{y}(0) = \mathbf{y}(T),$$

which fixes the period T. Periodic solutions $\mathbf{y}$ with periods T can be calculated by solving the augmented boundary-value problem

$$\begin{pmatrix} \dot{\mathbf{y}} \\ \dot{T} \end{pmatrix} = \begin{pmatrix} \mathbf{f}(\mathbf{y},\lambda) \\ 0 \end{pmatrix}, \quad \begin{pmatrix} \mathbf{y}(0) - \mathbf{y}(T) \\ p(\mathbf{y}(0),\lambda) \end{pmatrix} = \mathbf{0}.$$

For technical reasons (see the treatment of the Duffing equation in Section 6.1), it is advisable to normalize the interval to have unit length. Hence we rewrite the above boundary-value problem as

$$\begin{pmatrix} \mathbf{y} \\ T \end{pmatrix}' = \begin{pmatrix} T\mathbf{f}(\mathbf{y},\lambda) \\ 0 \end{pmatrix}, \quad \begin{pmatrix} \mathbf{y}(0) - \mathbf{y}(1) \\ p(\mathbf{y}(0),\lambda) \end{pmatrix} = \mathbf{0}. \tag{7.5}$$

The prime in Eq. (7.5) refers to a normalized time $0 \leq t \leq 1$. Defining $\tilde{n} = n + 1$ and

$$\tilde{\mathbf{y}} = \begin{pmatrix} \mathbf{y} \\ T \end{pmatrix}, \quad \tilde{\mathbf{f}}(\tilde{\mathbf{y}},\lambda) = \begin{pmatrix} T\mathbf{f}(\mathbf{y},\lambda) \\ 0 \end{pmatrix}, \quad \tilde{\mathbf{r}} = \begin{pmatrix} \mathbf{y}(0) - \mathbf{y}(1) \\ p(\mathbf{y}(0),\lambda) \end{pmatrix}, \tag{7.6a}$$

Eq. (7.5) is equivalent to

$$\tilde{\mathbf{y}}' = \tilde{\mathbf{f}}(\tilde{\mathbf{y}},\lambda), \quad \tilde{\mathbf{r}}(\tilde{\mathbf{y}}(0),\tilde{\mathbf{y}}(1)) = \mathbf{0}. \tag{7.6b}$$

Thus Eq. (7.5) is of the general form of Eqs. (6.1) and (6.2). All the tools available for boundary-value problems can be applied to periodic solutions of autonomous ODEs. Periodic solutions can be calculated by calling standard software for two-point boundary-value problems, and the branching techniques of the previous chapter can be applied to investigate the branching of periodic solutions (see also Section 7.6).

To illustrate, we show how previous methods can be adapted to tracing branches of periodic solutions. A local parameterization strategy for solutions of Eq. (7.5) imposes an additional boundary condition together with the trivial differential equation $\lambda' = 0$ for the parameter. With phase condition Eq. (7.3), a boundary-value problem of dimension $n + 2$ results:

$$\begin{pmatrix} \mathbf{y} \\ T \\ \lambda \end{pmatrix}' = \begin{pmatrix} T\mathbf{f}(\mathbf{y},\lambda) \\ 0 \\ 0 \end{pmatrix}, \quad \begin{pmatrix} \mathbf{y}(0) - \mathbf{y}(1) \\ y_k(0) - \eta_1 \\ y_j(0) - \eta_2 \end{pmatrix} = \mathbf{0}. \tag{7.6c}$$

Here $y_k(0) = \eta_1$ can be interpreted as a parameterizing equation, and $y_j(0) = \eta_2$ as a phase condition, or vice versa. Evidently, $k \neq j$ must hold. In case of the phase condition Eq. (7.4), the boundary-value problem is

$$\begin{pmatrix} \mathbf{y} \\ T \\ \lambda \end{pmatrix}' = \begin{pmatrix} T\mathbf{f}(\mathbf{y},\lambda) \\ 0 \\ 0 \end{pmatrix}, \quad \begin{pmatrix} \mathbf{y}(0) - \mathbf{y}(1) \\ y_k(0) - \eta \\ f_j(\mathbf{y}(0),\lambda) \end{pmatrix} = \mathbf{0}. \tag{7.6d}$$

This latter version of the boundary conditions does not require $k \neq j$; k and j are independent of each other, $j = 1$ works.

Example 7.1. Nerve Impulses.
The equations

$$\dot{y}_1 = 3(y_1 + y_2 - \tfrac{1}{3}y_1^3 + \lambda)$$

$$\dot{y}_2 = -(y_1 - 0.7 + 0.8y_2)/3$$

were investigated in Exercises 1.9 and 2.8. In order to calculate periodic solutions, choose, for instance, the phase condition Eq. (7.4) with $k = 2$. This choice of k (j in Eq. (7.6d)) leads to simpler boundary conditions than $k = 1$; $k = 1$ works too. We set $y_3 = T$

and obtain the boundary-value problem of the type in Eq. (6.1),

$$
\begin{aligned}
y_1' &= 3y_3(y_1 + y_2 - \tfrac{1}{3}y_1^3 + \lambda) \\
y_2' &= -\tfrac{1}{3}y_3(y_1 - 0.7 + 0.8y_2) \\
y_3' &= 0 \\
y_1(0) - y_1(1) &= 0 \\
y_2(0) - y_2(1) &= 0 \\
y_1(0) + 0.8y_2(0) - 0.7 &= 0.
\end{aligned}
$$

Solving this system for fixed λ is a method for calculating a periodic solution. In case of continuation with local parameterization, replace λ by y_4 and attach $y_4' = 0$ and $y_k(0) - \eta = 0$. Then, with appropriate values of k and η, a branch of periodic orbits can be traced by calling standard software. The branching diagram in Figure 2.34 has been established in this way. Figure 7.2 shows spike behavior of periodic solutions for $\lambda = -1.3$.

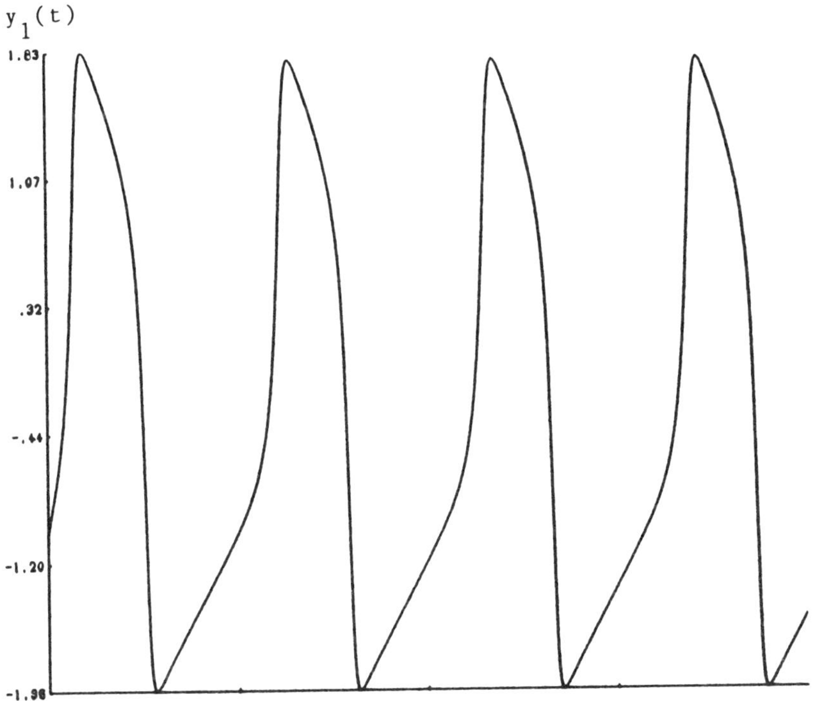

Figure 7.2

We note in passing that Eq. (7.6) does not exclude stationary solutions. Hence, especially in cases when an unstable orbit close to a stable equilibrium is to be calculated, solving Eq. (7.6) requires an appropriate initial guess. In the context of branch tracing or branch switching, such a guess is usually available. In critical cases, Eq. (7.6) can be modified so that it parameterizes with respect to amplitude [311]. Fixing a certain generalized amplitude leads to a three-point boundary-value problem that excludes stationary solutions.

7.2 THE MONODROMY MATRIX

When tracing a branch of periodic solutions, the question arises whether the periodic solutions are stable and where and in which way stability is lost. To analyze stability of periodic solutions, one needs the basic tools: the monodromy matrix and the Poincaré map. In what follows, we investigate stability of one particular periodic solution $\mathbf{y}^*(t)$ with period T and defer the dependence on λ for later study (in Section 7.4).

Stability of $\mathbf{y}^*$ manifests itself in the way neighboring trajectories behave (Figure 7.3). Trajectories of the differential equation are denoted by φ,

$$\varphi(t;\mathbf{z}) \text{ solves } \dot{\mathbf{y}} = \mathbf{f}(\mathbf{y},\lambda) \text{ with } \mathbf{y}(0) = \mathbf{z}. \tag{7.7}$$

A trajectory that starts from the perturbed initial vector $\mathbf{z}^* + \mathbf{d}_0$ progresses with the distance

$$\mathbf{d}(t) = \varphi(t;\mathbf{z}^* + \mathbf{d}_0) - \varphi(t;\mathbf{z}^*)$$

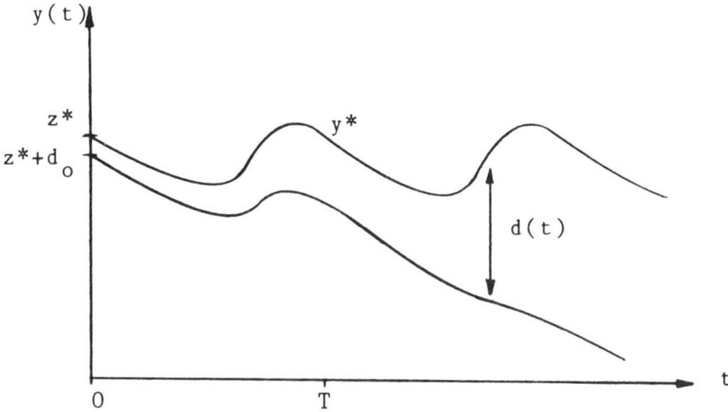

Figure 7.3

to the periodic orbit $\mathbf{y}^*$. Here $\mathbf{z}^* = \mathbf{y}^*(0)$ is taken. Measuring the distance after one period T gives

$$\mathbf{d}(T) = \varphi(T;\mathbf{z}^* + \mathbf{d}_0) - \varphi(T;\mathbf{z}^*).$$

Taylor expansion yields

$$\mathbf{d}(T) = \frac{\partial\varphi(T;\mathbf{z}^*)}{\partial\mathbf{z}}\,\mathbf{d}_0 + \text{terms of higher order.}$$

Apparently, the matrix

$$\frac{\partial\varphi(T;\mathbf{z}^*)}{\partial\mathbf{z}} \tag{7.8}$$

plays a role in deciding whether the initial perturbation $\mathbf{d}_0$ decays or grows. The matrix (7.8) is called the *monodromy matrix.*

Some properties of φ help to find another representation of the monodromy matrix. Note that φ from Eq. (7.7) obeys the differential equation in Eq. (7.1)

$$\frac{d\varphi(t;\mathbf{z})}{dt} = \mathbf{f}(\varphi(t;\mathbf{z}),\lambda) \qquad \text{for all } t.$$

Differentiating this identity with respect to $\mathbf{z}$ yields

$$\frac{d}{dt}\,\frac{\partial\varphi(t;\mathbf{z})}{\partial\mathbf{z}} = \frac{\partial\mathbf{f}(\varphi,\lambda)}{\partial\varphi}\,\frac{\partial\varphi(t;\mathbf{z})}{\partial\mathbf{z}}\,.$$

From $\varphi(0;\mathbf{z}) = \mathbf{z}$ we infer

$$\frac{\partial\varphi(0;\mathbf{z})}{\partial\mathbf{z}} = \mathbf{I}.$$

Consequently, the monodromy matrix (7.8) is identical to $\Phi(T)$, when $\Phi(t)$ solves the matrix initial-value problem

$$\dot{\Phi} = \mathbf{f}_\mathbf{y}(\mathbf{y}^*,\lambda)\Phi, \quad \Phi(0) = \mathbf{I}. \tag{7.9}$$

This specific *fundamental solution matrix* $\mathbf{\Phi}$ gives rise to the following definition:

Definition 7.1. *The n^2 monodromy matrix $\boldsymbol{M}$ of the periodic solution $\mathbf{y}^*(t)$ with period T and initial vector z^* is defined by*

$$\boldsymbol{M} = \Phi(T) = \frac{\partial\varphi(T;z^*)}{\partial z}\,.$$

φ and Φ are defined in Eqs. (7.7) and (7.9).

Local stability of $\mathbf{y}^*$ can be established by means of Floquet theory, based on the monodromy matrix [64, 128, 131, 132, 159]. We defer results to the next section and close this section with two further properties of $\mathbf{M}$, namely

$$\Phi(jT) = \mathbf{M}^j,$$

$\mathbf{M}$ has unity as eigenvalue.

This can be seen as follows. The Jacobian $\mathbf{J}(t) = \mathbf{f_y}(\mathbf{y}^*,\lambda)$ is periodic, $\mathbf{J}(t + T) = \mathbf{J}(t)$. Hence any matrix $\mathbf{Z}(t)$ that solves the matrix differential equation $\dot{\mathbf{Z}} = \mathbf{JZ}$ meets the property:

$\mathbf{Z}(t)$ is solution implies $\mathbf{Z}(t + T)$ is solution.

Thus there is a constant nonsingular matrix $\mathbf{Q}$ such that

$$\mathbf{Z}(t + T) = \mathbf{Z}(t)\mathbf{Q}.$$

For the special solution Φ from Eq. (7.9), we have

$$\Phi(0 + T) = \Phi(0)\mathbf{Q} = \mathbf{Q},$$

which shows $\mathbf{Q} = \mathbf{M}$ and

$$\Phi(2T) = \Phi(T + T) = \Phi(T)\mathbf{M} = \mathbf{M}^2.$$

This extends to $\Phi(jT) = \mathbf{M}^j$, which implies

$$\mathbf{d}(jT) = \mathbf{M}^j\mathbf{d}_0 + \text{terms of higher order}.$$

Because solutions of the autonomous differential equation in Eq. (7.1) can be shifted in t direction, there are periodic perturbations

$$\mathbf{d}(t) = \mathbf{y}^*(t + \Delta) - \mathbf{y}^*(t)$$

satisfying $\mathbf{d}(t + T) = \mathbf{d}(t)$. This leads to the conclusion that the linearized problem

$$\dot{\mathbf{h}} = \mathbf{J}(t)\mathbf{h}$$

has a periodic solution $\mathbf{h}(t)$. Every solution of this linearized problem satisfies $\mathbf{h}(t) = \Phi(t)\mathbf{h}(0)$. The relation

$$\mathbf{h}(0) = \mathbf{h}(T) = \Phi(T)\mathbf{h}(0) = \mathbf{Mh}(0)$$

shows that $\mathbf{M}$ has unity as an eigenvalue.

7.3 THE POINCARÉ MAP

The Poincaré map is extremely useful for describing dynamics of different types of oscillation [123]. In particular, arguments based on the Poincaré map establish stability results for periodic orbits.

Example 7.2 (cf. Exercise 1.12 and Section 2.6).
The system

$$\dot{y}_1 = y_1 - y_2 - y_1(y_1^2 + y_2^2)$$
$$\dot{y}_2 = y_1 + y_2 - y_2(y_1^2 + y_2^2)$$

can be written in polar coordinates as

$$\dot{\rho} = \rho(1 - \rho^2)$$
$$\dot{\vartheta} = 1.$$

A stable limit cycle is $\rho = 1$, $\vartheta = t$ (see the phase plane in Figure 7.4). Neighboring trajectories approach the limit cycle. Measuring the distance to the limit cycle is difficult when the observer travels along the trajectories. As in other sports, it is easier by far to wait at some finish line Ω and measure how the racetracks have varied after each lap. In this example we choose as the start and finish line the positive y_1 axis,

$$\Omega = \{\rho > 0, \quad \vartheta = 0\}.$$

The general solution to the differential equation is given by

$$\rho = [1 + (\rho_0^{-2} - 1)e^{-2t}]^{-1/2}, \quad \vartheta = t + \vartheta_0.$$

Consequently, a trajectory that starts from position $q \in \Omega$ requires the time 2π to complete one orbit and passes Ω at the radius

$$P(q) = [1 + (q^{-2} - 1)e^{-4\pi}]^{-1/2}.$$

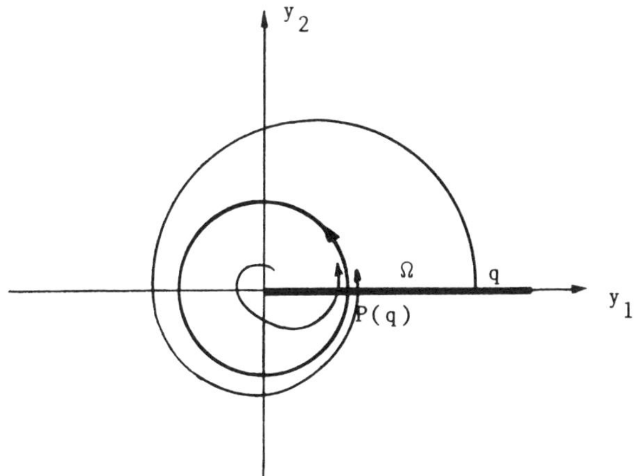

Figure 7.4

This is a simple example of a Poincaré map. The value $q^* = 1$ at which the periodic orbit passes Ω is a *fixed point* of the Poincaré map, $P(1) = 1$. A trajectory that starts, for instance, at $q = 0.0001$ shows the intermediate results

$$P(q) = 0.05347\ldots$$

$$P^2(q) = P(P(q)) = 0.99939\ldots$$

during its rapid approach to the stable limit cycle.

This concept of a Poincaré map can be generalized to the differential equation in (7.1) with n components. The set Ω must be an $(n-1)$-dimensional hypersurface (see Figure 7.5). The hypersurface Ω must be chosen so that all trajectories that cross Ω meet two requirements:

(a) the trajectories intersect Ω transversally

(b) the trajectories cross Ω in the same direction.

The hypersurface Ω is also called the *Poincaré section.*

The periodic orbit is the specific trajectory that passes Ω in $\mathbf{q}^*$ with

$$\mathbf{q}^* = \varphi(T;\mathbf{q}^*).$$

Let $T_\Omega(\mathbf{q})$ be the time taken for a trajectory $\varphi(t;\mathbf{q})$ to first return to Ω,

$$\varphi(T_\Omega(\mathbf{q});\mathbf{q}) \in \Omega.$$

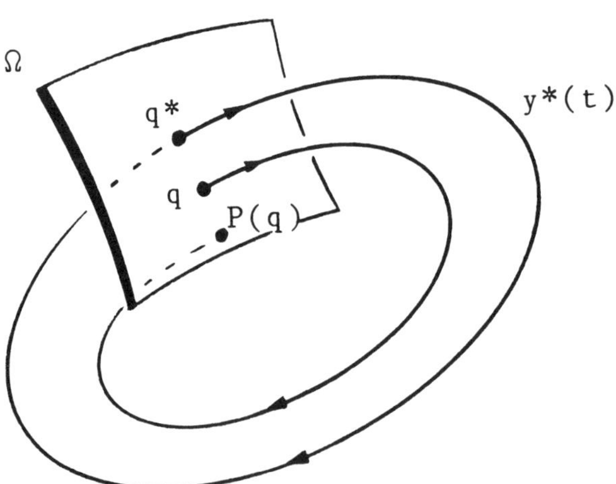

Figure 7.5

The *Poincaré map* or *return map* $\mathbf{P}(\mathbf{q})$ is defined by

$$\mathbf{P}(\mathbf{q}) = \varphi(T_\Omega(\mathbf{q});\mathbf{q}). \tag{7.10}$$

The geometrical meaning of $\mathbf{P}(\mathbf{q})$ is illustrated in Figure 7.5. $\mathbf{P}$ takes values in Ω, which is $(n - 1)$-dimensional. Because $\mathbf{q}$ is allowed to vary only in Ω, we consider $\mathbf{P}$ as an $(n - 1)$-dimensional map—that is, both $\mathbf{q}$ and $\mathbf{P}(\mathbf{q})$ can be interpreted to have $(n - 1)$ components. The Poincaré map satisfies

$$\mathbf{P}(\mathbf{q}^*) = \mathbf{q}^*;$$

$\mathbf{q}^*$ is a fixed point of $\mathbf{P}$. The return time $T_\Omega(\mathbf{q})$ is close to the period T for $\mathbf{q}$ close to $\mathbf{q}^*$,

$$\mathbf{q} \to \mathbf{q}^* \text{ implies } T_\Omega(\mathbf{q}) \to T.$$

The stability of $\mathbf{y}^*(t)$ is reduced to the behavior of the Poincaré map near its fixed point $\mathbf{q}^*$. Hence the desired information on stability is obtained by checking to see whether this fixed point is repelling or attracting. Let $\mu_1, \ldots, \mu_{n-1}$ be the eigenvalues of the linearization of $\mathbf{P}$ around the fixed point $\mathbf{q}^*$,

$$\mu_j \text{ eigenvalue of } \frac{\partial \mathbf{P}(\mathbf{q}^*)}{\partial \mathbf{q}}, \qquad j = 1, \ldots, n - 1.$$

Stability of fixed points is established by the following theorem.

Theorem 7.1.

(a) *If the moduli of all eigenvalues are smaller than* 1, *then* $\mathbf{q}^*$ *is stable (attracting).*

(b) *If the modulus of at least one eigenvalue is larger than* 1, *then* $\mathbf{q}^*$ *is unstable (repelling).*

The dynamic behavior of the sequence $\mathbf{P}^j(\mathbf{p})$ on Ω depends on the eigenvalues μ_j and the corresponding eigenvectors in an analogous manner, as noted in Section 1.2. Possible paths of the sequence

$$\mathbf{q} \to \mathbf{P}(\mathbf{q}) \to \mathbf{P}^2(\mathbf{q}) \to \ldots$$

are illustrated in Figure 7.6 by means of two-dimensional hypersurfaces Ω. Trajectories intersect these parts of a plane. The sequences of points $\mathbf{P}^j(\mathbf{q})$ of one trajectory do not generate continuous curves, like those depicted in Figure 7.6, but hit them at discrete points (Exercise 7.2). Depending on the dynamic behavior on Ω, the periodic orbit related to $\mathbf{q}^*$ is called the *saddle cycle* (Figure 7.6a), the *spiral cycle* (Figure 7.6b),

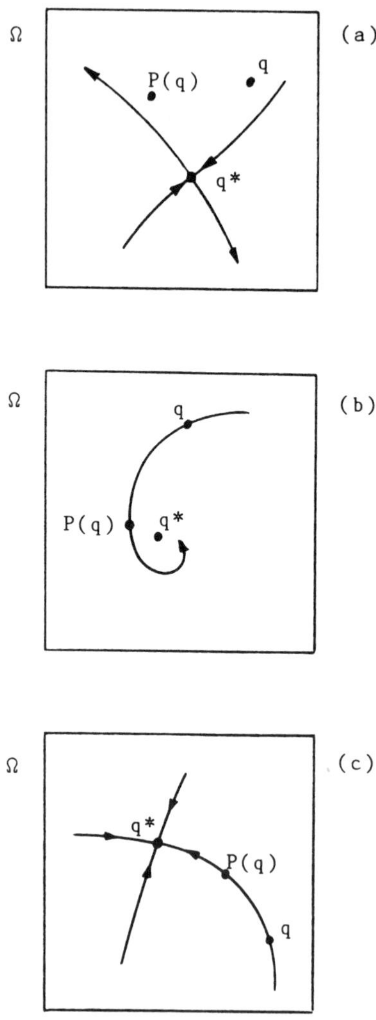

Figure 7.6

or the *nodal cycle* (Figure 7.6c). By reversing the arrows in (b) and (c), one obtains illustrations of repelling cycles.

The curves in Figure 7.6a and c are *invariant curves*, cross sections of *invariant manifolds* with Ω. For our purposes it is sufficient to imagine a *manifold* as a surface or its higher-dimensional analog; for a detailed discussion of manifolds in the context of ODEs, see [15]. An invariant manifold M of a system $\dot{\mathbf{y}} = \mathbf{f}(\mathbf{y})$ is a hypersurface in the state space with the property that for any initial vector $\mathbf{z}$ in M the associated trajectory $\varphi(t;\mathbf{z})$ stays in M for all t—that is, an invariant manifold is composed of flow lines. A saddle cycle has an unstable and a stable invariant manifold intersecting in the cycle. This can be visualized by means of Figure 7.5. Imagine $\mathbf{q}$ situated on an unstable invariant manifold of a saddle-type orbit. Then a closed band "spanned" by both trajectories of Figure 7.5 is part of this manifold. The reader is encouraged to draw an illustration that also includes part of a stable invariant manifold. Instructive illustrations of this kind can be found in [4, part 2].

We now apply the stability result for the fixed-point equation $\mathbf{P}(\mathbf{q}) = \mathbf{q}$ to periodic solutions of the differential equation in Eq. (7.1). In order to obtain the eigenvalues mentioned in the above theorem, the linearization of the Poincaré map in Eq. (7.10) around the fixed point $\mathbf{q}^*$ is calculated. The result is

$$\frac{\partial \mathbf{P}(\mathbf{q}^*)}{\partial \mathbf{q}} = \frac{\partial \varphi(T;\mathbf{q}^*)}{\partial \mathbf{q}} ;$$

in this equation, φ is restricted to the $(n-1)$-dimensional Ω. As pointed out in the previous section, the n^2 monodromy matrix

$$\frac{\partial \varphi(T;\mathbf{z}^*)}{\partial \mathbf{z}} = \mathbf{M}$$

has unity as an eigenvalue. Choosing an appropriate basis for the n-dimensional space, one can show that the remaining $n-1$ eigenvalues are those of

$$\frac{\partial \mathbf{P}(\mathbf{q}^*)}{\partial \mathbf{q}} .$$

Consequently, by the above theorem, the periodic orbit is stable if the remaining $n-1$ eigenvalues of $\mathbf{M}$ are smaller than unity in modulus. The orbit $\mathbf{y}^*$ is unstable if $\mathbf{M}$ has an eigenvalue μ with $|\mu| > 1$. The eigenvalue 1 corresponds to a perturbation in $\mathbf{y}^*$-direction leading out of Ω; the other $n-1$ eigenvalues of $\mathbf{M}$ determine what happens to small perturbations within Ω. The eigenvalues of the monodromy matrix are called the *Floquet multipliers* or *(characteristic) multipliers.*

Exercise 7.2.
A linear two-dimensional map is defined by

$$\mathbf{Pq} = \mathbf{P}\begin{pmatrix} q_1 \\ q_2 \end{pmatrix} = \begin{pmatrix} 1 & 1 \\ 1 & 2 \end{pmatrix}\begin{pmatrix} q_1 \\ q_2 \end{pmatrix} .$$

Calculate eigenvalues and eigenvectors and decide whether the fixed point is attractive or not. Draw the eigenvectors and indicate motion along these two lines by arrows. For

$$\mathbf{q} = \begin{pmatrix} 3 \\ -1.85 \end{pmatrix}$$

calculate the sequence $\mathbf{P}(\mathbf{q}), \ldots, \mathbf{P}^8(\mathbf{q})$ and enter the iterates in your sketch. (Remark: In [123] this map is discussed on a torus rather than on the plane.)

7.4 MECHANISMS OF LOSING STABILITY

So far we have discussed how (local) stability of a particular periodic solution manifests itself through Floquet multipliers. In general, the multipliers and hence the stability vary with λ. First we summarize the previous results on stability. Then we shall discuss mechanisms of losing stability. References are, for example, [16], chapter 34, and [123].

Summary 7.1. *For a value of λ, let $\mathbf{y}(t)$ be a periodic solution to $\dot{\mathbf{y}} = \mathbf{f}(\mathbf{y},\lambda)$ with period T. The monodromy matrix $\mathbf{M}(\lambda)$ is defined by $\mathbf{M}(\lambda) = \Phi(T)$, where $\Phi(t)$ solves the matrix initial-value problem*

$$\dot{\Phi} = \mathbf{f}_y(\mathbf{y},\lambda)\Phi, \ \Phi(0) = I.$$

The matrix $\mathbf{M}(\lambda)$ has n eigenvalues $\mu_1(\lambda), \ldots, \mu_n(\lambda)$. One of them is equal to unity, say $\mu_n = 1$. The other $n - 1$ eigenvalues determine (local) stability by the following rule:

$\mathbf{y}(t)$ is stable if $|\mu_j| < 1$ for $j = 1, \ldots, n - 1$.
$\mathbf{y}(t)$ is unstable if $|\mu_j| > 1$ for some j.

Sometimes stability is described by means of *characteristic or Floquet exponents* σ, defined by $\mu = \exp(\sigma T)$. Because the exponential function maps the halfplane $\mathrm{Re}(\sigma) < 0$ onto the interior of the unit circle, stability is also signaled by all characteristic exponents having a negative real part.

Figure 7.7 shows for three values of λ the multipliers μ of fictive periodic solutions that belong to the same branch. The circle is the unit circle, and one eigenvalue is unity for all λ. The left sketch ($\lambda = \lambda_1$) represents a stable solution because all eigenvalues μ_j lie inside the unit circle ("all" in the sense $j = 1, \ldots, n - 1$). The right sketch shows one multiplier outside the unit circle. Consequently, the periodic orbit for $\lambda = \lambda_2$ is unstable. Obviously, for some value λ_0 between λ_1 and λ_2 one multiplier crosses the unit circle and the stability is lost (or gained

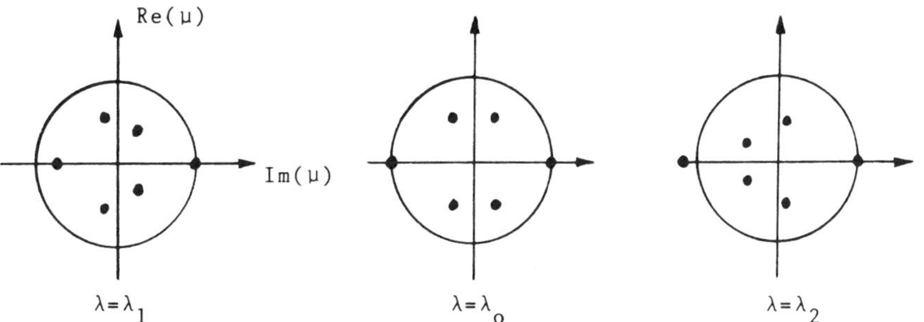

Figure 7.7

if we proceed from λ_2 to λ_1). Figure 7.7 assumes that the critical multiplier crosses the unit circle at -1. Depending on where the critical multiplier or pair of multipliers crosses the unit circle, different types of branching occur [16]. One distinguishes three ways of crossing the unit circle, with three associated types of branching. Figure 7.8 shows the path of the critical multiplier only—that is, the eigenvalue with $|\mu(\lambda_0)| = 1$. In Figure 7.8A, the eigenvalue gets unity in addition to μ_n, $\mu(\lambda_0) = 1$. In Figure 7.8B the multiplier crosses the unit circle at the negative real axis, $\mu(\lambda_0) = -1$. In Figure 7.8C the crossing is with a nonzero imaginary part—that is, a pair of complex conjugate eigenvalues crosses the unit circle. All three sketches refer to a loss of stability when λ passes λ_0; for an illustration of gain of stability, change the arrows to point to the opposite direction. In what follows, we discuss the three kinds of losing stability,

(A) $\mu(\lambda_0) = 1$

(B) $\mu(\lambda_0) = -1$

(C) $\mathrm{Im}(\mu(\lambda_0)) \neq 0$.

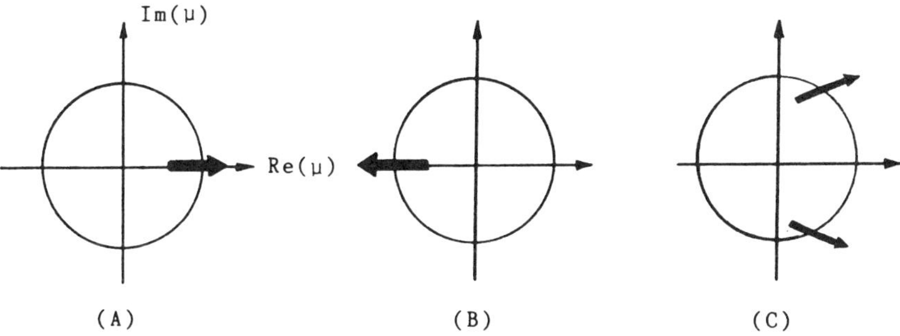

Figure 7.8

7.4.1 Branch Points of Periodic Solutions

Let $\mathbf{P}$ be the Poincaré map associated with the differential equation in Eq. (7.1). Because we are interested in periodic solutions, we study fixed points of the Poincaré map,

$$\mathbf{P}(\mathbf{q},\lambda) = \mathbf{q},$$

for $\mathbf{q} \in \Omega$. Assume situation (A) of the above three cases—that is,

$$\frac{\partial \mathbf{P}}{\partial \mathbf{q}}$$

has unity as an eigenvalue for $\lambda = \lambda_0$. The fixed point equation is equivalent to

$$\tilde{\mathbf{f}}(\mathbf{q},\lambda) = \mathbf{P}(\mathbf{q},\lambda) - \mathbf{q} = \mathbf{0}, \tag{7.11a}$$

which forms a system of $n - 1$ scalar equations. For this system of equations, the branching results of Chapter 2 apply. The Jacobian of $\tilde{\mathbf{f}}$ satisfies

$$\frac{\partial \tilde{\mathbf{f}}}{\partial \mathbf{q}} = \frac{\partial \mathbf{P}}{\partial \mathbf{q}} - \mathbf{I}. \tag{7.11b}$$

Hence, $\mu(\lambda_0) = 1$ implies that for λ_0 the Jacobian of $\tilde{\mathbf{f}}$ has an eigenvalue of zero. That is to say, for $\mu(\lambda_0) = 1$ we encounter the same kind of branching scenario we discussed in Chapter 2. In particular, assuming the standard situation of a simple eigenvalue (cf. Definitions 2.3 and 2.4), one has bifurcation points and turning points of the Poincaré map. This multiplicity of fixed points of $\mathbf{P}$ on Ω is illustrated by Figure 7.9.

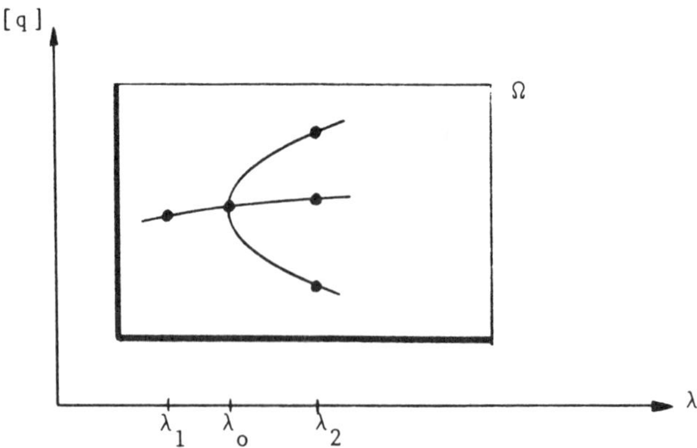

Figure 7.9

This figure shows schematically fixed points of a Poincaré map **P** for varying λ. For λ_1 we have one fixed point only, for λ_2 there are three fixed points. The splitting of fixed points at λ_0 (pitchfork bifurcation in Figure 7.9) relates to a bifurcation of the corresponding periodic orbits (cf. Figure 7.10). In this illustration, we assume the situation λ_2 of Figure 7.9 with two attracting fixed points and one repelling fixed point. The stable orbits that correspond to the attracting fixed points are emphasized by heavy curves; some approaching trajectories that start from neighboring points on Ω are indicated too.

Turning points can be illustrated analogously (see Figure 7.11). Figure 7.11 depicts one stable periodic orbit (heavy curve), one unstable periodic orbit, and one neighboring trajectory. The situation of Figure 7.11 (for some λ_1) can be regarded as being close to a turning point with $\lambda = \lambda_0$. If we approach λ_1 to λ_0, the two fixed points eventually collapse. At $\lambda = \lambda_0$ the resulting single periodic orbit is *semistable*; it dies when λ passes λ_0. This scenario is depicted in Figure 7.12, which shows three two-dimensional phase planes for λ_1, λ_0, and λ_2; for λ_2, no periodic solution exists. In the case of a planar situation ($n = 2$), inside any closed orbit there must be at least one equilibrium. Figure 7.12 is supposed to represent a two-dimensional projection of trajectories that move in a higher-dimensional space, so no equilibrium is indicated. Varying λ from λ_2 to λ_1 provokes birth of a limit cycle at

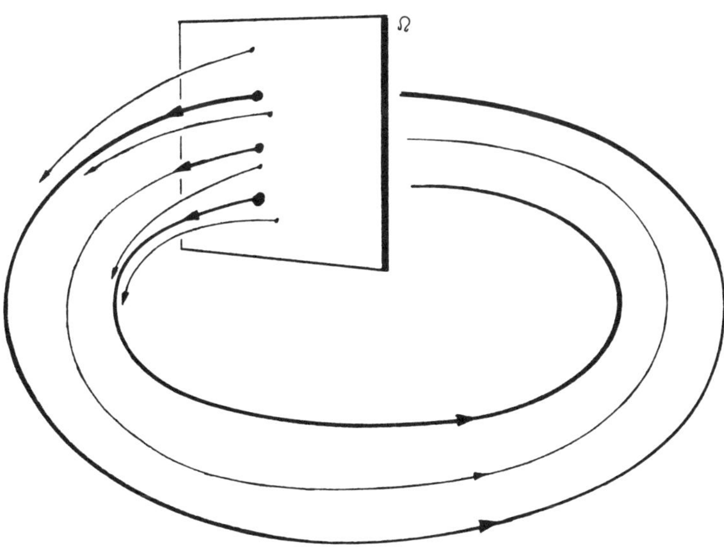

Figure 7.10

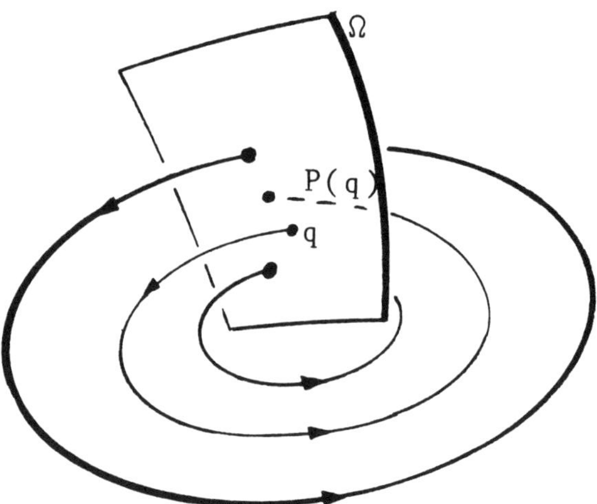

Figure 7.11

$\lambda = \lambda_0$, which immediately splits into two periodic orbits. An example is furnished by the FitzHugh model of nerve impulses (cf. Figure 2.34).

The case $\mu(\lambda_0) = 1$ can be summarized as follows. Typically, turning points occur, accompanied by the birth or death of limit cycles. In case **f** satisfies symmetry or other regularity properties, pitchfork or transcritical bifurcation points may occur.

Example 7.3. Brusselator with Two Coupled Cells.
We consider a third model derived from the Brusselator Eq. (5.42)—namely, the time-dependent behavior of two identical cells with mass exchange. The system is described by four differential equations:

$$\begin{aligned} \dot{y}_1 &= A - (B + 1)y_1 + y_1^2 y_2 + \delta_1(y_3 - y_1) \\ \dot{y}_2 &= By_1 - y_1^2 y_2 + \delta_2(y_4 - y_2) \\ \dot{y}_3 &= A - (B + 1)y_3 + y_3^2 y_4 - \delta_1(y_3 - y_1) \\ \dot{y}_4 &= By_3 - y_3^2 y_4 - \delta_2(y_4 - y_2). \end{aligned} \tag{7.12}$$

The variables $y_1(t)$ and $y_2(t)$ represent the substances in the first cell, and $y_3(t)$ and $y_4(t)$ represent those of the second cell. The coupling constants δ_1 and δ_2 control the mass exchange between the cells. The equations reflect the Brusselator kinetics in each cell. We choose the constants

$$A = 2, \quad \delta_1 = 1, \quad \delta_2 = 4$$

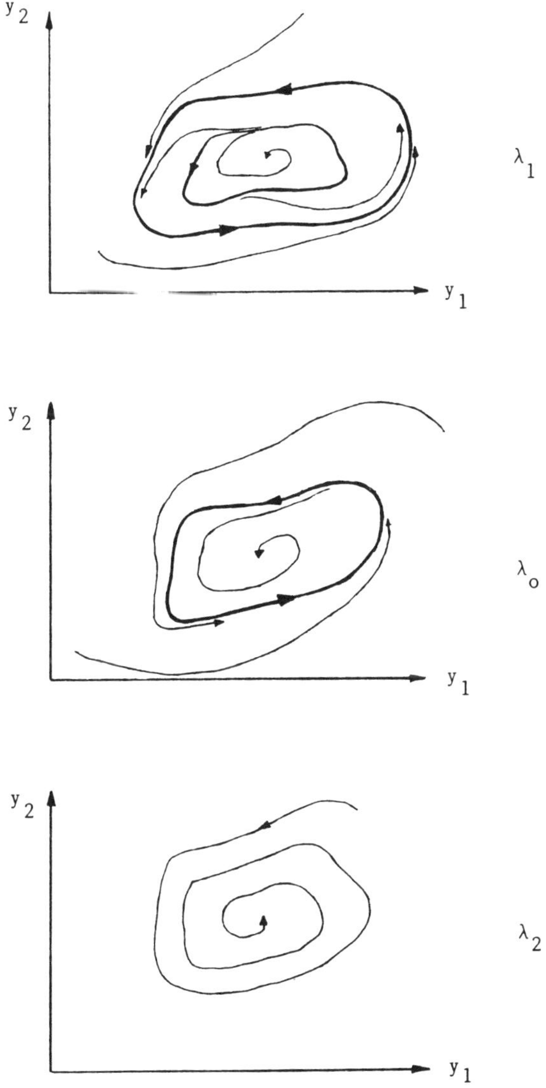

Figure 7.12

and take $\lambda = B$ as the branching parameter. There are two kinds of reactions. For a homogeneous reaction, both cells exhibit identical behavior,

$$y_1(t) = y_3(t), \quad y_2(t) = y_4(t) \qquad \text{for all } t.$$

The reaction is said to be nonhomogeneous if this identity does not hold.

The branching behavior of Eq. (7.12) is depicted in Figure 7.13. Stationary solutions are indicated by full lines (stable equilibria) and dashed lines (unstable equilibria). The dots refer to periodic orbits, the solid dots signal stability. The branch $y_1 = 2$ consists of homogeneous stationary solutions, which lose stability at $\lambda_0 = 4.5$. For the value $\lambda_0 = 5$, there is a Hopf bifurcation into periodic homogeneous orbits. First, the emerging branch is unstable; it gains stability at $\lambda_0 = 5.6369$ ($T_0 = 4.42624$) via a subcritical pitchfork bifurcation. That is, for $\lambda > 5.6369$ there are stable homogeneous oscillations, which are "encircled" by unstable nonhomogeneous orbits. The two branches of stationary nonhomogeneous solutions, which exist for $\lambda > 4.5$, lose their stability at the two Hopf bifurcation points with $\lambda_0 = 7.1454$, $T_0 = 2.8888$.

The remarkable part of the solutions of Eq. (7.12) is the Hopf bifurcation at $\lambda_0 = 5$. Because all the solutions merging in the bifurcation point

$$y_1 = y_3 = 2, \quad y_2 = y_4 = 2.5, \quad \lambda = 5$$

are homogeneous, the coupling terms in Eq. (7.12) vanish; the equations of the two cells decouple. Consequently, the same solutions can be calculated by solving only the subsystem of the first two equations ($n = 2$). In this way, both the homogeneous equilibria and the homogeneous oscillations are obtained. However,

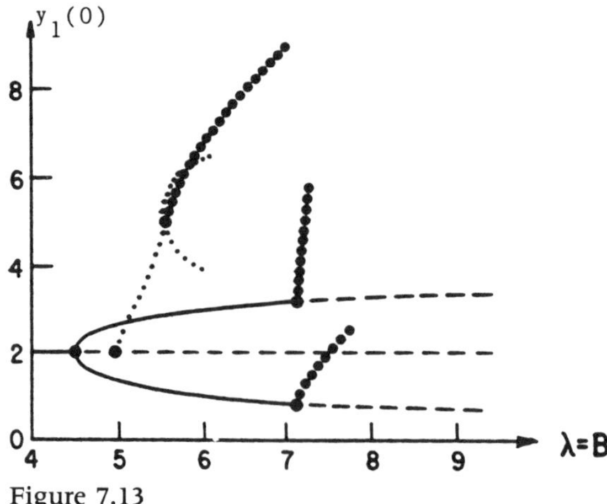

Figure 7.13

the stability behavior is different. The bifurcations into nonhomogeneous solutions do not exist, and there is an exchange of stability at $\lambda_0 = 5$ (Exercise 7.4). This shows that by the process of coupling two identical cells the stability behavior changes drastically. In particular, the homogeneous orbits in the range

$$5 < \lambda < 5.6369$$

lose stability, caused by the onset of additional bifurcations. This serves as another example for the phenomenon that relaxing restrictions can enrich the bifurcation behavior (see the example in Section 6.4). Concerning the underlying "spaces" of admissible functions, distinguish between the space of smooth vector functions with independent components and the small subspace of homogeneous functions satisfying $y_1 = y_3$, $y_2 = y_4$.

7.4.2 Period Doubling

We now discuss the second case of losing stability, $\mu(\lambda_0) = -1$. From Eq. (7.11) we expect that the Jacobian of

$$\tilde{\mathbf{f}}(\mathbf{q},\lambda_0) = \mathbf{P}(\mathbf{q},\lambda_0) - \mathbf{q}$$

is nonsingular (eigenvalue $-2 \neq 0$). Consequently, at λ_0 there is no branch point of the kind discussed above. Rather, a smooth branch $\mathbf{q}(\lambda)$ passes through $\mathbf{q}(\lambda_0)$ without bifurcating itself.

Example 7.4.
We study the fixed points of the real-valued function

$$P(q,\lambda) = \lambda(1 - q)q \qquad \text{for } q > 0. \tag{7.13}$$

From the fixed-point equation we obtain the two fixed points $q(\lambda)$,

$$q(\lambda) = 0, \text{ and}$$

$$q(\lambda) = \frac{\lambda - 1}{\lambda}.$$

Consequently, there is a transcritical bifurcation at

$$(q_0,\lambda_0) = (0,1).$$

Stability is determined by the eigenvalues of the Jacobian of $P(q)$, which reduces here to the single value

$$\mu = \frac{dP}{dq} = \lambda(1 - 2q).$$

The bifurcation with exchange of stability at $\lambda_0 = 1$ is confirmed by $\mu(1) = +1$. Evaluated at the nontrivial branch, the expression for μ is

$$\mu = 2 - \lambda.$$

The range of stability is determined by $|\mu| = |2 - \lambda| < 1$, which leads to

$$1 < \lambda < 3.$$

Consequently, the nontrivial branch loses stability at $\lambda = 3$. This point is no bifurcation of fixed points, because $\mu = -1$. In order to check where the fixed-point iteration converges for $\lambda > 3$, we choose as an example $\lambda = 3.1$ and $q = 0.1$ and iterate

$$P(q),\ P(P(q)), \ldots$$

After about 30 iterations the values settle down to

0.7644 . . .
0.5583
0.7645
0.5582
0.7645.

Apparently every other iterate converges to a limit—that is, the iteration exhibits a periodicity with period two. For the composed *double-period map*

$$P^2(q,\lambda) = P(P(q,\lambda),\lambda),$$

both the values that occur as limits of the above sequence for $\lambda = 3.1$ are fixed points. For $\lambda > 3$, $\lambda \approx 3$, there is no attracting fixed point of P, but there are two attracting fixed points of P^2. Consequently, at $\lambda_0 = 3$ there is an exchange of stability of period-one fixed points to period-two fixed points. This phenomenon is called *period doubling*. Period doubling is a bifurcation of the P^2 fixed-point equation, because all fixed points of P are also fixed points of P^2. The eigenvalues of the branch

$$q(\lambda) = \frac{\lambda - 1}{\lambda}$$

of solutions to

$$P^2(q,\lambda) - q = 0$$

indicate the bifurcation at $\lambda_0 = 3$ by $\mu(\lambda_0) = +1$ (Exercise 7.5). We shall meet this example again in another context (Section 9.2).

We are now prepared to understand what happens to periodic oscillations, when for $\lambda = \lambda_0$ the monodromy matrix has an eigenvalue $\mu = -1$. Because for the related Poincaré map the chain rule implies

$$\frac{\partial}{\partial q}\,[\mathbf{P}(\mathbf{P}(\mathbf{q}))] = \left(\frac{\partial \mathbf{P}(\mathbf{q})}{\partial \mathbf{q}}\right)^2 ,$$

the fixed point $\mathbf{q}$ of $\mathbf{P}(\mathbf{q}) = \mathbf{q}$ leads to the eigenvalue $\mu = +1$ for $\mathbf{P}^2(\mathbf{q}) = \mathbf{P}(\mathbf{P}(\mathbf{q}))$. This bifurcation of the double period is illustrated by Figure 7.14. The dashed curve represents an unstable oscillation with the simple period; the heavy curve is the stable orbit of the double period. The depicted situation can be seen as representing solutions for a λ close to the value λ_0 of a period doubling. A band spanned by the trajectories of Figure 7.14 forms a Möbius strip. For λ approaching λ_0, the curve that winds twice shrinks to the curve that winds once. Shifting λ in the other direction leads to the phenomenon that at λ_0 the stable single-period oscillation splits into stable double-period oscillations.

The term "period" has a slightly different meaning for the Poincaré map $\mathbf{P}$ and the periodic oscillation $\mathbf{y}(t)$. In the case of the Poincaré map, the period is the integer that reflects how many iterations of $\mathbf{P}$ are required to reproduce the fixed point from which the iteration starts. For varying λ, this

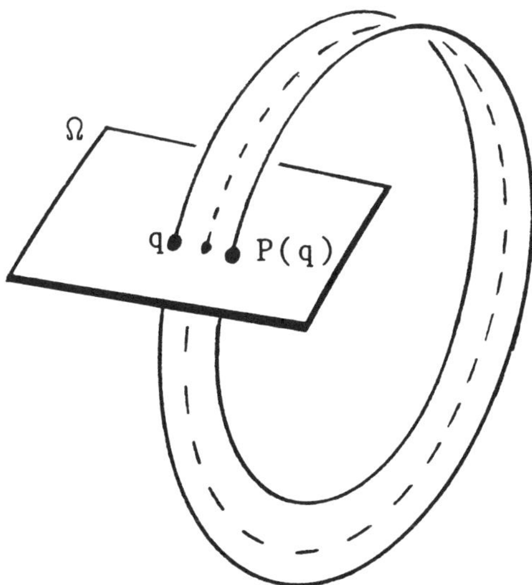

Figure 7.14

integer remains constant if no further period doubling occurs. In contrast, the periods T of the oscillations vary with λ. In general, this variation is different for the single-period oscillations and the double-period oscillations. Consequently, for a given $\lambda \neq \lambda_0$ the periods do not differ exactly by a factor of 2. For λ tending to λ_0, this factor is attained as limit. This will be illustrated in Figure 7.18.

Period doubling is also called *flip bifurcation* or *subharmonic bifurcation*. This phenomenon is observed in many applications—for example, in chemical reactions [165, 168, 325], in nerve models [279], or in Navier-Stokes equations [99]. Period doubling also occurs in the Brusselator [199] and Lorenz equations [329]. In experiments, for increasing λ, often a sequence of period doublings occurs. As pointed out in [67], after a first period doubling a second period doubling is more likely than other bifurcations. Accordingly, sequences of period doublings are typical for a class of bifurcation problems. A characteristic branching diagram is shown in Figure 7.15a. After a Hopf bifurcation at λ_H is passed, a series of period doublings at λ_{PD1} occur. In a corresponding experiment with increasing λ (Figure 7.15b), a period doubling manifests itself by a lowering of every other maximum (or minimum). Period doubling does not exist in fewer than three dimensions; $n \geq 3$ is necessary. Figure 7.15 shows a soft loss of stability. Period doubling can also occur subcritically (see Figure 7.16 and Section 2.8).

In a number of real-valued mappings, the sequence of repeated period doubling obeys a certain law concerning the distribution of the period doubling parameters λ_{PD1} [90, 121]. In the limit, these values satisfy

$$\lim_{i \to \infty} (\lambda_{i+1} - \lambda_i)/(\lambda_i - \lambda_{i-1}) = 0.214\ldots \qquad (7.14)$$

Frequently, the inverse value is quoted, 4.6692 . . . This limit does not only occur in iterations of simple maps; it has also been observed with periodic solutions to ODEs. The *scaling law* in Eq. (7.14) appears to have a universal character.

Example 7.5.
The ODE system from [325]

$$\begin{aligned} \dot{y}_1 &= y_1(30 - 0.25y_1 - y_2 - y_3) + 0.25y_2^2 + 0.1 \\ \dot{y}_2 &= y_2(y_1 - 0.001y_2 - \lambda) + 0.1 \\ \dot{y}_3 &= y_3(16.5 - y_1 - 0.5y_3) + 0.1 \end{aligned} \qquad (7.15)$$

describes an isothermal chemical reaction. A sequence of period

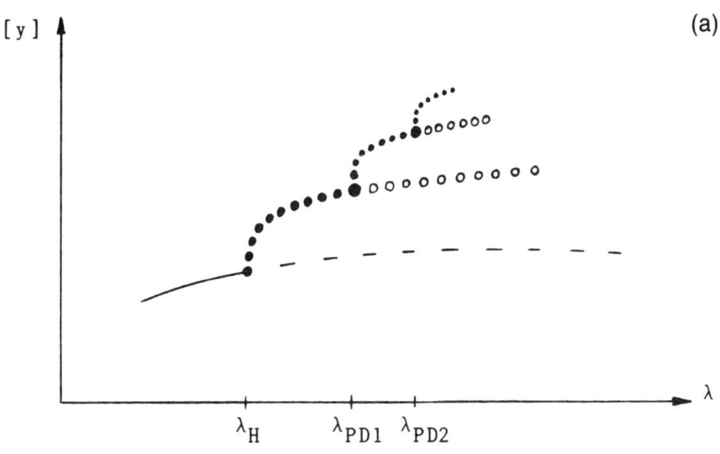

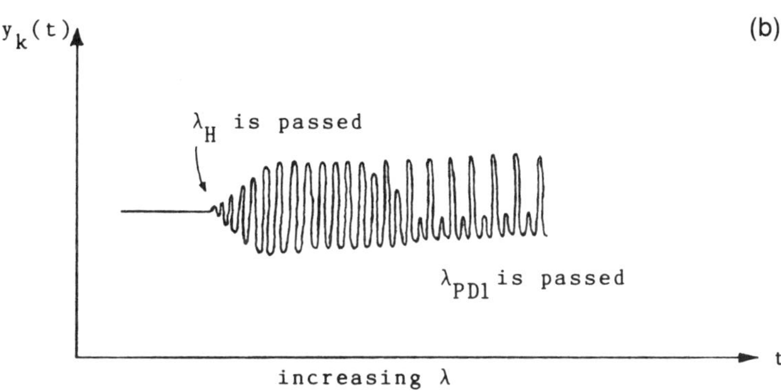

Figure 7.15

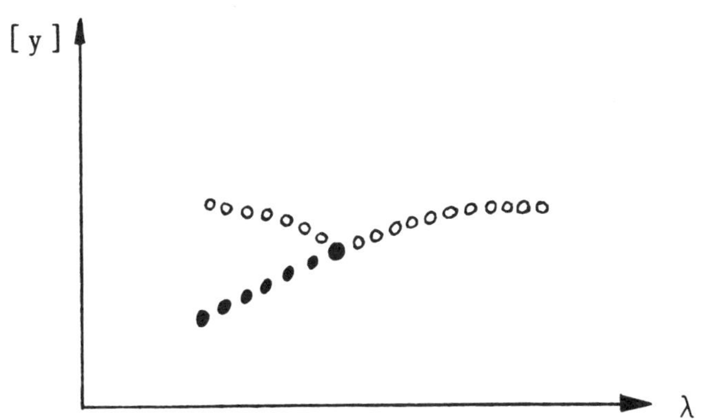

Figure 7.16

doubling points has been revealed, with values of the parameter

$$\lambda_1 = 10.58$$
$$\lambda_2 = 10.19$$
$$\lambda_3 = 10.137$$
$$\lambda_4 = 10.126.$$

These four values allow calculation of two of the quotients in Eq. (7.14). The values are

0.136
0.208,

coming close to the limit in Eq. (7.14).

Example 7.6. Lorenz's Fourth-Order System.
In [218], Lorenz introduced the system of four ordinary differential equations:

$$\begin{aligned}
\dot{y}_1 &= -y_1 + 2\lambda - y_2^2 + (y_3^2 + y_4^2)/2 \\
\dot{y}_2 &= -y_2 + (y_1 y_2 - y_3 y_4) + (y_4^2 - y_3^2)/2 \\
\dot{y}_3 &= -y_3 + (y_2 - y_1)(y_3 + y_4)/2 \\
\dot{y}_4 &= -y_4 + (y_2 + y_1)(y_3 - y_4)/2.
\end{aligned} \tag{7.16}$$

A rigorous bifurcation analysis is given in [174]. We summarize part of the results in the branching and stability diagram in Figure 7.17. There are two stationary bifurcations, one Hopf bifurcation (for $\lambda_0 = 3.8531$) and a sequence of period doublings. Five period doubling points have been calculated, with the values of λ

14.4722
17.0105
17.3607
17.4175
17.4290

The resolution of Figure 7.17 does not show all the period doublings; the way the stability is exchanged remains the same for the further period doublings. As indicated in Figure 7.18, the periods T decrease as λ increases. The five calculated period doubling

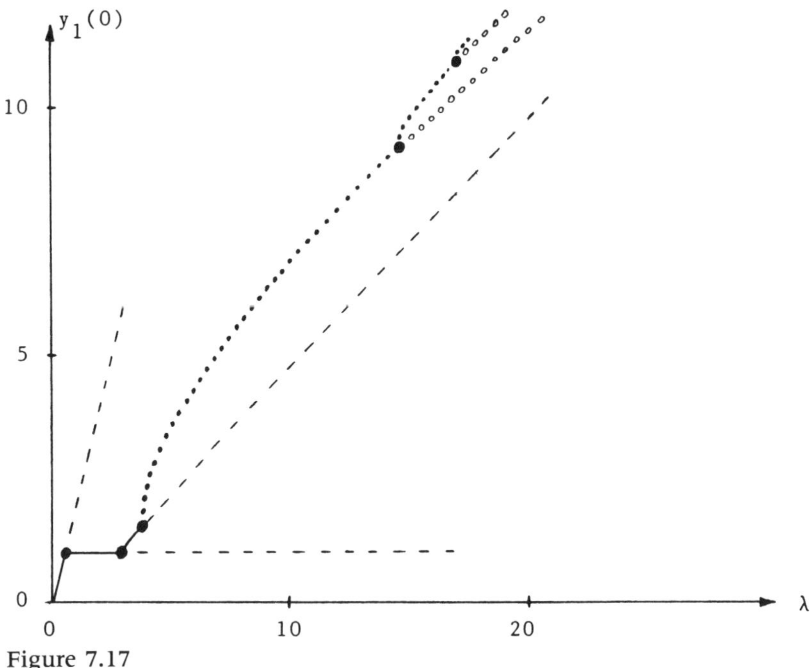

Figure 7.17

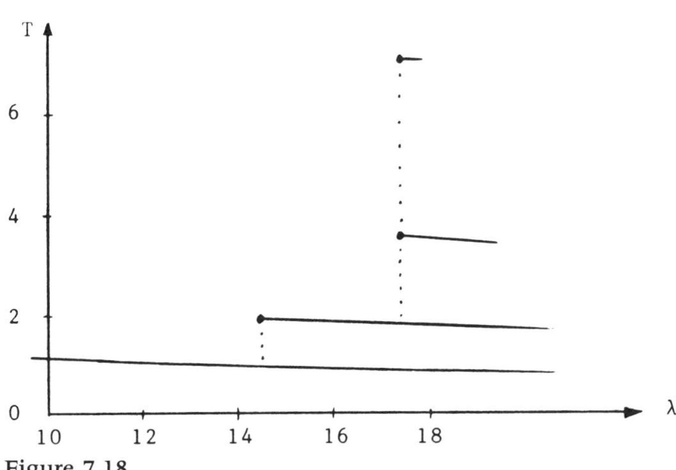

Figure 7.18

points allow us to determine three values of the sequence in Eq. (7.14),

0.138
0.162
0.203.

Assuming a distribution according to Eq. (7.14) enables us to calculate a guess for the next period doubling point, which is expected for

$$\lambda \approx 17.4314.$$

The small relative deviation (10^{-4}) between these points, and the increasingly rich structure of the oscillations, make it difficult to calculate further period doubling points.

It is instructive to study the path of the multipliers $\mu(\lambda)$ during the successive period doublings. In what follows, we illuminate the schematic diagram of Figure 7.7 by concrete results obtained for Eq. (7.16). The monodromy matrices $\mathbf{M}(\lambda)$ that correspond to Eq. (7.16) have four eigenvalues. One is always unity, another eigenvalue remains close to zero. We suppress these two eigenvalues in Figure 7.19 and only depict in a heavy line and a dashed line the paths $\mu(\lambda)$ of the two remaining multipliers. Figure 7.19 shows the multipliers for the part of the periodic branch between the Hopf point and the first period doubling. The numbers report on values of the branching parameter λ. Inspecting Figure 7.19, we realize that in addition to the multiplier that is unity for all λ,

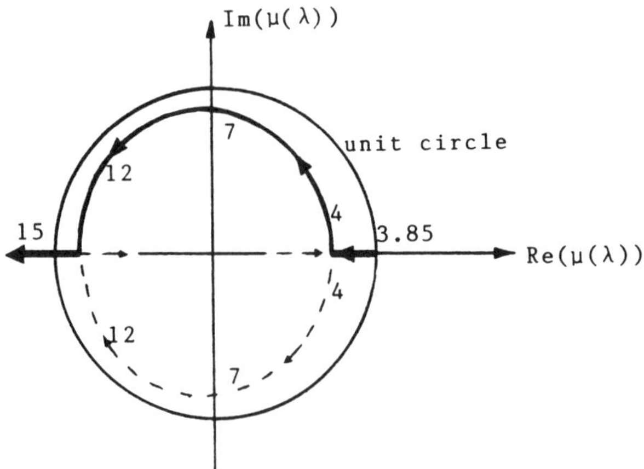

Figure 7.19

one multiplier is unity at the Hopf bifurcation ($\lambda = 3.85\ldots$). For increasing λ, this multiplier moves inside the unit circle along the real axis. For $\lambda \approx 4$ the multiplier meets the second multiplier, and both become a complex-conjugate pair. For increasing λ, the pair of multipliers moves along curved paths (roughly circular) toward the negative real axis. Here, for $\lambda \approx 14.3$, the two multipliers divorce and move in opposite directions. We follow the multiplier with the maximal modulus because this one determines stability. For $\lambda = 14.47\ldots$ this critical multiplier leaves the unit circle at -1, indicating period doubling.

Following the multipliers for any stable branch between two period doublings, one obtains an analog picture. In Example 7.16, the curved paths of conjugate-complex multipliers decrease in diameter after each period doubling. For instance, for the stable periodic orbits in the range $17.4175 < \lambda < 17.429$, the diameter of the curved part of the path is smaller than 0.05.

7.4.3 Bifurcation into Torus

The third type of losing stability is characterized by a pair of complex-conjugate multipliers crossing the unit circle at

$$\mu(\lambda_0) = e^{\pm i\vartheta}$$

(i: imaginary unit). For simplicity, we assume that the angle ϑ is an irrational multiple of 2π (Exercise 7.6). Then the Poincaré map has an *invariant curve* C in Ω (cf. Figure 7.20)—that is to say, take any point $\mathbf{q}$ on C and all iterates of the Poincaré map stay on that curve. Consequently, trajectories spiral around a torus-like object. The invariant curve is the cross section of the torus with the hypersurface Ω. The periodic orbit that has lost its stability to the torus can be seen as the central axis of the torus. For λ tending to λ_0, the diameter of the torus shrinks to zero and eventually reduces to the periodic orbit—that is, at λ_0 a bifurcation from a periodic orbit to a torus takes place. Because this scenario requires complex multipliers in addition to the eigenvalue that is unity, a bifurcation into a torus can take place only for $n \geq 3$.

Tori can be attractive or repelling, as with periodic orbits. For $n = 3$, trajectories approach an attractive torus from the inside domain and the outside domain. Trajectories projected to the hyersurface Ω resemble the dynamics close to a Hopf bifurcation, where a stable limit cycle encircles an unstable equilibrium. Accordingly, the phenomenon of bifurcation into a torus is sometimes called *Hopf bifurcation of periodic orbits, secondary Hopf bifurcation,* or *generalized Hopf bifurcation.* A re-

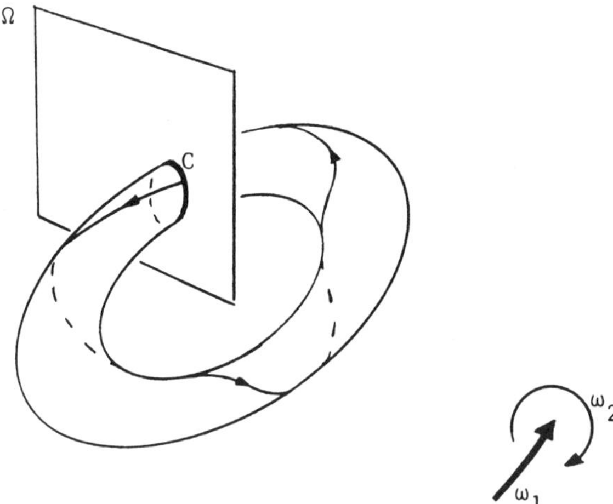

Figure 7.20

pelling torus can be visualized as a tube surrounding a stable periodic orbit ($n = 3$). For λ tending to λ_0, the domain of attraction of the limit cycle vanishes and the torus dies. This is the scenario of a subcritical torus bifurcation.

For ϑ being an irrational multiple of 2π, a trajectory that starts on the invariant curve C will never arrive in the same position. After the bifurcation from periodic orbit (with frequency $\omega_1(\lambda)$) into torus, there are two frequencies $\omega_1(\lambda)$, $\omega_2(\lambda)$. One frequency describes the component of motion along the axis within the torus, the other frequency is measured along the cross section (see the inset in Figure 7.20). The related flow is called *quasiperiodic* if the ratio ω_1/ω_2 is irrational. For some λ, this ratio may be rational ("locked state," the trajectory on the torus is closed). For results on quasiperiodic flow on tori, see [131, 243]. Higher-dimensional tori ("m-torus" for m-dimensional torus; 1-torus is a periodic orbit) have more frequencies; for bifurcation from 2-torus to 3-torus see Section 9.3.

The assumption that $\vartheta/2\pi$ is irrational seems reasonable, because in "nature" almost all numbers are irrational. (On a computer, however, no irrational numbers exist.) For rational fractions, additional resonant terms exist. In particular, the "strongly resonant" cases $\mu^j(\lambda_0) = 1$ for $j = 1, 2, 3, 4$ require specific analysis. We discussed the cases $j = 1$ and $j = 2$ earlier in this chapter; for $j = 3$ and $j = 4$, see [96, 97, 159, 360]. Weak resonances can manifest themselves in that trajectories on the torus concentrate in bands. Graphical illustrations of a related example are shown in [211]; other plots of tori can be found in [11] (Exercise 7.7). Numerical experiments were also conducted in [40, 185]. Interactions of Hopf and stationary bifurcations are investigated in [98, 209, 210, 300, 332].

Example 7.7. A Model Related to Nerve Impulses.
The ODE system

$$\begin{aligned} \dot{y}_1 &= y_1 - \tfrac{1}{3}y_1^3 - y_2 + y_3 + \lambda \\ \dot{y}_2 &= 0.08(y_1 + 0.7 - 0.8y_2) \\ \dot{y}_3 &= 0.0006(-y_1 - 0.775 - y_3) \end{aligned} \tag{7.17}$$

from [277] is an extended version of the FitzHugh model of nerve impulses [95]. A branching diagram is shown in Figure 7.21. There is a Hopf bifurcation at the value of the electric current $\lambda_0 = 0.13825$ and initial period $T_0 = 22.7115$. The emerging branch of periodic orbits loses its stability at $\lambda_0 = 0.1401$ via bifurcation into torus. The system in Eq. (7.17) exhibits further phenomena—for example, a period doubling has been found at $\lambda_0 = 0.31415$. Characteristic solutions to Eq. (7.17) show a double-spike pattern.

Literature on *nerve models* that exhibit branching phenomena has become richer. Apart from the simple FitzHugh model, the more refined Hodgkin-Huxley model has been investigated repeatedly [151, 134, 137, 138, 279]. Other nerve models that show nonlinear phenomena are the Plant model [156] and the model attributed to Beeler and Reuter [29] (cf. Section 2.9). A general discussion of nerve activities requires studying the cable equation—that is, PDEs must be solved [278, 319]. A compromise between the

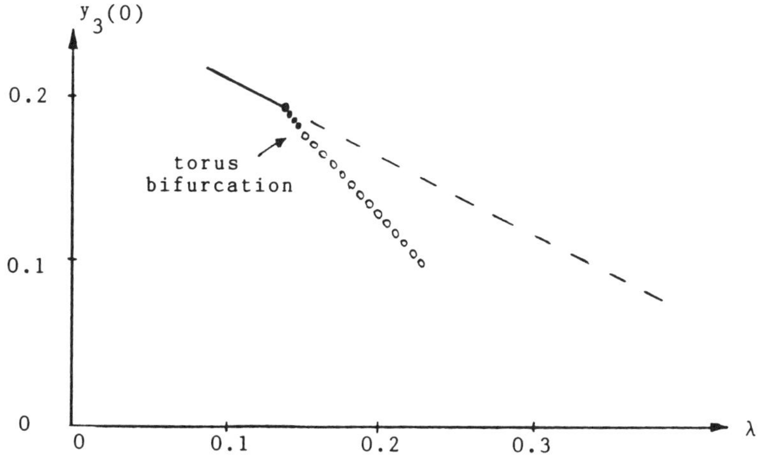

Figure 7.21

demanding Hodgkin-Huxley equations and the space-independent FitzHugh model are the FitzHugh-Nagumo equations

$$u_t = u_{xx} + u(u - a)(1 - u) - v$$

$$v_t = \epsilon(u - \gamma v)$$

$$0 < a < 1, \epsilon > 0, \gamma > 0, |\epsilon|, |\gamma| \text{ small},$$

x is the distance along the axon. Traveling waves have been reported for $a < \frac{1}{2}$. In a mixed initial boundary-value problem, solutions are investigated on the quarter plane $x > 0$, $t > 0$. Nerves that are initially at rest ($u = v = 0$ for $t = 0$) may be stimulated at an endpoint (boundary conditions for $x = 0$). In case the stimulus is stronger than a critical level, a pulse travels along the axon. For stimuli lower than the threshold, the pulse decays to zero. Apart from a single traveling pulse, wave trains have been observed and calculated. For a discussion of traveling waves in nerves, and particularly in the FitzHugh-Nagumo model, see [137, 138, 177, 232, 265, 278, 280] and the references therein.

Exercise 7.3.
In a manner similar to that in Figure 7.12, draw two-dimensional phase planes illustrating the dynamic behavior indicated in Figure 7.9 and Figure 7.10.

Exercise 7.4.
Consider the branching diagram in Figure 7.13.

(a) Draw a diagram depicting only those solutions that can be observed in biochemical reality. (Assume that Eq. (7.12) represents reality.)
(b) Draw a branching diagram of the solutions to Eq. (7.12) in the decoupled case ($\delta_1 = \delta_2 = 0$).

Exercise 7.5.
Consider the fixed-point equation that corresponds to (7.13).

(a) Establish explicitly the fixed-point equation that belongs to P^2.
(b) To which values does the fixed-point iteration $q^{(j)} = P(q^{(j-1)}, \lambda)$ settle for $\lambda = 3.5$, with $q^{(0)} = 0.1$?
(c) Sketch a branching diagram. (Hint: Something "happens" at $\lambda = 3.45$.)

Exercise 7.6.
Consider the linear map in the plane,

$$\mathbf{P}(\vartheta)\mathbf{q} = \begin{pmatrix} \cos\vartheta & -\sin\vartheta \\ \sin\vartheta & \cos\vartheta \end{pmatrix} \mathbf{q}.$$

(a) Show that the eigenvalues of $\mathbf{P}$ lie on the unit circle of the complex plane.
(b) Show that the map establishes a rotation.
(c) Show that $(\mathbf{P}(\vartheta))^m = \mathbf{P}(m\vartheta)$.
(d) For which values of ϑ is $(\mathbf{P}(\vartheta))^m = \mathbf{I}$ (identity matrix)?

Exercise 7.7.
In a diagram $y_1(t)$ versus t, sketch the time history that is characteristic for a quasiperiodic motion with two frequencies.

7.5 CALCULATING THE MONODROMY MATRIX

We learned about the outstanding role of the monodromy matrix $\mathbf{M}$ from Definition 7.1. In this section, we discuss methods for calculating $\mathbf{M}$. In the first subsection, we assume that a periodic solution $\mathbf{y}^*(t)$ to Eq. (7.1) with period T has been calculated and that $\mathbf{M}$ is calculated subsequently.

7.5.1 A Posteriori Calculation

Integrating the n^2 scalar differential equations in Eq. (7.9) for $0 \le t \le T$ yields the monodromy matrix $\mathbf{M}$ from Definition 7.1. Because the right-hand side

$$\mathbf{f_y}(\mathbf{y}^*,\lambda)\Phi$$

varies with $\mathbf{y}^*(t)$ and hence with t, the Jacobian matrix must be available for each t. A convenient way of providing the periodic oscillation $\mathbf{y}^*(t)$ for all t is to attach the original ODE system in Eq. (7.1) as an initial-value problem,

$$\mathbf{y}^*(t) = \varphi(t;\mathbf{y}^*(0));$$

see Eq. (7.7) for a definition of φ. A first method for calculating $\mathbf{M}$ is as follows: Integrate the initial-value problem

$$\begin{pmatrix} \dot{\mathbf{y}} \\ \dot{\Phi} \end{pmatrix} = \begin{pmatrix} \mathbf{f}(\mathbf{y},\lambda) \\ \mathbf{f}_\mathbf{y}(\mathbf{y},\lambda)\Phi \end{pmatrix}, \quad \begin{pmatrix} \mathbf{y}(0) \\ \Phi(0) \end{pmatrix} = \begin{pmatrix} \mathbf{y}^*(0) \\ \mathbf{I} \end{pmatrix} \tag{7.18}$$

until $t = T$, then $\mathbf{M}$ is given by $\Phi(T)$. This method requires that $n + n^2$ scalar differential equations be integrated simultaneously. The enormous storage requirements usually restrict the use of Eq. (7.18) to differential equations with small n. The differential equation of Eq. (7.18) is incorporated in [155] as part of the solution procedure to calculate $\mathbf{y}^*(t)$.

As an alternative, the monodromy matrix can be calculated columnwise [317], reducing storage requirements significantly. For each column of $\mathbf{M}$ a separate system is integrated, consisting of essentially $2n$ scalar differential equations. For practical reasons that will soon become apparent, one may include trivial differential equations for λ and T and integrate

$$\begin{pmatrix} \mathbf{y} \\ T \\ \lambda \\ \mathbf{h} \end{pmatrix}' = \begin{pmatrix} T\mathbf{f}(\mathbf{y},\lambda) \\ 0 \\ 0 \\ T\mathbf{f}_\mathbf{y}(\mathbf{y},\lambda)\mathbf{h} \end{pmatrix}. \tag{7.19}$$

The vector $\mathbf{h}$ with n components serves as an auxiliary variable for generating a column of $\mathbf{M}$. The calculation of $\mathbf{M}$ by integrating Eq. (7.18) is equivalent to the following algorithm:

Algorithm 7.1. Calculation of M. For $j = 1, \ldots, n$:

> Integrate Eq. (7.19) for $0 \le t \le 1$ with initial values $\mathbf{y}(0) = \mathbf{y}^*(0)$, $\mathbf{h}(0) = \mathbf{e}_j$. Then $\mathbf{h}(1)$ is the j-th column of $\mathbf{M}$.

Carrying out Algorithm 7.1, one integrates n times a $(2n + 2)$ system. By doing this, storage problems are not nearly as stringent as they are in integrating Eq. (7.18). It must be noted, however, that Algorithm 7.1 is about twice as slow as integrating Eq. (7.18). This disadvantage can be removed with the following modification: We do not keep $\mathbf{y}^*(t)$ available as the solution of the initial-value problem, but replace $\mathbf{y}^*(t)$ by an approximation $\bar{\mathbf{y}}(t)$. For $\bar{\mathbf{y}}(t)$ we choose, for instance, a spline or a Hermite polynomial that interpolates $\mathbf{y}^*(t)$. The modified algorithm then reads:

Algorithm 7.2. Modified algorithm for calculating M. For $j = 1, \ldots, n$:

Integrate $\mathbf{h}' = T\mathbf{f}_{\mathbf{y}}(\bar{\mathbf{y}},\lambda)\mathbf{h}$ for $0 \leq t \leq 1$
with initial value $\mathbf{h}(0) = \mathbf{e}_j$. Then $\mathbf{h}(1)$ approximates the j-th column of $\mathbf{M}$.

Algorithm 7.2 is efficient with regard to both storage and computing time. The drawbacks of Algorithm 7.2 are technical: Storaging $\bar{\mathbf{y}}$ is more tedious than saving the vector $\mathbf{y}^*(0)$; in addition, an error control takes more effort. Hence Algorithm 7.2 is much more complex than carrying out Algorithm 7.1 or integrating Eq. (7.18). Note that $\mathbf{M}$ is still a n^2 matrix; the storage efficiency of Algorithm 7.1 and Algorithm 7.2 refers to the calculation only.

7.5.2 Monodromy Matrix as a By-product of Shooting

The amount of work required for an a posteriori calculation of the monodromy matrix $\mathbf{M}$ is significant, so it is worthwhile to think about alternatives that provide $\mathbf{M}$ as a side result during the calculation of a periodic orbit. In what follows, we show how to obtain $\mathbf{M}$ without any cost as a by-product of the (multiple) shooting approach. As in Section 6.3, the Jacobian $\mathbf{f}_{\mathbf{y}}$ is not explicitly required.

The iteration matrix of a shooting algorithm in condensed form was introduced in Eqs. (6.21) and (6.22). The definitions of $\mathbf{M} = \Phi(T)$ and $\mathbf{G}_j$ imply that

$$\mathbf{M} = \mathbf{G}_{m-1} \cdot \ldots \cdot \mathbf{G}_1 .$$

From the definitions of $\mathbf{A}$ and $\mathbf{B}$ (cf. Eq. (6.12)), we infer that for periodic boundary conditions

$$\mathbf{r}(\mathbf{y}(0),\mathbf{y}(1)) = \mathbf{y}(0) - \mathbf{y}(1) = \mathbf{0}$$

the relations $\mathbf{A} = \mathbf{I}$ and $\mathbf{B} = -\mathbf{I}$ hold. Consequently,

$$\mathbf{E} = \mathbf{A} + \mathbf{B}\mathbf{G}_{m-1} \cdot \ldots \cdot \mathbf{G}_1 = \mathbf{I} - \mathbf{M}. \tag{7.20}$$

Because shooting is applied to the extended system in Eq. (7.5) or Eq. (7.6) rather than Eq. (7.1), the approximation of the $(n + 1)$-vector $\tilde{\mathbf{y}} = (\mathbf{y}^*,T)$ terminates with an iteration marix $\tilde{\mathbf{E}}$, which is an $(n + 1)^2$ matrix or $(n + 2)^2$ matrix. By investigating the structure of Eqs. (7.5) and (7.6), one realizes that $\mathbf{E}$ is the leading n^2 submatrix of $\tilde{\mathbf{E}}$.

This gives another explanation of why standard shooting cannot be applied directly to the n-system in Eq. (7.1) with boundary conditions $\mathbf{y}(0) = \mathbf{y}(T)$:

Eq. (7.20) implies that the iteration matrix **E** is singular because **M** has unity as eigenvalue. However, the extended iteration matrix $\tilde{\mathbf{E}}$ is mostly nonsingular. $\tilde{\mathbf{E}}$ may get singular at bifurcations of the type at which **M** has unity as an additional eigenvalue (turning point, Hopf bifurcation, stationary bifurcation point).

We summarize the above strategy as follows [317]:

Algorithm 7.3. Simultaneous calculation of y^* and **M**.

Solve Eq. (7.5) or Eq. (7.6) by multiple shooting [334].
E is the leading n^2 submatrix of the final iteration matrix. Then $\mathbf{M} = \mathbf{I} - \mathbf{E}$.

Because $\tilde{\mathbf{E}}$ is obtained by numerical differentiation, the accuracy of the monodromy matrix calculated by Algorithm 7.3 is strongly affected by the approximation strategies in a particular implementation of shooting. Of course, the essence of Algorithm 7.3 for calculating **M** can be carried out independently of the shooting algorithm—that is, the version of **M** in Eq. (7.8) is approximated by numerical differentiation.

7.5.3 Numerical Aspects

The calculated monodromy matrix **M** has a feature that must be appraised highly—namely, a measure of its accuracy is accessible. Let $\mathbf{M}'$ be a numerically approximated monodromy matrix with eigenvalues μ_j' $(j = 1, \ldots, n)$. If $\mathbf{M}'$ is obtained accurately and the underlying solution is periodic, then one eigenvalue (say, μ_n') is close to unity. Consequently, the distance

$$\delta = | 1 - \mu_n' | \tag{7.21}$$

serves as a measure of the quality of approximation of $\mathbf{M}'$. If δ is small (say, $\delta < 0.005$), we accept the other $n - 1$ multipliers $\mu_1', \ldots, \mu_{n-1}'$ as reasonable approximations to the exact multipliers. Because all the methods in Eq. (7.18) and Algorithms 7.1 and 7.3 are essentially based on numerically integrating initial-value problems, the measure δ reflects how accurately this "shooting" is carried out.

Numerical experience suggests the following conclusions. Broadly speaking, shooting methods work well for stable orbits and weakly unstable orbits. Here we call a periodic orbit *weakly unstable* if the multipliers are not large in modulus, say,

$$\max | \mu_j | < 10^3.$$

Shooting encounters difficulties in providing enough accuracy when there are large multipliers.

The above bound of 10^3 is artificial; it depends on the chosen numerical integrator and the prescribed error tolerance. Hence, the conclusions above do not necessarily apply to all problems and all integrators in that extent. In a small number of problems with strongly unstable orbits, backward shooting may help. Because all changes in stability occur in the stable and weakly unstable range, shooting can be expected to behave well in a bifurcation and stability analysis. For example, all results shown in the figures of this chapter have been calculated by shooting methods. With inaccurate values of $\mathbf{M}'$ (δ large), there can be difficulty deciding safely which of the eigenvalues approximates unity; this frequently happens when $\mathbf{M}$ has a multiplier close to unity.

The number δ from Eq. (7.21) allows us to compare the algorithms for approximating $\mathbf{M}'$. Clearly, Algorithm 7.1 and the integration of Eq. (7.18) yield the same results, because both methods differ only in their organization. Algorithm 7.3, which exploits the iteration matrix of multiple shooting, is strongly affected by the strategies of both continuation and approximating the shooting matrix (compare Section 5.1). As a rule, it is worthwhile to calculate the eigenvalues of $\mathbf{M}'$ from Algorithm 7.3. In the following cases, $\mathbf{M}'$ must be recalculated by a more accurate method, such as Algorithm 7.1:

when δ based on Algorithm 7.3 is too large

after every five (say) continuation steps, in order to check the accuracy of Algorithm 7.3 by comparing it with Algorithm 7.1

when close to a loss or gain of stability—that is, when a multiplier approaches unity in modulus.

Such a combination of the cheap Algorithm 7.3 with the accurate Algorithm 7.1 appears to be a practical compromise. If the underlying system $\dot{\mathbf{y}} = \mathbf{f}(\mathbf{y},\lambda)$ results from a discretization of PDEs—that is, if the dimension n is large—a calculation of $\mathbf{M}$ and its eigenvalues may be too costly. In several of these cases, it is advisable to test stability by means of a coarser discretization, thereby assuming that the stability result is not severely affected by discretization errors. That is, smaller monodromy matrices that reflect the stability behavior of the large-scale system are to be generated. This has been done for finite-element equations modeling a wind energy converter [230].

7.6 CALCULATING BRANCHING BEHAVIOR

In the two preceding chapters we discussed methods for handling branch points, with emphasis on systems of nonlinear equations and boundary-value problems of ODEs. The main ideas underlying these methods apply also to periodic solutions. Recall the correspondence between the problem of calculating periodic solutions and the boundary-value problem in Eq. (7.6a and b). Instead of closing this section with that, it is worthwhile to discuss some specific aspects of periodic orbits.

The standard approach for locating branch points is to monitor an appropriate test function. Because branching is related to a multiplier μ_j that escapes or enters the unit disk of the complex plane, we can choose the modulus for test function: The function $\tau(\lambda)$ defined by

$$\tau = |\mu_j| - 1, \text{ with } |\mu_j| = \max\{|\mu_1|, \ldots, |\mu_{n-1}|\} \quad (7.22)$$

signals a change in stability by a change of sign. This assumes that the multipliers are ordered such that $\mu_n = 1$. In order to locate branch points within an unstable range, choose the j of the term that minimizes the sequence

$$||\mu_1| - 1|, \ldots, ||\mu_{n-1}| - 1|.$$

These test functions correspond to Eqs. (5.5) and (5.6).

Indirect methods based on the branching test function Eq. (7.22) have proved successful also in the context of periodic solutions; strategies like that outlined in Section 5.3.1 work. Apart from indirect methods, direct methods have also been applied. The branching system for periodic solutions consist of $2n + 3$ components (Exercise 7.8),

$$\begin{pmatrix} \mathbf{y} \\ T \\ \lambda \\ \mathbf{h} \\ \zeta \end{pmatrix}' = \begin{pmatrix} T\mathbf{f}(\mathbf{y},\lambda) \\ 0 \\ 0 \\ T\mathbf{f}_\mathbf{y}(\mathbf{y},\lambda)\mathbf{h} + \zeta\mathbf{f}(\mathbf{y},\lambda) \\ 0 \end{pmatrix}, \quad \begin{pmatrix} \mathbf{y}(0) - \mathbf{y}(1) \\ f_1(\mathbf{y}(0),\lambda) \\ \mathbf{h}(0) - \mathbf{h}(1) \\ \sum_i h_i \partial f_1/\partial y_i \\ h_1(0) - 1 \end{pmatrix} = \mathbf{0}. \quad (7.23)$$

In Eq. (7.23) the boundary conditions are based on the phase condition in Eq. (7.4) with $k = 1$. For other phase conditions, one obtains analog branching systems. Hopf bifurcation, pitchfork bifurcation, and period doubling are among the solutions of Eq. (7.23). For the latter case, the double period must be entered in order to obtain a standard bifurcation with a multiplier $\mu = 1$. Because numerical results obtained so far in applying Eq. (7.23) have shown poor convergence, this system cannot be recommended. In the special case of Hopf bifurcation, however, Eq. (7.23) can be reduced favorably. Observing $\mathbf{f}(\mathbf{y},\lambda) = \mathbf{0}$, the specific branching system

$$\begin{pmatrix} \mathbf{y} \\ T \\ \lambda \\ \mathbf{h} \end{pmatrix}' = \begin{pmatrix} T\mathbf{f}(\mathbf{y},\lambda) \\ 0 \\ 0 \\ T\mathbf{f_y}(\mathbf{y},\lambda)\mathbf{h} \end{pmatrix}, \quad \begin{pmatrix} \mathbf{y}(0) - \mathbf{y}(1) \\ \mathbf{h}(0) - \mathbf{h}(1) \\ \sum_i h_i \partial f_1/\partial y_i \\ h_1(0) - 1 \end{pmatrix} = \mathbf{0} \tag{7.24}$$

results [311]. The summation term in the boundary conditions of Eqs. (7.23) and (7.24) is evaluated at $t = 0$,

$$\sum_i h_i(0) \frac{\partial f_1(\mathbf{y}(0),\lambda)}{\partial y_i} = 0.$$

The convergence of an iteration procedure (shooting combined with quasi-Newton) that solves Eq. (7.24) has shown satisfactory results in a wide range of examples. Concerning singularity and convergence speed, the remarks in Section 5.3.2 apply to Eq. (7.24). Notice that the right-hand side of Eq. (7.24) is the same one used in Algorithm 7.1 for calculating the monodromy matrix. Solving Eq. (7.24) is an alternative to the direct method of solving Eq. (5.59). The two methods approach the Hopf point from different sides: solving Eq. (5.59) is a steady-state method, while solving Eq. (7.24) is a transient method. A further reduction of Eq. (7.24) toward a differential-algebraic system is possible. Solving ODEs in Eq. (7.24) is more expensive than solving nonlinear equations in Eq. (5.59), but it compares favorably with the average cost of calculating periodic orbits.

After a Hopf point ($\mathbf{y}_0$, T_0, λ_0, $\mathbf{h}_0$) is calculated, an initial guess for small-amplitude periodic solutions is given by

$$\bar{\mathbf{y}} = \mathbf{y}_0 + \delta\mathbf{h}_0, \quad \bar{T} = T_0, \quad \bar{\lambda} = \lambda_0. \tag{7.25}$$

These data (for small $|\delta|$) serve as an initial approximation for Eq. (6.3), which is based on the boundary-value problem in Eqs. (7.6a) and (7.6b). Consequently, the $(n + 2)$-dimensional boundary-value problem in Eqs. (7.6c) and (7.6d) is solved, starting from Eq. (7.25). The boundary conditions Eq. (7.6d) require the value

$$\eta = \bar{y}_k(0).$$

In case the Hopf point is calculated by the steady-state method of solving Eq. (5.59), the vector function $\mathbf{h}_0(t)$ is not immediately available. Instead, the complex-conjugate eigenvector $\mathbf{w}$ is part of the solution. Hence, in order to use Eq. (7.25), $\mathbf{h}_0(t)$ must be calculated based on the real part $\mathbf{h}$ and the imaginary part $\mathbf{g}$ of $\mathbf{w}$. (Here the constant vector $\mathbf{h}$ in Eq. (5.59) is not to be confused with $\mathbf{h}_0(t)$.) The derivation of $\mathbf{h}_0(t)$ is a standard analysis in ODE textbooks. Let the eigenvalues of $\mathbf{f_y}(\mathbf{y}_0,\lambda_0)$ be $\mu_1 = \beta$, $\mu_2 = -\beta$, μ_3, μ_4, . . . , μ_n. Then solutions $\mathbf{h}_0(t)$ to

$$\mathbf{h}' = T\mathbf{f_y}(\mathbf{y}_0,\lambda_0)\mathbf{h}$$

are given by

$$\mathbf{h}_0(t) = (c_1\mathbf{h} + c_2\mathbf{g}) \cos T\beta t + (c_2\mathbf{h} - c_1\mathbf{g}) \sin T\beta t + \sum_{i=3}^{n} c_i\mathbf{z}^i \exp(T\mu_i t). \quad (7.26)$$

In Eq. (7.26) we assumed that the Jacobian $\mathbf{f_y}(\mathbf{y}_0,\lambda_0)$ has n linear independent eigenvectors $\mathbf{z}^i$ corresponding to μ_i. The constants c_i are determined by the boundary conditions imposed on $\mathbf{h}_0(t)$ (see Eq. (7.24)). The resulting vector function $\mathbf{h}_0(t)$ is

$$\mathbf{h}_0(t) = \mathbf{h}\cdot\cos 2\pi t - \mathbf{g}\cdot\sin 2\pi t. \quad (7.27)$$

This function is based on the normalization $w_1 = 1$ and matches the normalizations chosen in Eq. (7.24) (Exercise 7.9). Other normalizations in Eq. (5.59) or Eq. (7.24) lead to expressions similar to that in Eq. (7.27). The solutions to Eqs. (7.24) and (5.59) are related via Eq. (7.27).

Note that, with the chosen normalization, the first component of $\mathbf{h}_0(t)$ is known without solving Eqs. (7.24) or (5.59),

$$h_{01}(t) = \cos 2\pi t.$$

In the special case $n = 2$, the other component can be precalculated as well (Exercise 7.10). Thus, in the case of $n = 2$ the initial guess Eq. (7.25) can be calculated without requiring any eigenvectors. The approach Eq. (7.24) has been modified to parabolic PDEs [317].

Other direct methods for calculating period doubling points were proposed in [199]. For instance, the equation

$$(\mathbf{M} + \mathbf{I})\mathbf{z} = \mathbf{0}$$

exploits the fact that -1 is the eigenvalue of the monodromy matrix $\mathbf{M}$. One component of $\mathbf{z}$ is prescribed as normalization. Hence, this system constitutes n equations for $(n - 1)$ unknown components of $\mathbf{z}$ and the parameter value λ of a period doubling. In addition, equations that constitute $\mathbf{M}$ must be attached. At the present state, direct methods do not appear to be efficient enough; indirect methods are recommended for practical use.

Concerning branch switching, the methods of Section 6.4 apply also in the case of periodic solutions. The kind of regularity breaking that is appropriate for period doubling was introduced by Eqs. (6.34) and (6.30d).

Simulation (Algorithm 1.1) is successfully applied in a practical bifurcation and stability analysis of periodic solutions. Sometimes, because it often requires long integration times until the dynamic behavior becomes clear, simulation is discredited. The uncertainty about where to choose the starting vector and to choose the termination time t_f gives simulation a random character. On the other hand, compared with solving boundary-value problems like Eq. (7.5), simulation has strong advantages:

The computational overhead is small (only an integrator is used).

Results are more global and include information on stability.

In many instances, a simulation yields a limit cycle even faster than a solution of the boundary-value problem in Eq. (7.5) does—not counting the effort of checking the stability, which is also required when Eq. (7.5) is applied. Simulation often makes possible an easy branch switching from repellors to attractors; this includes equilibria, periodic orbits, and tori. Simulation can be tedious near Hopf bifurcations; here a solution is only weakly attracting or repelling, because the real part of the critical eigenvalue is close to zero. In case of a hard loss of stability, however, simulation can rapidly generate a periodic orbit. If simulation is avoided in instances where it works slowly, it is highly recommended. The advantages of simulation seem to grow with the complexity of the underlying model.

Exercise 7.8.
Derive Eq. (7.23) from Eq. (6.14).

Exercise 7.9.
Confirm Eq. (7.26). Determine the constants c_i such that $\mathbf{h}_0(t)$ obeys the boundary conditions in Eq. (7.24). Assume hereby that $\mathbf{h}$ and $\mathbf{g}$ are normalized by $h_1 = 1, g_1 = 0$.

Exercise 7.10.
Assume that $n = 2$ and $\partial f_1/\partial y_2 \neq 0$ in the Hopf bifurcation point. Establish a formula for the second component of $\mathbf{h}_0(t)$.

Example 7.8. Chemical Reaction with Autocatalytic Step.
This example is attributed to [193]. Dimensionless variables are

$y_i(t)$: concentrations of chemicals

ψ: flow ($\psi = 3$)

λ: velocity coefficient of the autocatalytic step.

The reaction is defined by the differential equations

$$\begin{aligned} \dot{y}_1 &= \psi - y_1 - \lambda y_1 y_3 \\ \dot{y}_2 &= y_1 - y_2 y_3 \\ \dot{y}_3 &= y_2 y_3 - \lambda y_1 y_3. \end{aligned} \tag{7.28}$$

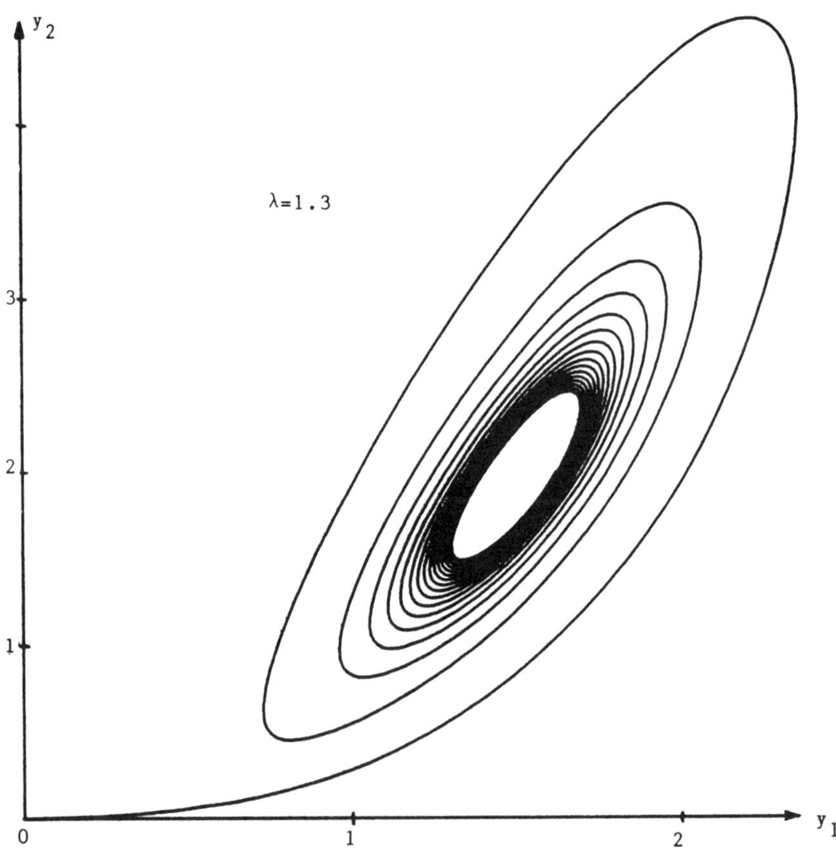

Figure 7.22a

There is a Hopf bifurcation with

$$\lambda_0 = 1.3017,\ T_0 = 6.0355,$$

calculated by means of Eq. (7.24). A branch of stable periodic orbits branches off for $\lambda \geq \lambda_0$. Figure 7.22 shows simulations starting from $\mathbf{z} = \mathbf{0}$ for three parameter values; the phase space is projected to the (y_1,y_2)-plane. In Figure 7.22a, the trajectory is approaching the stable equilibrium close to the Hopf point so slowly that the simulation had to be terminated before the underlying dynamics became evident. Figure 7.22b shows a trajectory approaching a stable limit cycle, still at a dilatory rate. This illustrates that a simulation too close to the Hopf bifurcation can be a waste of time, especially in instances where there is a soft

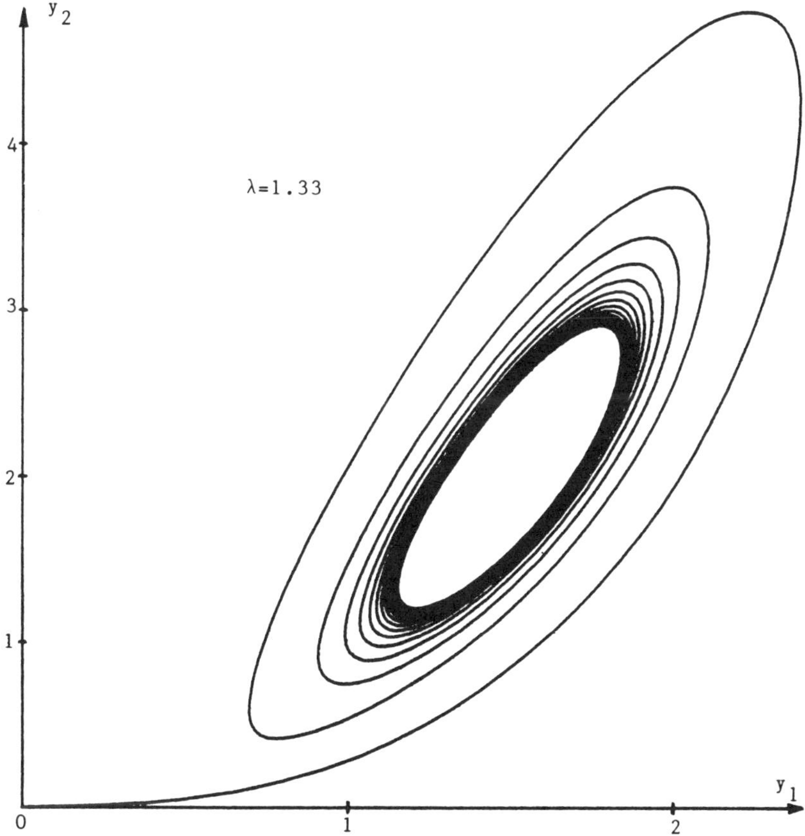

Figure 7.22b

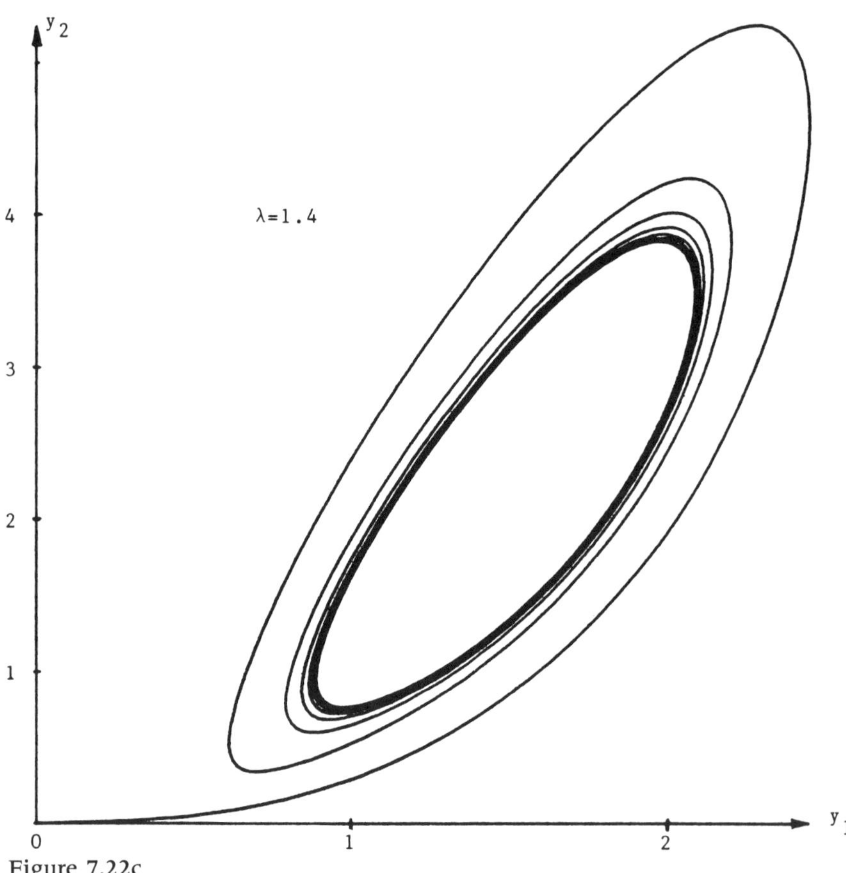

Figure 7.22c

loss of stability. In such situations, simulation is no alternative to Newton-like procedures for calculating equilibria and period orbits. If we try simulation again for $\lambda = 1.4$ (Figure 7.22c), it works very well: The trajectory encircles the limit cycle only a few times before being absorbed. Generating the limit cycle in this way compares favorably with calculating the limit cycle by solving the boundary-value problem in Eq. (7.5). Note that each numerical differentiation involved in solving Eq. (7.5) requires calculation of n trajectories. In Eq. (7.28) with $n = 3$, one numerical differentiation corresponds to three turnings of the spiraling trajectory. Hence, establishing the limit cycle and its stability in Figure 7.22c by simulation requires less effort than solving a boundary-value problem such as that of Eq. (7.5).

The periodic oscillations of Eq. (7.28) become a relaxation oscillation for $\lambda \approx 5$. Such an oscillation is depicted in Figure 7.23 for $\lambda = 6$. Close to a Hopf bifurcation, oscillations have a sinusoidal shape. This sinusoidal function is approximated by Eq. (7.25), which is the function in Eq. (7.27) added to a constant vector. In Eq. (7.28), the initial smoothness diminishes gradually with increasing λ and gives way to the saw-toothed shape of Figure 7.23.

It is not necessary to use positive values of the terminating time t_f of a simulation. A negative value of t_f means backward integration, $0 \geq t \geq t_f$. In this way, in some examples, unstable periodic orbits can be generated. In particular, this is possible for all examples with $n = 2$. For instance, consider the unstable periodic orbit for $\lambda = -0.337$ in the FitzHugh nerve model (see the branching diagram Figure 2.34). This orbit was obtained by simulation using backward integration (see the phase diagram in Figure 7.24). Here the convergence is again slow, because of the nearby Hopf bifurcation. On the other side of this bifurcation, however, there is a hard loss of stability and the quick success that a simulation can bring.

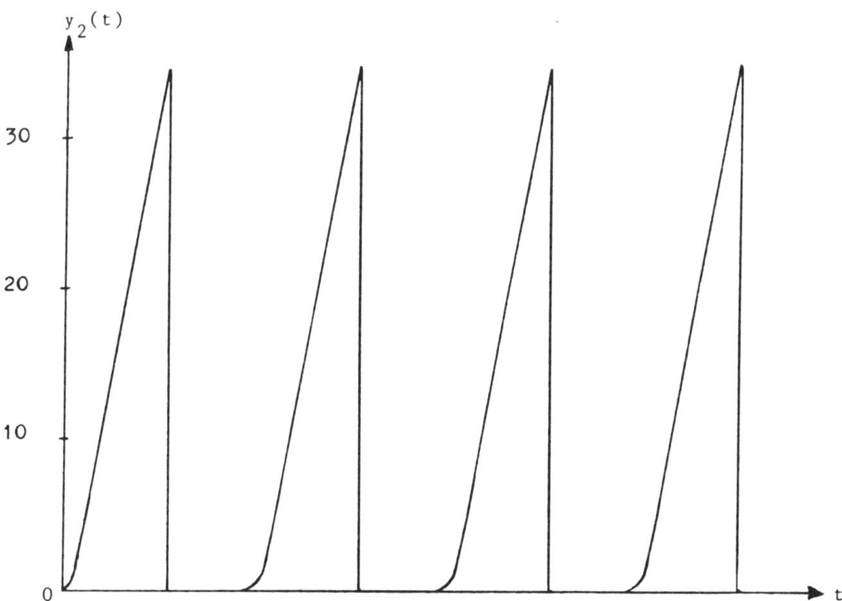

Figure 7.23

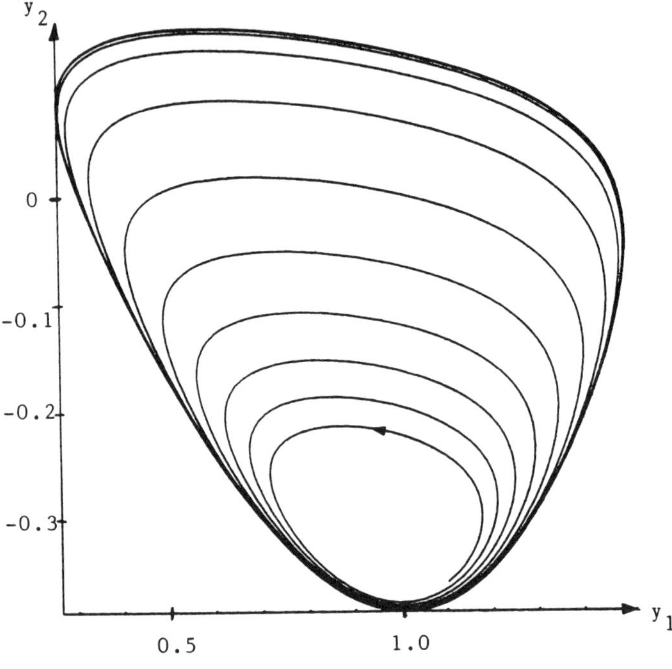

Figure 7.24

7.7 FURTHER EXAMPLES AND PHENOMENA

Oscillatory phenomena are richer and more widespread than indicated so far in this chapter. We complete the preceding material with hints about further examples and investigations, including phenomena that are nonoscillatory. Faced with the great variety of nonlinear phenomena that have been reported, we confine ourselves to relatively few hints. The reader will know which journals to consult for recent reports on nonlinear phenomena in his or her own field. A general reference is the journal *Physica D*, which is devoted specifically to nonlinear phenomena.

In chemistry and chemical engineering, much attention has been given to flames and combustion (see, for example, [140, 224, 225, 226, 229, 255, 328]). Some papers on chemical kinetics have already been quoted; more hints can be found in [85, 266]. Most popular is the Belousov-Zhabotinskii reaction, modeled in [91]. This model (often called "Oregonator") has been investigated, for instance, in [139, 281]. Feedback is related to bifurcation [107]; it was even introduced artificially to discriminate among different dynamic behaviors [204, 220]. For biochemical phenomena, see [145], for example.

As examples of physical bifurcation, we mention Hopf bifurcation in a model describing glaciation cycles [102], Hopf bifurcation in lasers [269], equations describing meteorological effects more accurately than the Lorenz system in Eq. (2.18) does [70], and the study of the earth's magnetic field [51]. A great variety of bifurcation problems occur in fluid dynamics. Reviews and handbooks are [166, 186, 336, 338, 372]. In fluid dynamics, much depends on the underlying geometry. Accordingly, different studies for different geometries have been conducted. Flows in channels or tubes [41, 356] and the flow between two rotating concentric cylinders (the Taylor problem) [341, 32, 187, 237, 357] have found specific interest. Convection driven by buoyancy or surface tension is mentioned in Section 2.7 and Section 3.4; for further references, see [338, 372].

In biology, most current investigations related to bifurcation appear to be devoted to nerve models. Some of these models are designed to model heartbeat. References concerning nerve models were given in Section 7.4.3. Interesting nonlinear effects occur in predator-prey models, which describe oscillations of populations of certain species that struggle for existence [42, 173]. Related phenomena occur, for example, with fish [358] and starfish [14], in immune response [30], or with the Canadian lynx (a cycle with a period of 10 years) [297]. Hopf bifurcation also occurs in animal orientation [126].

Classical applications for bifurcation occur in mechanics. In particular, the bending and buckling of structures have found much attention. We mention mode jumping of rectangular plates, which are modeled by the von Kármán equations [150, 337], the bending of tubes of elliptic cross sections [362], and the buckling of spherical shells (see [114] and the references therein). General references for nonlinear phenomena in mechanical engineering are [344, 346]. Specific experiments are a double pendulum (motivated by vibrations of the Trans-Arabian pipeline in 1950) [152, 298], snap-through of imperfect shallow arches [158], and the tractor-semitrailer, the straight motion of which loses stability for certain critical speeds [348]. The Hopf bifurcation of the Watt steam engine is analyzed in [6].

Another aspect of nonlinear oscillations is synchronization (entrainment). Consider a self-oscillating system affected by a time-periodic perturbation. Synchronization has two aspects: resonance occurs (amplitude increases), or the phases of the excitation and the system become locked [238]. Entrainment takes place in all fields—for example, biology [256], electrical engineering [350], and chemistry [18, 301].

Apart from periodic perturbations, effects of single stimuli are studied. In particular, a single pulse of proper timing and amplitude serves as a neural and cardiac pacemaker [60].

The extent of frequency locking may vary with parameters. Let the applied excitation to an oscillator be

$$K \cos(\omega t),$$

with frequency ω, and let the response frequency of the system be

$$\omega_s = \frac{2\pi}{T}.$$

In Section 6.1 (Example 2.1, the forced Duffing equation), we artificially imposed $\omega_s = \omega$ by means of appropriate boundary conditions. Free oscillations may take other ratios between the exciting frequency ω and the response frequency ω_s. Whenever this ratio is a rational number, we say that the phase is locked. Upon plotting the ratio versus an underlying parameter λ (which may be ω itself), locking manifests itself by horizontal "steps." For weak coupling (K small), the phase-locked intervals are narrow and the motion is mostly quasiperiodic (ratio irrational). For increasing coupling (K larger), the phase-locked portions increase. The resulting step function, called *devil's staircase* [222], has been observed in many applications (see Marek, Schreiber in [205] and Bak in *Physics Today,* 1986, 12).

Exercise 7.11 (requires the use of a computer).
Consider the *circle map* [302]

$$\vartheta_{j+1} = \vartheta_j + \omega + \frac{K}{2\pi} \sin(2\pi\vartheta_j)$$

and the *winding number*

$$\lim_{j\to\infty} \frac{1}{j} (\vartheta_{j+1} - \vartheta_1).$$

For at least three values of the coupling K ($K = 0$, $K = 0.5$, $K = 0.9$), calculate and draw the winding number versus ω. (For nonzero K, you must resort to approximations.) For increasing K, increasing ω-intervals of phase locking become visible, leading to a devil's staircase. For continuous K, in a (ω,K)-plane, these intervals extend to *Arnold tongues.*

8 Qualitative Instruments

8.1 SIGNIFICANCE

Numerical calculations are primarily quantitative in nature. Appropriate interpretation transforms quantitative information into meaningful results. The diversity of nonlinear phenomena requires more than just the ability to understand numerical output. The preceding chapters have concentrated on one-parameter problems, with only a brief excursion (in Section 5.7) to two-parameter models, mainly to show how one-parameter studies can be applied to investigation of certain aspects of two-parameter models. Reducing a parameter space to lower-dimensional subsets will remain a standard approach for obtaining quantitative insight for a long time. However, a full qualitative interpretation of multiparameter models requires instruments that have not yet been introduced, instruments provided by singularity theory and catastrophe theory. Both these fields are qualitative in nature; in no way do they replace numerical parameter studies. Knowledge of singularity theory and catastrophe theory helps organize a series of partial results into a global picture.

Before numerical methods became widely used, analytical methods, such as the Liapunov-Schmidt method [333, 354] and the center manifold theory (to be discussed briefly in Section 8.5), were the standard means for carrying out bifurcation analysis. Today, and presumably to an even larger extent in the future, the analysis is accomplished by numerical methods. This is due to the high level of sophistication that numerical methods have reached, as well as to the limitations of analytical methods. These limitations are twofold. First, numerical methods are required for evaluating analytical expressions anyway. Second, analytical results are generally local—that is, they consist of assertions of the type "holds for a sufficiently small distance," where "sufficiently small" is left unclarified. Although remarkable results have been obtained by analytical methods (for example, [224, 289, 298]), the extensive application of numerical methods is indispensable for practical bifurcation and stability analysis. With a black box of numerical methods at hand, it is easy—sometimes too easy—to take advantage of the skills of others. Unfortunately, the use of black boxes occasionally results in a shallower understanding. Nevertheless, the former rule, "Analysis must come ahead of numerical analysis," has lost its validity.

8.2 SINGULARITY THEORY FOR ONE SCALAR EQUATION

Singularity theory provides a framework for classifying branching phenomena. Many of the related results are attributable to Golubitsky and Schaeffer [112, 113] (see also [110, 111]). In this section we give a brief account of those parts of singularity theory. Some of the material was touched upon in Section 2.9. A review of Exercise 2.12 (finite-element analog of a beam) will help in understanding the present section. Here we restrict ourselves to one scalar equation

$$0 = f(y,\lambda)$$

with a scalar variable y. A singularity is a solution (y,λ) with $f_y(y,\lambda) = 0$. Both bifurcation points and turning points are singular points. For convenience, assume that the singularity under investigation is situated in the origin—that is, in particular, $f(0,0) = 0$.

A first step toward a classification of singularities and branching phenomena is to ignore differences that are not essential. For example, a branching problem does not change its quality when the coordinates are changed. This leads to the following local definition:

Definition 8.1. $f(y,\lambda) = 0$ *and* $\tilde{f}(\tilde{y},\tilde{\lambda}) = 0$ *are said to be contact equivalent when the qualitative features are the same. Or, more formally, there is a scaling factor* $\zeta(\tilde{y},\tilde{\lambda})$ *such that for the transformation*

$$(\tilde{y},\tilde{\lambda}) \rightarrow y(\tilde{y},\tilde{\lambda}),\ \lambda(\tilde{\lambda})$$

the following holds:

$$\tilde{f}(\tilde{y},\tilde{\lambda}) = \zeta(\tilde{y},\tilde{\lambda}) f(y(\tilde{y},\tilde{\lambda}),\lambda(\tilde{\lambda})),$$

$$y(0,0) = 0,\ \lambda(0) = 0,\ \zeta(0,0) \neq 0,$$

$$\partial y(0,0)/\partial \tilde{y} > 0,\ \partial \lambda(0)/\partial \tilde{\lambda} > 0.$$

Example 8.1.
The two bifurcation problems

$$\tilde{f}(\tilde{y},\tilde{\lambda}) = 1 - \tilde{\lambda} - \cos(\tilde{y}) = 0$$

$$f(y,\lambda) = y^2 - \lambda = 0$$

are contact equivalent in a neighborhood of $y = 0$. This is prompted by the power series of $\cos(y)$,

$$1 - \tilde{\lambda} - \cos(\tilde{y}) = 1 - \tilde{\lambda} - (1 - \tfrac{1}{2}\tilde{y}^2 \pm \ldots) = \tfrac{1}{2}\tilde{y}^2 - \tilde{\lambda} \pm \ldots$$

The dots refer to terms of order $\tilde{y}^4$. The formal transformation requested in Definition 8.1 is easily found (Exercise 8.1).

The definition of contact equivalence allows us to identify branching problems that differ only by a change of coordinates. The second step toward a classification is to investigate how many perturbations are required to "unfold" a branching phenomenon.

Definition 8.2. An unfolding of $f(y,\lambda)$ is an m-parameter family

$$U(y,\lambda,\Lambda_1, \ldots, \Lambda_m)$$

with

$$U(y,\lambda,0, \ldots, 0) = f(y,\lambda).$$

Here $\Lambda_1, \ldots, \Lambda_m$ are m additional parameters. A simple example is furnished by Eq. (2.21): The expression

$$y^3 + \lambda y + \Lambda_1$$

unfolds

$$y^3 + \lambda y$$

($m = 1$). This unfolding is illustrated in Figure 2.36. Attaching an additional term with parameter Λ_2 is also an unfolding; hence, the above meaning of unfolding is still too general to be useful for a classification. This leads to another definition:

Definition 8.3. An universal unfolding is an unfolding with the two features:

(1) *It includes all possible small perturbations of $f(y,\lambda)$ (up to contact equivalence).*
(2) *It uses the minimum number of parameters $\Lambda_1, \ldots, \Lambda_m$.*

This number m is called the codimension of $f(y,\lambda)$.

Example 8.2.
The equation

$$f(y,\lambda) = y^3 - \lambda = 0 \tag{8.1}$$

possesses a singularity at $(y,\lambda) = (0,0)$ (see Figure 8.1a). Up to contact equivalence, there are two possible qualitative changes, (see Figure 8.1b and c). Both situations (b) and (c) are characterized by one equation

$$y^3 - \gamma y - \lambda = 0$$

with two signs of γ. The unfolding

$$U(y,\lambda,\Lambda_1) = y^3 + \Lambda_1^2 y - \lambda$$

is not universal because it does not allow for both signs of γ; perturbation (c) is ruled out, and hence feature (1) is not satisfied.

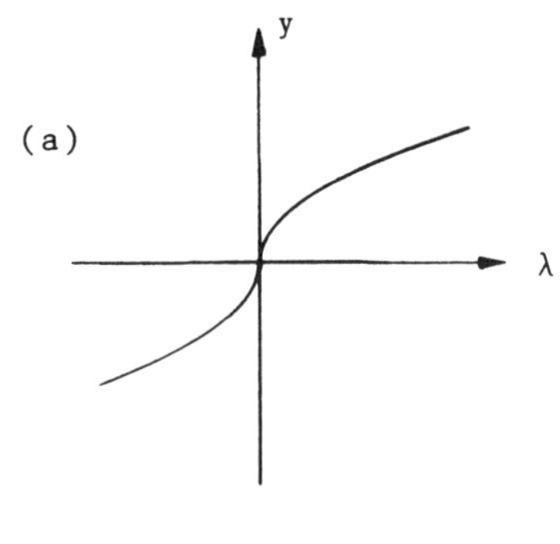

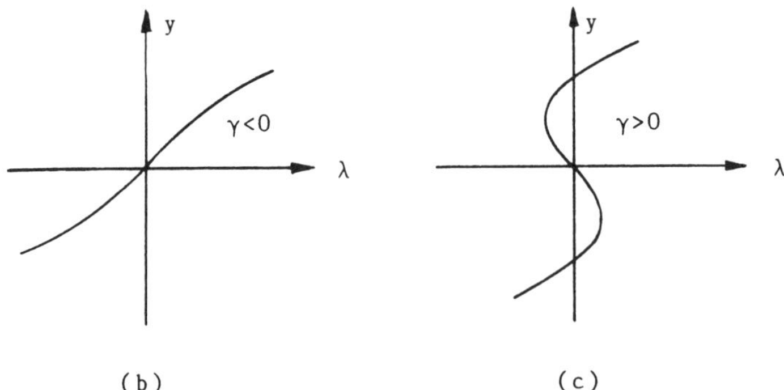

Figure 8.1

Also, the unfolding

$$U(y,\lambda,\Lambda_1,\Lambda_2) = y^3 - \Lambda_1 y^2 - \Lambda_2 y - \lambda$$

is not universal, since the two parameters Λ_1,Λ_2 are not the minimum; feature (2) is violated. Apparently the universal unfolding is given by

$$U(y,\lambda,\Lambda_1) = y^3 - \Lambda_1 y - \lambda,$$

allowing for all perturbations and requiring only $m = 1$ unfolding parameter—that is, the codimension of (8.1) is one.

As a sample of the type of assertions provided by singularity theory, we give a special case of a theorem in [112] (see also [113, 25]):

Theorem 8.1. Suppose that $f(y,\lambda)$ has a singular point $(y,\lambda) = (0,0)$ that obeys

$$f(0,0) = 0$$

$$\frac{\partial^i f(0,0)}{\partial y^i} = 0 \qquad \textit{for } i = 1, 2, \ldots, j$$

$$\xi \overset{\text{def}}{=} \frac{\partial^{j+1} f(0,0)}{\partial y^{j+1}} \frac{\partial f(0,0)}{\partial \lambda} \neq 0.$$

Then, locally, $f(y,\lambda)$ is contact equivalent to

$$y^{j+1} + \lambda \qquad \textit{in case } \xi > 0$$

$$y^{j+1} - \lambda \qquad \textit{in case } \xi < 0.$$

The universal unfolding is

$$U(y,\lambda,\Lambda) = y^{j+1} - \Lambda_{j-1} y^{j-1} - \cdots - \Lambda_1 y \mp \lambda.$$

The theorem states that bifurcation problems can be represented by polynomials, which can be seen as *normal forms.* The application is as follows. Calculate the order j of a singularity. The higher the integer j, the more degenerate a singularity is. The singularity of highest order is the solution (y,λ) at which the maximum number of derivatives of f with respect to y vanish. After j is determined, check the sign of ξ. In case $\xi \neq 0$, $j + 1$ different solutions can be found locally. In the above example in Eq. (8.1), $(y,\lambda) = (0,0)$ is the only singular point; we find $j = 2$, $\xi = -6$. As illustrated in Figure 8.1, there are parameter combinations for which $j + 1 = 3$ different solutions y exist.

Generalizations of the above theorem are established (see reference [110], p. 453, which includes a table of related results). For instance, in case $f = f_y = f_{yy} = f_\lambda = 0$, $f_{yyy}f_{y\lambda} \neq 0$, the normal form is

$$y^3 \pm \lambda y.$$

Theorems of this kind focus on the "worst case"—that is, the solution with a singularity of maximal order. Turning points are the least degenerate singular points ($f = f_y = 0$, $f_{yy} \neq 0$, $f_\lambda \neq 0$). Bifurcation points ($f_\lambda = 0$) and hysteresis points ($f_{yy} = 0$) are singularities of higher order. In what follows, the parameter vector $\mathbf{\Lambda}$ stands for the "given" problem-inherent control parameters instead of unfolding parameters. Suppose the singularity of highest order takes place for the specific parameter vector $\mathbf{\Lambda}^0$. Then for some parameters $\mathbf{\Lambda}$ close to $\mathbf{\Lambda}^0$, lower-order singularities arise. The lower-order singularities are organized by the singularity of highest order, which prompts the name "organizing center." It is tempting to calculate organizing centers beforehand and then to fill the parameter space with other solutions. This works in a few scalar examples, but in practice it is usually illusive. First, because higher-order derivatives of f can be evaluated in only the simplest examples, a direct calculation of high-order organizing centers is too laborious. Second, even if an organizing center is calculated, its actual structure becomes apparent only *after* global results are calculated. Because the entire formation, not only the location of the organizing center, is of interest, the approximation of the manifolds defined by $f(y,\lambda,\mathbf{\Lambda}) = 0$ will remain a subject for further research (see Section 5.7).

In order to obtain a global picture, the parameter space can be subdivided in a systematic way, and samples can be taken. This is best described for the scalar equation

$$f(y,\lambda,\Lambda_1, \ldots, \Lambda_m) = 0. \tag{8.2}$$

In singularity theory, λ is seen as a distinguished branching parameter that is not mixed with the other m parameters $\Lambda_1, \ldots, \Lambda_m$. Accordingly, we interpret the parameter space here as being m-dimensional. (The advantage of this will become apparent later.) Singularities of Eq. (8.2) are given by

$$\frac{\partial f(y,\lambda,\mathbf{\Lambda})}{\partial y} = 0.$$

This equation and Eq. (8.2) define an m-dimensional surface in the $(y,\lambda,\mathbf{\Lambda})$-space, called the *singular set*. The projection of the singular set onto the $(\lambda,\mathbf{\Lambda})$-space is the branching set or catastrophe set. For instance, in the equation

$$f(y,\lambda,\alpha,\beta) = y - \alpha - 2\lambda \sin(y) + \beta \cos(y) = 0 \qquad (8.3)$$

of Exercise 2.12, we have $m = 2$ ($\mathbf{\Lambda}_1 = \alpha$, $\Lambda_2 = \beta$), and the parameters for which there are singular points form a two-dimensional surface in the (λ,α,β)-space. This branching set is illustrated in Figure 8.2.

For some purposes, the branching set is too large to distinguish between different branching phenomena. The branching set includes all parameters $(\lambda,\mathbf{\Lambda})$ for which turning points arise, and these least degenerate singular points are abundant. It is a question on a "higher" level to ask for higher-order singularities, which separate different kinds of branching diagrams. For example, a hysteresis point (Figure 8.1a) stands between a region without a turning point (Figure 8.1b) and a region with two turning points (Figure 8.1c). Similarly, bifurcation points separate different branching diagrams (see Figure 2.39). Higher-order singularities of the scalar equation in Eq. (8.2) are defined in what follows.

Definition 8.4. A hysteresis variety V_{HY} *is defined by the three equations*

$$\begin{aligned} f(y,\lambda,\mathbf{\Lambda}) &= 0 \\ f_y(y,\lambda,\mathbf{\Lambda}) &= 0 \\ f_{yy}(y,\lambda,\mathbf{\Lambda}) &= 0. \end{aligned} \qquad (8.4)$$

After eliminating y and λ (not always possible explicitly), there remains one equation in $\mathbf{\Lambda}$, defining a $(m - 1)$-dimensional hypersurface in the $\mathbf{\Lambda}$-space. Obviously, every hysteresis point satisfies Eq. (8.4). Accordingly, the hypersurface defined by Eq. (8.4) separates parameter combinations $\mathbf{\Lambda}$ with branching diagrams like that in Figure 8.1b from

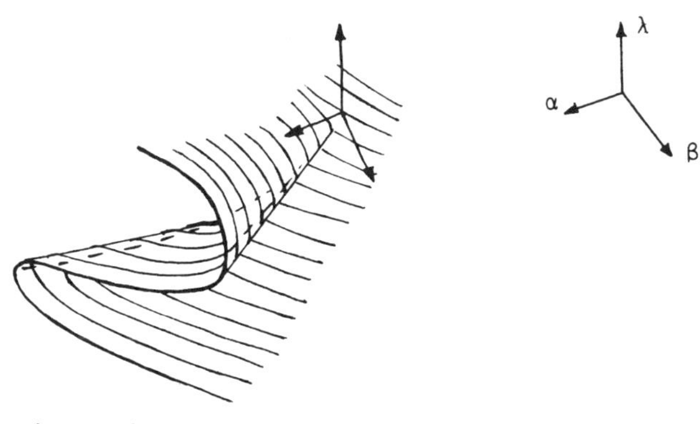

Figure 8.2

parameter combinations $\mathbf{\Lambda}'$ with branching diagrams like that in Figure 8.1c.

In Eq. (8.3), the two additional equations defining a hysteresis variety are

$$f_y = 1 - 2\lambda \cos(y) - \beta \sin(y) = 0$$

$$f_{yy} = 2\lambda \sin(y) - \beta \cos(y) = 0.$$

A straightforward analysis (Exercise 8.2) reveals that the hysteresis variety forms a curve (one-dimensional manifold) in the (α,β)-plane. This curve is given by

$$\beta = \sin \alpha.$$

Another hypersurface of higher-order singularities is defined by the following definition:

Definition 8.5. A bifurcation and isola variety V_{BI} *is defined by the three equations*

$$\begin{aligned} f(y,\lambda,\mathbf{\Lambda}) &= 0 \\ f_y(y,\lambda,\mathbf{\Lambda}) &= 0 \\ f_\lambda(y,\lambda,\mathbf{\Lambda}) &= 0. \end{aligned} \tag{8.5}$$

Both bifurcation points and isola centers solve Eq. (8.5). To see this, consider $u = f(y,\lambda)$ to be a surface in the three-dimensional (y,λ,u)-space. Solutions of $f(y,\lambda) = 0$ are given as the intersection of the surface with the plane $u = 0$, which in view of Eq. (8.5) are critical points of $u = f(y,\lambda)$—that is, the gradient vanishes. Bifurcation points correspond to saddles of u, and isola centers correspond to maxima or minima. (The discriminant of the second-order derivatives decides which case occurs; see calculus textbooks. This can be seen as a simple special case related to the more general analysis outlined in Section 5.4.1.) The varieties V_{BI} and V_{HY} are hypersurfaces separating the branching diagrams of the two right columns in Figure 2.39; in addition, they separate a closed branch from a situation with no solution locally. A further exception occurs when two turning points exist for the same parameter value. This hypersurface in the Λ-space is defined by this definition:

Definition 8.6. A double limit variety V_{DL} *satisfies the four equations*

$$\begin{aligned} f(y_1,\lambda,\mathbf{\Lambda}) &= 0 \\ f(y_2,\lambda,\mathbf{\Lambda}) &= 0, \quad y_1 \neq y_2 \\ f_y(y_1,\lambda,\mathbf{\Lambda}) &= 0 \\ f_y(y_2,\lambda,\mathbf{\Lambda}) &= 0. \end{aligned} \tag{8.6}$$

In Eq. (8.3), the curves of V_{HY} and V_{BI} separate different regions in the (α,β)-parameter plane (see Figure 8.3). Five insets in Figure 8.3 illustrate sample branching diagrams that are characteristic of the three regions and the two separating curves. Each branching diagram depicts y versus λ. For example, for any parameter combination in the shaded region, the branching diagram has three turning points. If we move a sample parameter combination $\mathbf{\Lambda} = (\alpha,\beta)$ toward the bifurcation and isola variety V_{BI}, two of the turning points approach each other. When $\mathbf{\Lambda}$ reaches and crosses V_{BI}, the two turning points collapse into a bifurcation point and detach immediately, as indicated by two further sample branching diagrams in Figure 8.3. On the other hand, if we start with Λ again in the shaded region and move toward the hysteresis variety V_{HY}, the two turning points that form a hysteresis approach each other and join. After V_{HY} is crossed, the hysteresis has disappeared. This is illustrated by the two remaining sample branching diagrams in Figure 8.3. For any point (α,β) in Figure 8.3, there is at least one singularity. Figure 8.2 ilustrates the two-dimensional branching set in the (λ,α,β)-space that represents turning points. The dashed lines in Figure 8.2 correspond to V_{HY} and V_{BI}. The collapsing of turning points at the dashed lines of higher-order singularities becomes apparent.

Diagrams like that in Figure 8.3 are used frequently to summarize the variety of nonlinear phenomena. These multidiagrams are especially useful in case $m = 2$, making possible a two-dimensional representation of the branching behavior in the higher-dimensional $(y,\lambda,\mathbf{\Lambda})$-space. Modifications

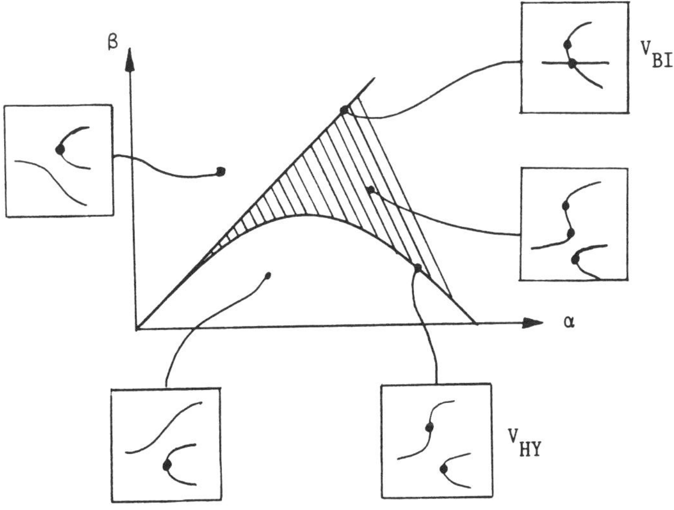

Figure 8.3

are used frequently. For instance, with $m = 1$ and $n = 2$ it is instructive to draw phase diagrams in insets surrounding a (λ,Λ_1)-parameter chart.

The hypersurfaces V_{HY}, V_{BI}, V_{DL} separate the $\boldsymbol{\Lambda}$-parameter space into connected components (Figure 8.3). Let us denote the interior of any such connected component by C (for example, the shaded area in Figure 8.3, excluding V_{BI} and V_{HY}). For each of these subsets C of the parameter space $\boldsymbol{\Lambda}$, the following holds:

Theorem 8.2. Suppose the two parameter vectors $\boldsymbol{\Lambda}$ *and* $\boldsymbol{\Lambda}'$ *are in the same region C. Then* $f(y,\lambda,\boldsymbol{\Lambda})$ *and* $f(y,\lambda,\boldsymbol{\Lambda}')$ *are contact equivalent.*

This statement justifies choosing only one sample $\boldsymbol{\Lambda}$ in each C for the purpose of illustrating characteristic branching behavior. The resulting branching diagram is representative of the entire class C and stable under variations of $\boldsymbol{\Lambda}$ that remain inside C.

The above remarks illuminate what was introduced in Section 2.9 under the name "generic branching." Inside any component C there are only turning points, which are stable and ubiquitous. Hopf bifurcations cannot occur in our scalar setting; for $n \geq 2$ they are generic too. Higher-order singularities are restricted to the boundaries of the components C; these hypersurfaces (curves in Figure 8.3) may be crossed or touched when $\boldsymbol{\Lambda}$ is varied. The higher the order of a singularity, the more unlikely it is to pass the singularity. There is a hierarchy of singularities. Singularities of a higher order are escorted on their flanks by singularities of a lower order. The least degenerate turning points may coalesce in singularities of the next higher order—namely, bifurcation and hysteresis points. On their part, the hypersurfaces V_{HY} or V_{BI} may collapse in singularities of an even higher order ($\alpha = \beta = 0$ in Figure 8.3). In the following section we shall find out what high-order singularities look like.

Exercise 8.1.
Verify the criterion for contact equivalence for Example 8.1.

Exercise 8.2.
Calculate the hysteresis variety and the bifurcation and isola variety of Eq. (8.3). Is there a double limit variety? Compare your results with Figure 8.3 and Exercise 2.12.

Exercise 8.3.
Consider Eq. (2.21). What are V_{HY} and V_{BI}? Draw a sketch that corresponds to Figure 8.3 ($\Lambda_1 = \gamma$).

8.3 THE ELEMENTARY CATASTROPHES

Catastrophe theory has found a remarkable resonance among engineers and scientists. The popular work of Thom [343] and Zeeman [371] has both offered explanations to various phenomena and provoked a controversy about its value [17, 122]. For textbooks on catastrophe theory other than those mentioned, refer to [105, 262]. The aims of catastrophe theory resemble those of singularity theory in that qualitative explanations are given together with a classification of some of the nonlinear phenomena. A central role in catastrophe theory is played by the *elementary catastrophes,* which correspond to the singularities of ascending order mentioned in the previous section. Because catastrophe theory does not distinguish between a branching parameter λ and other control parameters assembled in $\mathbf{\Lambda}$, we use new notations to minimize confusion. Throughout this section, the parameters will be γ, δ, ζ, and ξ. The state variable y is a scalar again; it may be a scalar representative of more complex variables. In what follows, the elementary catastrophes are ordered according to the number of involved parameters.

8.3.1 The Fold (Requires One Parameter, γ)

The *fold* is the name of a manifold, the simplest representative of which is defined by

$$0 = y^2 + \gamma. \tag{8.7}$$

This equation is a parabola in the two-dimensional (y,γ)-space (see Figure 8.4). The singularity at $(y_0,\gamma_0) = (0,0)$ is a turning point. Because there are two solutions for $\gamma < \gamma_0$ and no solution for $\gamma > \gamma_0$,

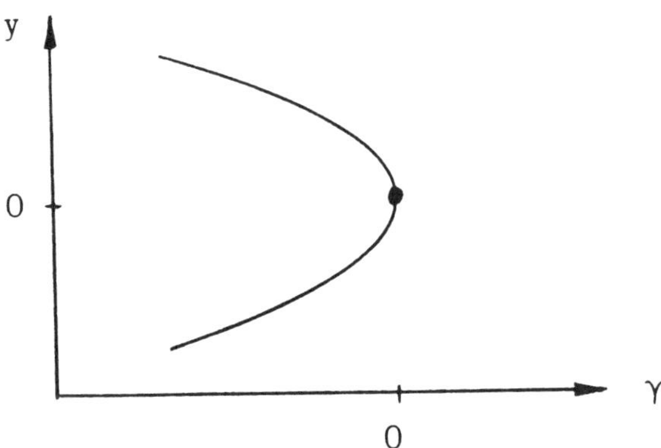

Figure 8.4

the multiplicity changes at γ_0. This point γ_0 establishes the branching set of the parameter "space." For later reference, we point out that the branching set of a fold is zero-dimensional in the one-dimensional parameter space. The least degenerate singularity represented by a fold is well known to us from the examples that exhibit turning points.

8.3.2 The Cusp (Requires Two Parameters, δ, ζ)

The model equation that represents a *cusp* is defined by

$$0 = y^3 + \zeta y + \delta. \tag{8.8}$$

This example was discussed in detail in Section 2.9. The manifold of a cusp is illustrated in Figure 8.5. The branching set given by

$$4\zeta^3 + 27\delta^2 = 0$$

(cf. Eq. 2.22) bounds a region in the parameter space (shaded in Figure 8.5). Equation (8.8) has three solutions for parameter values inside the hatched area, and one solution for outside parameter values. The branching set is one-dimensional in the two-dimensional parameter space; the cusp manifold is a two-dimensional surface in the (y,ζ,δ)-space. The singularity of highest order occurs at $(\zeta,\delta) = (0,0)$. Cutting

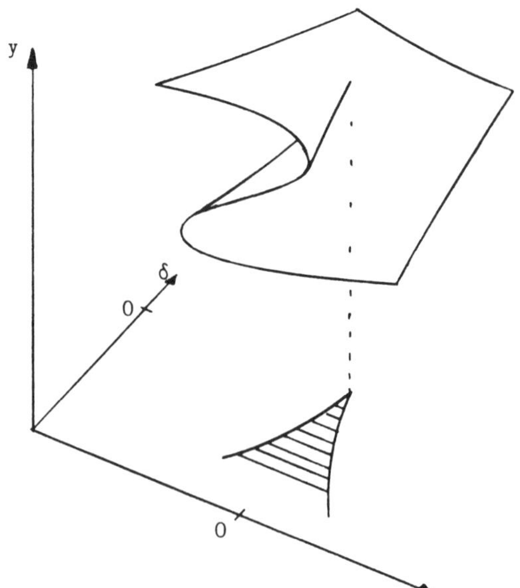

Figure 8.5

one-dimensional slices into the parameter plane (vertical planes in Figure 8.5), one obtains folds. These folds escort the cusp center—that is, singularities of low order form the flanks of the singularity of higher order. Cross sections of the cusp are depicted in Figure 2.36 and Figure 2.38. On one "side" of the cusp there is hysteresis. Another example of a cusp occurs in the model of a catalytic reaction (see Section 6.2.2 and Figures 5.24 and 5.25).

8.3.3 The Swallowtail (Requires Three Parameters, γ, δ, ζ)

The *swallowtail* is described by

$$0 = y^4 + \gamma y^2 + \delta y + \zeta. \tag{8.9}$$

This equation defines a three-dimensional manifold in the four-dimensional (y,γ,δ,ζ)-space. In order to obtain the branching set, differentiate Eq. (8.9) with respect to y,

$$0 = 4y^3 + 2\gamma y + \delta.$$

This equation and Eq. (8.9) define a surface in the (γ,δ,ζ)-space. For a parameterization of the branching set, two parameters ρ and σ are needed; we choose $\rho = y$ and $\sigma = \gamma$ and obtain

$$\begin{aligned} \gamma &= \sigma \\ \delta &= -4\rho^3 - 2\sigma\rho \\ \zeta &= 3\rho^4 + \sigma\rho^2. \end{aligned} \tag{8.10}$$

To visualize the branching set given by Eq. (8.10), we take cross sections. For instance, discussing the relation between γ and ζ for the three cases $\sigma = 0$, $\sigma > 0$, and $\sigma < 0$ provides enough information to draw an illustration of the branching set (Figure 8.6; see Exercise 8.4). The heavy curve is a cross section of the branching set with a plane defined by $\sigma = \gamma =$ constant, $\gamma < 0$. The branching set of Figure 8.6 separates the (γ,δ,ζ)-space into three parts. A core part for $\gamma < 0$ extends to the curves 1, 2, and 3, reminding one of a tent-like structure. The two other parts are "above" and "below" the manifold sketched in Figure 8.6. In the center of the swallowtail, the three curves 1, 2, and 3 collapse. This center is the singularity of highest order, organizing the hierarchy of lower-order singularities. Curve 1 is a curve of self-intersection, curves 2 and 3 are formed by cusp ridges, which are accompanied by the lowest-order singularities (Exercise 8.4).

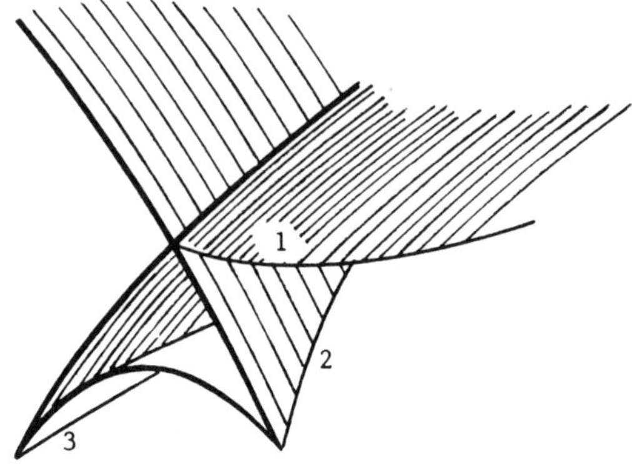

Figure 8.6

The normal form of a *butterfly* is given by

$$0 = y^5 + \gamma y^3 + \delta y^2 + \zeta y + \xi. \tag{8.11}$$

Because its branching set extends in the four-dimensional $(\gamma,\delta,\zeta,\xi)$-space, an illustration of a butterfly is complicated. We show typical slices, holding two parameters fixed (γ and δ). Then the cross sections are (ζ,ξ)-planes, drawn in the insets in Figure 8.7—that is, Figure 8.7

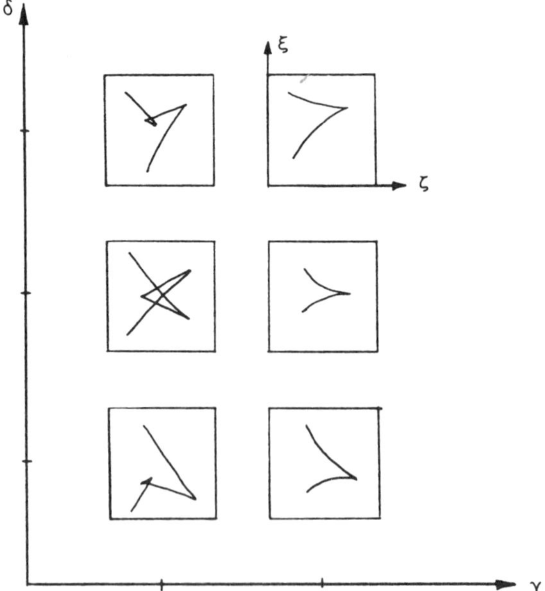

Figure 8.7

represents a two-dimensional family of two-dimensional cross sections [262].

There is another group of elementary catastrophes—the *umbilics*. These are described by two scalar equations involving either three or four parameters. For details and illustrations, see the special literature.

We close this section with a disillusioning remark: The higher the order of a singularity, the more unlikely we are to discover the singularity in a practical parameter study, and the more costly any attempt to calculate it.

Exercise 8.4.
In Eq. (8.10), assume three fixed values of σ ($\sigma > 0$, $\sigma = 0$, $\sigma < 0$) and discuss the curves in the (δ,ζ)-plane. Draw a diagram for each case. Try to find your curve for $\sigma < 0$ in Figure 8.6. Which parts of the surface in Figure 8.6 are the loci of the lowest-order singularities?

Exercise 8.5.
Extend Figure 8.7 (that is, vary γ and δ further) to illustrate the birth and death of swallowtails.

8.4 ZEROTH-ORDER REACTION IN A CSTR

The example of this section follows [25]. A simple zeroth-order reaction in a continuous stirred tank reactor (CSTR) is described by the scalar equation

$$f(y,\lambda,\gamma,\beta) = y - \lambda \exp\left(\frac{\gamma\beta y}{1 + \beta y}\right) = 0. \tag{8.12}$$

The meaning of the variables is the same as with example Eq. (6.15). The feasible range for the concentration is $0 < y < 1$. In the spirit of singularity theory, λ is a distinguished branching parameter. Accordingly, we ask how (y,λ)-branching diagrams vary with the two parameters β and γ. The example is simple enough to evaluate singularities. The derivative

$$f_y = 1 - \lambda\gamma\beta(1 + \beta y)^{-2} \exp\left(\frac{\gamma\beta y}{1 + \beta y}\right) = 1 - \gamma\beta y(1 + \beta y)^{-2} \tag{8.13}$$

vanishes for

$$\gamma\beta y = (1 + \beta y)^2,$$

which establishes a quadratic equation in y. This gives y explicitly in terms of β and γ, $y = u(\beta,\gamma)$,

$$y_{1,2} = [\gamma - 2 \pm (\gamma(\gamma - 4))^{1/2}]/(2\beta) = u(\beta,\gamma).$$

Real solutions are given for $\gamma \geq 4$, which implies that there are no singularities and hence no branching phenomena for $0 < \gamma < 4$. Substituting $y_{1,2}$ for $\gamma \geq 4$ into the equation $f = 0$ yields the branching set in implicit form,

$$0 = u(\beta,\gamma) - \lambda \exp\left(\frac{\gamma\beta u(\beta,\gamma)}{1 + \beta u(\beta,\gamma)}\right). \tag{8.14}$$

Equation (8.14) involves only the three parameters; it defines a two-dimensional surface in the (λ,β,γ)-parameter space.

Next we investigate the subset formed by singularities of higher order. To this end, we shall discuss the varieties introduced in Section 8.2. Clearly $f_\lambda \neq 0$, since the exponential term in Eq. (8.12) cannot vanish. Hence, there is no bifurcation and isola variety V_{BI} in this example (cf. Eq. (8.5)). In order to find a hysteresis variety V_{HY} (cf. Eq. (8.4)), we must evaluate the second-order derivative with respect to y,

$$f_{yy} = -\gamma\beta(1 - \beta y)(1 + \beta y)^{-3},$$

which vanishes for

$$y = \frac{1}{\beta}.$$

Inserting this expression into the equation $f_y = 0$ gives

$$\gamma = 4.$$

This establishes V_{HY} as a straight line in the (β,γ)-plane. Because of $y < 1$, V_{HY} is defined only for $\beta > 1$. Corresponding values of λ are given by Eq. (8.12),

$$\lambda = y \cdot \exp\left(-\frac{\gamma\beta y}{1 + \beta y}\right) = \frac{1}{\beta}\exp(-2).$$

The loci of hysteresis points are now known; parameterized by β, the hysteresis points are given by

$$y = \frac{1}{\beta}, \quad \lambda = \frac{1}{\beta e^2}, \quad \gamma = 4 \qquad \text{for } \beta > 1. \tag{8.15}$$

The (λ,β,γ)-values of Eq. (8.15) form a subset of the branching set defined by Eq. (8.14). Differentiating f further shows

$$f_{yyy}f_{\lambda} \neq 0.$$

Hence, in this example, the hysteresis points in Eq. (8.15) are the singularities of highest order. We summarize the preliminary results in Figure 8.8. It can easily be shown that the curve in the inset branching diagram for $\gamma < 4$ must be monotonically increasing (Exercise 8.6).

The restriction $\beta > 1$ is essential only for the existence of the hysteresis points with $y < 1$. It is still unclear what happens for $\beta \leq 1$. The role of β in Eq. (8.12) is artificial; β can be removed by introducing new variables, but this would conceal the feasible range of y. What we can do is establish the curve in the (β,γ)-plane, for which the y-value of a singularity reaches unity. A straightforward calculation establishes the curve

$$\gamma = (1 + \beta)^2/\beta.$$

This curve is shown in Figure 8.9. Above that curve, only one of the turning points satisfies $y < 1$, both for $\beta > 1$ and $\beta \leq 1$. It remains to discuss the region between the curve and the straight line $\gamma = 4$. Here, for $\beta > 1$ both turning points satisfy $y < 1$, whereas for $\beta \leq 1$ none of the turning points satisfies $y < 1$. A characteristic branching behavior of this latter part resembles the behavior for $\gamma < 4$; accordingly, we

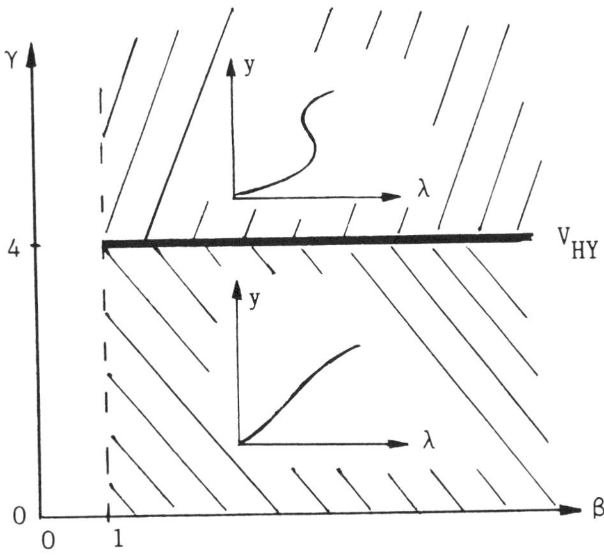

Figure 8.8

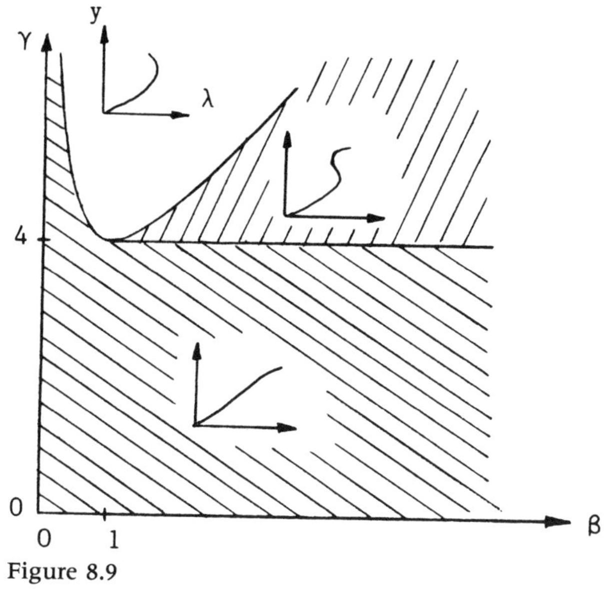

Figure 8.9

join these two components of the (β,γ)-plane. Hence, the (β,γ)-quadrant is subdivided into three regions, with branching diagrams ($y < 1$) as indicated in Figure 8.9.

Exercise 8.6.
Consider Eq. (8.12). Show that for $\gamma < 4$ the solution increases monotonically with λ.

8.5 CENTER MANIFOLDS

In Section 1.2 we investigated trajectories close to stationary solutions by means of linearizations. In nondegenerate cases, the linearized problem describes the flow correctly in the neighborhood of the equilibrium. Because most of the nonlinear phenomena have their origin in solutions that are characterized by a zero real part of an eigenvalue of the Jacobian, it is natural to ask whether there are analytical tools for analyzing the flow in the degenerate situation. The answer is yes, and the proper framework is provided by center manifold theory. A detailed study of how to apply center manifold theory is given in [55], and specific applications can be found, for instance, in [152, 298]. Troubles with linearization techniques and center manifold procedures are reported in [23].

Consider an autonomous system of ODEs, with a stationary solution assumed in the coordinate origin,

$$\dot{\mathbf{y}} = \mathbf{f}(\mathbf{y}), \quad \mathbf{f}(\mathbf{0}) = \mathbf{0}.$$

Suppose further that the system can be written as

$$\begin{aligned} \dot{\mathbf{y}}^- &= \mathbf{A}^-\mathbf{y}^- + \mathbf{f}^-(\mathbf{y}^-,\mathbf{y}^+,\mathbf{y}^0) \\ \dot{\mathbf{y}}^+ &= \mathbf{A}^+\mathbf{y}^+ + \mathbf{f}^+(\mathbf{y}^-,\mathbf{y}^+,\mathbf{y}^0) \\ \dot{\mathbf{y}}^0 &= \mathbf{A}^0\mathbf{y}^0 + \mathbf{f}^0(\mathbf{y}^-,\mathbf{y}^+,\mathbf{y}^0) \end{aligned} \tag{8.16}$$

with square matrices $\mathbf{A}^-$, $\mathbf{A}^+$, and $\mathbf{A}^0$ of orders i, j, and k, whose eigenvalues have negative, positive, and zero real parts, respectively. In Eq. (8.16) the n-vector $\mathbf{y}$ is partitioned into three subvectors of lengths i, j, and k such that $i + j + k = n$. Not all the dimensions i, j, or k need to be nonzero. The nonlinear functions $\mathbf{f}^-$, $\mathbf{f}^+$, and $\mathbf{f}^0$ and their first-order derivatives are assumed to vanish at the equilibrium $(\mathbf{y}^-,\mathbf{y}^+,\mathbf{y}^0) = \mathbf{0}$. Decomposing the original system $\dot{\mathbf{y}} = \mathbf{f}(\mathbf{y})$ as in Eq. (8.16) enables splitting off stable and unstable parts; in this way a stability analysis can be reduced to a subsystem for the subvector $\mathbf{y}^0$. Note that the setup of $\dot{\mathbf{y}} = \mathbf{f}(\mathbf{y})$ into the decomposition in Eq. (8.16) is not guaranteed to be practical.

To gain more insight into the structure of the flow of Eq. (8.16), one can use the concept of invariant manifolds. Recall that an invariant manifold M is composed of solution curves; a trajectory that starts on M, stays on it. Center manifold theory says that for Eq. (8.16) there are five types of invariant manifolds (locally) that pass $(\mathbf{y}^-,\mathbf{y}^+,\mathbf{y}^0) = \mathbf{0}$ —namely, the stable manifold M_S, the unstable manifold M_U, the center manifold M_C, the center-stable manifold M_{CS}, and the center-unstable manifold M_{CU} [183]. Not all these manifolds need to exist simultaneously. This is pictured by two sketches in Figure 8.10, both illustrating the three-dimensional situation $n = 3$. In Figure 8.10a we depict the situation $i = 2, j = 1, k = 0$—that is, there is a two-dimensional stable manifold M_S (here with a focus), a one-dimensional unstable manifold M_U, and no center manifold. The two manifolds M_S and M_U intersect in the stationary solution. Any flow line starting on the invariant manifold M_S remains there and spirals toward the equilibrium, although it is unstable. Each trajectory not starting exactly on M_S spirals toward M_U, leaving the neighborhood of the stationary solution. Figure 8.10b shows a center on a two-dimensional center manifold M_C ($k = 2$), intersected by a stable manifold M_S ($i = 1$); here the

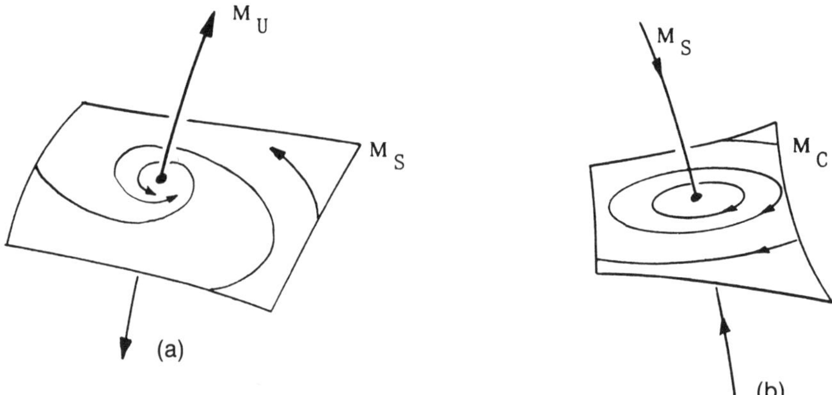

Figure 8.10

flow tends to M_C. In Figure 8.11 we illustrate an example $i = j = k = 1$. Here also a center-stable manifold M_{CS} (spanned by M_C and M_S) and a center-unstable manifold M_{CU} (spanned by M_C and M_U) are indicated.

To study the matter more closely, we simplify Eq. (8.16), assuming that the + components are absent ($j = 0$),

$$\begin{aligned} \dot{\mathbf{y}}^- &= \mathbf{A}^-\mathbf{y}^- + \mathbf{f}^-(\mathbf{y}^-,\mathbf{y}^0) \\ \dot{\mathbf{y}}^0 &= \mathbf{A}^0\mathbf{y}^0 + \mathbf{f}^0(\mathbf{y}^-,\mathbf{y}^0). \end{aligned} \tag{8.17}$$

Such a scenario is depicted in Figure 8.10b. Assume also temporarily that $\mathbf{f}^-$ and $\mathbf{f}^0$ are zero. Then this linear problem has two invariant manifolds—the stable manifold $\mathbf{y}^0 = 0$ and the center manifold $\mathbf{y}^- = 0$.

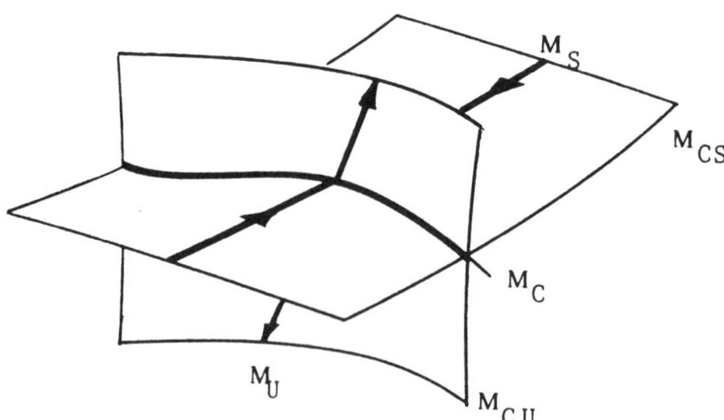

Figure 8.11

On the center manifold, the linear system is reduced to

$$\dot{\mathbf{y}}^0 = \mathbf{A}^0\mathbf{y}^0.$$

This k-dimensional equation on the center manifold determines whether $(\mathbf{y}^-,\mathbf{y}^0) = 0$ is attractive or not. The reduction also works in the nonlinear case in Eq. (8.17). Let the center manifold be described by a function $\mathbf{C}$ with

$$\mathbf{y}^- = \mathbf{C}(\mathbf{y}^0) \tag{8.18a}$$

$$\mathbf{C}(\mathbf{0}) = \mathbf{0}, \quad \mathbf{C}'(\mathbf{0}) = \mathbf{0}. \tag{8.18b}$$

Here $\mathbf{C}'(\mathbf{y}^0)$ denotes the rectangular matrix of the first-order partial derivatives of $\mathbf{C}$. The center manifold M_C satisfies Eq. (8.18b) because it passes the equilibrium tangentially to the center manifold $\mathbf{y}^- = 0$ of the linearized problem. Substituting Eq. (8.18a) into the second equation of Eq. (8.17) leads to the k-dimensional system

$$\dot{\mathbf{y}}^0 = \mathbf{A}^0\mathbf{y}^0 + \mathbf{f}^0(\mathbf{C}(\mathbf{y}^0),\mathbf{y}^0). \tag{8.19}$$

One of the main conclusions of center manifold theory states that the stability/instability of the equilibrium $\mathbf{y} = 0$ of Eq. (8.17) is reflected by the reduced system in Eq. (8.19).

In order to benefit from the above *reduction principle,* we need the center manifold $\mathbf{C}(\mathbf{y}^0)$, or at least an approximation. Substituting Eq. (8.18) into the first equation in Eq. (8.17) yields

$$\mathbf{C}'(\mathbf{y}^0)\dot{\mathbf{y}}^0 = \mathbf{A}^-\mathbf{C}(\mathbf{y}^0) + \mathbf{f}^-(\mathbf{C}(\mathbf{y}^0),\mathbf{y}^0).$$

Because $\dot{\mathbf{y}}^0$ can also be replaced using Eq. (8.19), the function $\mathbf{C}(\mathbf{y}^0)$ solves the differential equation

$$\mathbf{C}'(\mathbf{y}^0)[\mathbf{A}^0\mathbf{y}^0 + \mathbf{f}^0(\mathbf{C}(\mathbf{y}^0),\mathbf{y}^0)] - \mathbf{A}^-\mathbf{C}(\mathbf{y}^0) - \mathbf{f}^-(\mathbf{C}(\mathbf{y}^0),\mathbf{y}^0) = \mathbf{0}. \tag{8.20}$$

The solution to Eq. (8.20) is illusive, since it is equivalent to solving Eq. (8.17). However, an approximation $\overline{\mathbf{C}}(\mathbf{y}^0)$ of $\mathbf{C}(\mathbf{y}^0)$ can be found. This is established by the following theorem ([55], p. 5):

Theorem 8.3. Let $\mathbf{y}^- = \overline{\mathbf{C}}(\mathbf{y}^0)$ *be a smooth mapping with* $\overline{\mathbf{C}}(0) = 0$, $\overline{\mathbf{C}}'(0) = 0$. *Suppose that* $\overline{\mathbf{C}}(\mathbf{y}^0)$ *satisfies Eq. (8.20) to the order* $O(|\mathbf{y}^0|^q)$ *for* $q > 1$. *Then* $\overline{\mathbf{C}}(\mathbf{y}^0)$ *approximates the center manifold* $\mathbf{C}(\mathbf{y}^0)$ *with the same order* q.

Analog reduction and approximation principles hold for Eq. (8.16).

Example 8.3.
We apply the above ideas to the simple example

$$\begin{aligned}\dot{y}_1 &= -y_1 + y_2^2 + y_1 y_2^2 \\ \dot{y}_2 &= y_1 y_2 + y_2^3 + y_1^2 y_2.\end{aligned} \tag{8.21}$$

Because the Jacobian at the only equilibrium $\mathbf{y} = \mathbf{0}$ has zero as an eigenvalue, the principle of linearized stability does not apply. We try to obtain stability results by means of center manifold theory. The system in Eq. (8.21) is already written in the form of Eq. (8.17). Only the first equation has a linear term; hence, Eq. (8.21) corresponds to Eq. (8.17) as follows:

$$A^- = -1, \quad y_1 = y^-, \quad A^0 = 0, \quad y_2 = y^0,$$

$$f^-(y^-,y^0) = f_1(y_1,y_2) = (y_1 + 1)y_2^2,$$

$$f^0(y^-,y^0) = f_2(y_1,y_2) = y_2(y_1 + y_1^2 + y_2^2).$$

To approximate the center manifold $y_1 = C(y_2)$, we form Eq. (8.20),

$$C'(y_2)[y_2(C(y_2) + (C(y_2))^2 + y_2^2)] + C(y_2) - (C(y_2) + 1)y_2^2 = 0. \tag{8.22}$$

The simplest trial that satisfies Eq. (8.18b) is

$$\bar{C}(y_2) = \gamma y_2^2.$$

Substituting this expression into Eq. (8.22) yields

$$\gamma y_2^2 - y_2^2 + y_2^4(2\gamma^2 + \gamma) + 2\gamma^3 y_2^6 = 0.$$

For $\gamma = 1$ the differential equation Eq. (8.22) is satisfied to the order $O(|y_2|^4)$. Accordingly, by the above theorem, y_2^2 approximates the center manifold with the same order $q = 4$—that is,

$$C(y_2) = y_2^2 + O(y_2^4).$$

This approximation of C is substituted into the second equation in Eq. (8.21), leading to the scalar differential equation

$$\begin{aligned}\dot{u} &= u^2 u + u^3 + u^4 u + O(u^5) \\ &= 2u^3 + O(u^5).\end{aligned}$$

Because $u = 0$ is an unstable equilibrium of $\dot{u} = 2u^3$, the equilibrium $\mathbf{y} = \mathbf{0}$ of Eq. (8.21) is unstable.

9 Chaos

The oscillations we have met in previous chapters have been periodic. Periodicity reflects a high degree of regularity and order. Frequently, however, one encounters *irregular* oscillations, like those illustrated in Figure 9.1. Irregular oscillations are also called *chaotic, aperiodic,* or *erratic.* Irregularity does not imply that the oscillation is completely arbitrary. Notice that the amplitude in Figure 9.1 is bounded. In addition, the "frequency" seems to vary only slightly. These observations nurture the impression that erratic oscillation, or "chaos," may conceal an inherent structure. In this chapter, we briefly discuss how chaos arises and characterize the structure in chaos. Such questions are of immediate interest in many fields, especially because chaos is related to turbulence [292]. Some aspects of chaos have stimulated deep philosophical questions about determinism in nature, including the question of whether the outcome of an experiment is predictable.

Throughout this book, we have assumed that the equations modeling a physical situation are given. Calculations were based, for instance, on a known system of ODEs

$$\dot{\mathbf{y}} = \mathbf{f}(\mathbf{y},\lambda). \tag{9.1}$$

Aperiodic oscillations can arise not only from numerical integrations but also in measurements of experiments. It can be the case that only one scalar variable $y_1(t)$ is observable (Figure 9.1) and that the underlying equation is unknown. In the present chapter, we address this situation and accordingly do not require that $\mathbf{f}(\mathbf{y},\lambda)$ be known.

The field of chaos is developing rapidly. Illuminating surveys about fundamental results are presented in [84, 251, 291].

9.1 FLOWS AND ATTRACTORS

Consider the ODE system in Eq. (9.1), with $\mathbf{f}$ known or unknown. Every initial vector $\mathbf{z}$ defines a trajectory or flow line $\varphi(t;\mathbf{z})$. Flow lines

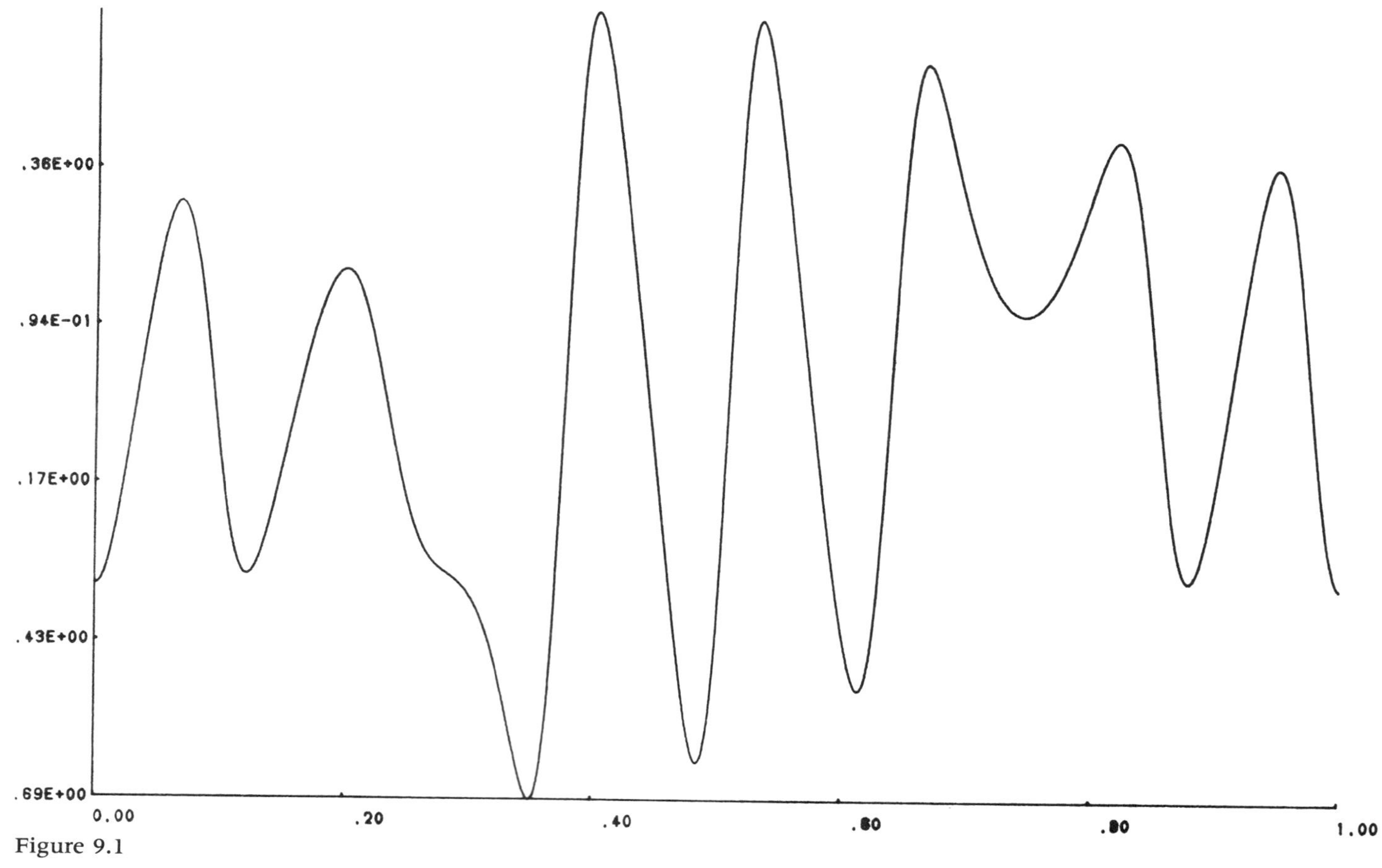

Figure 9.1

have the attributes

(1) $\varphi(0;\mathbf{z}) = \mathbf{z}$
(2) $\varphi(s;\varphi(t;\mathbf{z})) = \varphi(t + s;\mathbf{z})$.

Let $\mathbf{z}$ be variable in a set M_0. The corresponding flow lines can be pictured as densely filling part of the state space (cf. Figure 9.2). At time instant t_1, the original set M_0 has been moved or deformed to a set M_1. In what follows, we discuss how, for a fixed value of λ, a set M_0 is deformed by the flow governed by the ODE system in Eq. (9.1).

First, let the set M_0 consist of a single stationary point $\mathbf{y}^s$ of Eq. (9.1): $\mathbf{f}(\mathbf{y}^s,\lambda) = \mathbf{0}$. Then the flow line remains at this point,

$$\varphi(t;\mathbf{y}^s) = \mathbf{y}^s \text{ for all } t.$$

In other words, the stationary solution is a fixed point of the flow; the set M_0 is invariant under the flow. As a second example, consider the set M_0 consisting of a limit cycle of Eq. (9.1). Then, for any $\mathbf{z}$ in M_0, the flow line follows the limit cycle, and M_0 remains invariant under the flow:

$$\varphi(t;M_0) = M_0 \text{ for all } t.$$

Apart from these two exceptional cases, no flow line can intersect with itself. General sets M_0 are deformed by the flow.

Example 9.1. Damped Harmonic Oscillator $\ddot{u} + \dot{u} + u = 0$.
An associated first-order system is

$$\dot{y}_1 = -y_2$$
$$\dot{y}_2 = y_1 - y_2.$$

Because oscillations are damped and eventually decay to zero, any

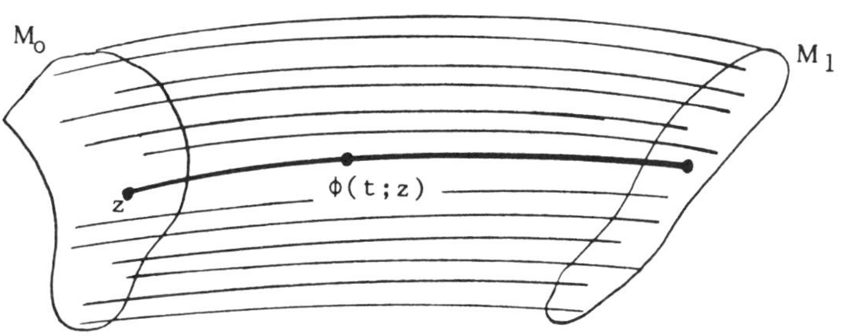

Figure 9.2

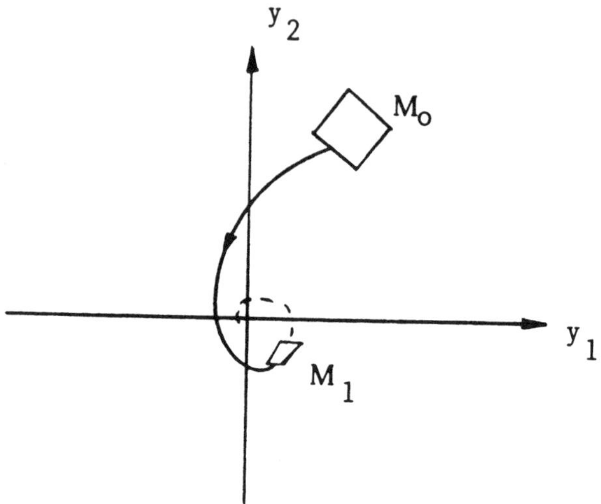

Figure 9.3

set M_0 of the two-dimensional (y_1,y_2)-space shrinks to zero,

$$M_\infty = \{0\}$$

(cf. Figure 9.3)—that is, the flow not only contracts volumes but even reduces the dimension (here to zero).

Another example for a contraction of volume is furnished by the van der Pol equation. As outlined in Section 1.3, the van der Pol equa-

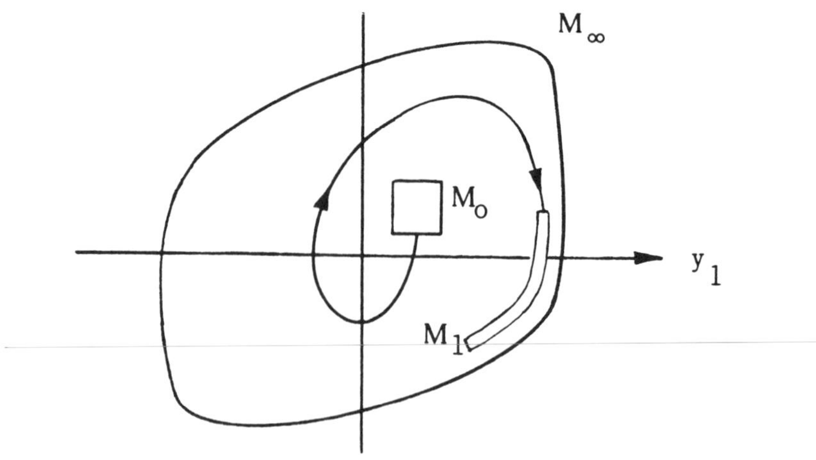

Figure 9.4

tion possesses an attracting closed orbit. Accordingly, we expect a deformation of an initial set M_0, as depicted schematically in Figure 9.4. The set M_0 is stretched in one dimension and contracted in the other dimension. For $t \rightarrow \infty$, the flow produces a set M_∞ identical to the limit cycle. As above, the flow of this example not only contracts volumes or areas but also reduces dimension. Because in the van der Pol example the final limit is a curve, the dimension is reduced from two to one. The limiting set is an attractor.

Next we imagine a flow that not only stretches into one dimension and contracts in another dimension, but also folds (see Figure 9.5). The question arises, Where does M finally arrive after an infinite number of stretchings and foldings? A related example, the "horseshoe map," was constructed by Smale [324]. Attractors of such flows look strange

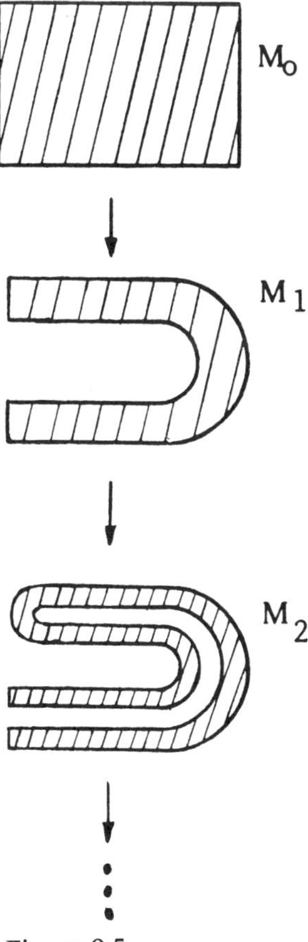

Figure 9.5

and cannot easily be compared to the elementary attractors formed by stable equilibria and limit cycles.

Attractors are formed by the sets to which trajectories tend for $t \to \infty$. Hence, attractors are closely related to stability; they describe the long-term behavior of physical systems. The invariant sets of attractors are generalizations of stable equilibria and limit cycles. In spite of the significance of attractors, there is no agreement on which definition of an attractor is most useful. Here we give a definition following [84]. (The reader may like to draw sketches of situations in which hypotheses (3) or (4) are not satisfied.)

Definition 9.1. A compact region A is an attractor of a flow $\varphi(t;\mathbf{z})$, if the following four hypotheses hold:

(1) *A is invariant under φ.*
(2) *A has a shrinking neighborhood.*
(3) *The flow is recurrent—that is, trajectories starting from any (open) subset of A hit that subset again and again for arbitrarily large values of t; the flow is nowhere transient.*
(4) *The flow cannot be decomposed—that is, A cannot be split into two nontrivial invariant parts.*

Attractors that evolve by the process of repeated stretching and folding (cf. Figure 9.5) have a peculiar feature that gives rise to the following definition:

Definition 9.2. An attractor is called a strange attractor if flow lines depend sensitively on the initial values.

In a strange attractor, initial points that are arbitrarily close to each other are macroscopically separated by the flow after sufficiently long time intervals. The sensitive dependence on initial conditions can be visualized by a turbulent flow of a fluid: Particles that are momentarily close to each other will not stay together. In terms of the ODE system in Eq. (9.1), compare two trajectories, one starting from $\mathbf{z}$, the other starting from the perturbed point $\mathbf{z} + \epsilon$. No matter how small $|\epsilon|$ is chosen, the two trajectories depart exponentially from each other. Strange attractors are aperiodic solutions.

The sensitive dependence on initial conditions has practical consequences. Because small deviations of initial conditions are always present, the position of a trajectory inside a strange attractor is not accurately predictable. The longer the time interval of a prediction, the less that can be said about the future state. This affects, for example, a long-term weather forecast. Integrating an ODE system that exhibits a strange attractor is to some extent a nondeterministic problem; limits of computability are reached. Note that the chaos represented by strange attractors is governed by deterministic laws, not by stochastic perturbations.

The contraction of volume is characteristic for dissipative systems. Conservative systems, which include the dynamic systems of classical mechanics [3], do not exhibit strange attractors. In conservative systems, other kinds of irregular or chaotic motions may appear [142, 215]. Here we focus on dissipative systems.

9.2 EXAMPLES OF STRANGE ATTRACTORS

Many examples exhibiting strange attractors have been found. Autonomous systems, Eq. (9.1), of ODEs must consist of at least three scalar equations ($n \geq 3$), otherwise no strange attractors exist. The most famous example of a strange attractor occurs for the Lorenz equations in Eq. (2.18). Duffing equations too are known to possess strange attractors [318, 345, 349]. Rössler has introduced the system

$$\begin{aligned} \dot{y}_1 &= -y_2 - y_3 \\ \dot{y}_2 &= y_1 + ay_2 \\ \dot{y}_3 &= b + y_1 y_3 - cy_3, \end{aligned}$$

which exhibits strange attractors, for instance, for $a = 0.55$, $b = 2$, $c = 4$ [282]. Other equations that show chaos are forced van der Pol oscillators [123].

Strange attractors have been discussed also for *maps*

$$\mathbf{y}(i + 1) = \mathbf{f}(\mathbf{y}(i)).$$

Recall the important class of Poincaré maps, which are generated by flows (Section 7.3). For maps, the restriction to dimensions greater than two does not hold; simple examples of dimension one or two can be studied. We mention the "logistic map" ($n = 1$)

$$y(i + 1) = \lambda y(i)(1 - y(i)), \tag{9.2}$$

which simulates growth of a population (think of y to represent the percentage of a population suffering from an epidemic; λ is then the infection rate). Equation (9.2) serves as an elementary example to illustrate period doubling; compare Eq. (7.13) and Exercise 7.5. Recall that we encountered period doublings at $\lambda = 3$ and $\lambda \approx 3.45$. For example, for $\lambda = 3.5$, there is an attractor that consists of four points

(Exercise 9.1). After further period doublings, chaos sets in for $\lambda \approx 3.57$. The map in Eq. (9.2) has been treated extensively in the literature [90, 121, 231, 251, 302]. Another popular map is the Hénon map,

$$\begin{aligned} y_1(i+1) &= 1 - a(y_1(i))^2 + y_2(i) \\ y_2(i+1) &= by_1(i) \end{aligned} \tag{9.3}$$

with, for instance, $a = 1.4$, $b = 0.3$ [143]. It is highly instructive to iterate this map (Exercises 9.2 and 9.3). Plots of the Hénon attractor and other strange attractors are included in [291]. The circle map (Exercise 7.11) exhibits chaos for $K \geq 1$.

In this section we study the strange attractors of the specific Duffing equation

$$\ddot{u} + \tfrac{1}{25}\dot{u} - \tfrac{1}{5}u + \tfrac{8}{15}u^3 = \tfrac{2}{5}\cos \omega t, \tag{9.4}$$

which was discussed in Sections 2.3 and 6.1. The flow of this equation in the $(u,\dot{u})$-plane is studied by means of the *stroboscopic map* [238] as follows. The state of the system is observed only at discrete time instances t_k, sampled at constant time intervals. The subharmonic driving force in Eq. (9.4) suggests "flashing the stroboscope" once per period,

$$t_k = k\frac{2\pi}{\omega}.$$

To calculate an attractor, choose initial values $y_1(0) = u(0)$, $y_2(0) = \dot{u}(0)$ and integrate the Duffing equation numerically. At the time instances t_k the integration is halted temporarily and the point $(y_1(t_k), y_2(t_k))$ is plotted. In this way, a sequence of points in the plane is obtained. Notice that the stroboscopic map is a special case of the Poincaré map. To see this, introduce a third variable for the frequency, $y_3 = \omega t$, with the differential equation $y_3' = \omega$. Now the system is autonomous with $n = 3$ scalar differential equations, the minimum number for chaos to occur in ODEs. Because y_3 can be seen as an angle, we interpret y_1, y_2, y_3 as cylindrical coordinates and imagine a phase plane rotating proportional to time (see Figure 9.6). Evaluating y_1 and y_2 at the discrete time instances t_k means to fix a (y_1,y_2)-plane Ω and measure the points in Ω where the trajectory intersects after each revolution. This evaluation of the *stroboscopic points* $(y_1(t_k), y_2(t_k))$ amounts to carrying out a Poincaré map. In Figure 9.6 a subharmonic orbit of order two is depicted, which has the double period compared with the reference period $2\pi/\omega$ of the excitation. A planar graphical illustration of a subharmonic attractor consists of two points, emphasized by the dots in Ω (Figure 9.6).

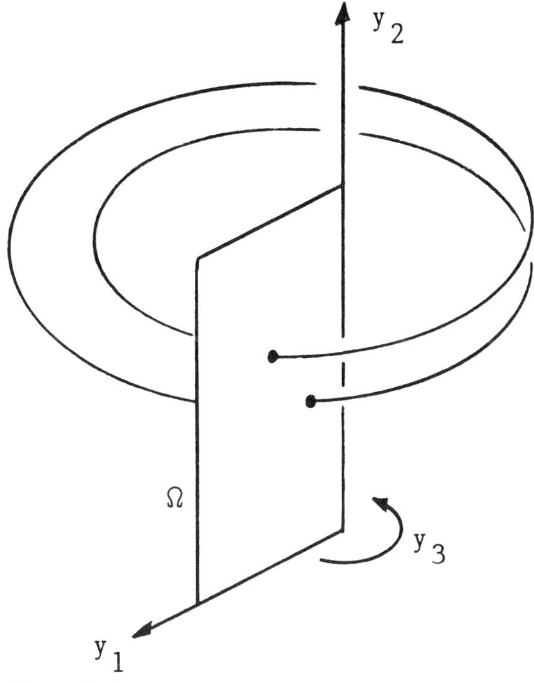

Figure 9.6

Figure 9.7 assembles 4,000 stroboscopic points that are generated by the specific trajectory, with $\omega = 0.2$ starting from (0,0). Apparently, the points indicate lines that are nested in a complicated structure. One can imagine that after calculation of an infinite number of points the gaps along the lines would be filled. Figure 9.7 illustrates a strange attractor or, rather, an approximation of a strange attractor. A numerical calculation of a strange attractor is never finished. Even the calculation of "only" 4,000 points was costly; the differential equation had to be integrated for the interval

$$0 \leq t \leq 125664.$$

Strange attractors for maps can be studied in a less expensive way (Exercise 9.3).

The impression that the "lines" in Figure 9.7 give is misleading. A strong enlargement of any of the "lines" reveals that they consist of a bunch of "sublines." Each magnification of sublines leads to the same observation—lines are resolved into more lines. Theoretically, there is an infinite number of lines, comparable to a *Cantor set*. Notice that the

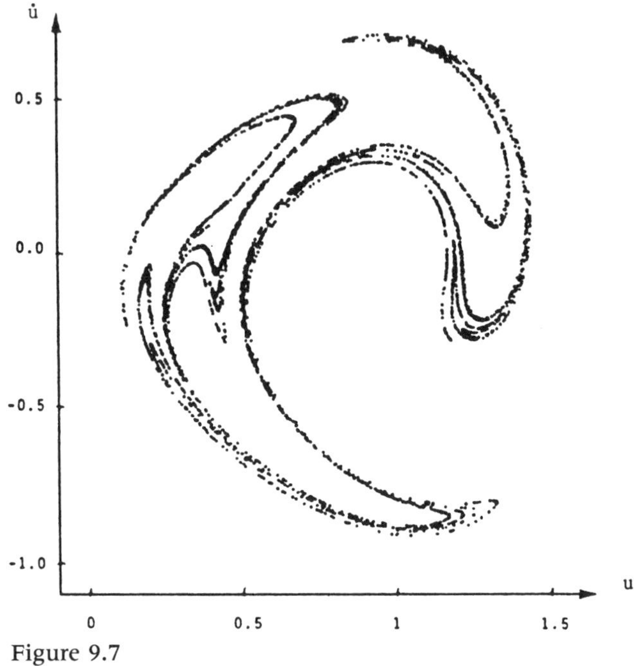

Figure 9.7

strange attractor illustrated by its Poincaré set in Figure 9.7 is actually a surface-type set in the three-dimensional space (cf. Figure 9.6). From the line structure indicated in Figure 9.7, we infer that this strange attractor has a leaved structure, with leaves densely folded and nested in a highly complex way.

A *Cantor set* can be constructed as follows: Take a straight line that fills the unit interval, and remove the inner third. Both the remaining short lines are treated in the same way, resulting together in four short intervals, each of length $\frac{1}{9}$ (cf. Figure 9.8). The process of removing thirds is repeated inifinitely many times; the remaining structure converges to the Cantor set. This way of constructing the Cantor set can be generalized to higher dimensions. The removed parts need not be just a third; similar structures are obtained when other portions are removed. Looking at Figure 9.8 from a large enough distance, the gaps are not visible and the structure resembles a dot. Now imagine that a cross section of one of the "lines" in Figure 9.7 is the fictitious "dot"

Figure 9.8

of a Cantor set. This explains how, with increasing magnification, points are resolved into more points, lines into more lines, leaves into more leaves. Only the outer leaves are visible. The structure repeats itself on finer and finer scales. This *self-similarity* becomes apparent with the Cantor set. For instance, each half of Figure 9.8 is similar to the first stage of the construction. "Self-similar" patterns are found frequently in the context of chaos (Exercise 9.3b). Theoretically, the fineness of the self-similar pattern reaches zero, but in practice uncertainties present in any system (noise, discretization errors) truncate Cantor-like sets, whole families of trajectories become identified, and the number of points (lines, leaves) is finite.

The order in which stroboscopic points are generated by trajectories is unpredictable. Recalculating Figure 9.7 with another accuracy or another integrator leads to a different arrangement of points along the same "lines" of the attractor. This reflects the attribute of strange attractors that trajectories, once inside the attractor, diverge from each other.

Observations on chaotic phenomena have been reported repeatedly—for example, investigations on nerve membranes [61], brain activity [24], biophysics (glycolysis) [227], Rayleigh-Bénard cells [34, 103], driven semiconductor oscillations [342], cardiac cells [124], and ecological systems [297]. Poincaré maps obtained experimentally are illustrated in [240]. Classical examples are reported in [84, 251, 302] and in the reprint selection [71].

Exercise 9.1.
Draw three parallel y-axes and illustrate the attractors of the map in Eq. (9.2) for the parameter values $\lambda = 2.7$, $\lambda = 3.1$, and $\lambda = 3.5$ (see Section 7.4.2 and Exercise 7.5).

Exercise 9.2.
Let $\mathbf{J}$ be the Jacobian of a map. A map is dissipative if $|\det \mathbf{J}| < 1$. For which values of the constants a and b is the Hénon map in Eq. (9.3) dissipative?

Exercise 9.3.
If you have access to a programmable computer with a plotting device:

(a) Iterate Eq. (9.3) as often as possible (according to how patient you are) and plot all iterates $\mathbf{y}(i)$ for $i > N$. The first N iterates are suppressed; they represent the transient initial phase until the iteration is trapped by the strange attractor. N depends on your choice of a starting vector $\mathbf{y}(0)$.

(b) If you have even more patience, it is instructive to enlarge a "line" of the attractor in order to see that the "line" consists of more lines. (The figures in [251] are based on more than 10,000 points.)

9.3 ROUTES TO CHAOS

Because our focus is on parameter-dependent equations, it is natural to ask how chaos depends on λ. In particular, we like to know for which values of λ one may expect chaotic behavior. As motivation, imagine that one wants to predict when a hitherto laminar flow becomes turbulent. Both theoretical and experimental investigations have revealed that there is no unique way in which chaos arises. Some of the different possible scenarios are discussed in this section. As we shall see, bifurcations play a crucial role in initiating chaos.

9.3.1 Route via Torus Bifurcation

An early conjecture attributed to Landau (1944, [207]) postulated that chaos is caused by an infinite sequence of secondary Hopf bifurcations. That is, after a first Hopf bifurcation from equilibrium to limit cycle a sequence of bifurcations into tori arises, each bifurcation adding another fundamental frequency. It suggests that, as more and more frequencies occur, the motion gets more and more turbulent. However, the Landau scenario has not been backed by experiments. Theoretically it was shown that the infinite sequence of Hopf bifurcations is not generic (see below).

Another route to chaos based on torus bifurcation was proposed by Newhouse, Ruelle, and Takens [246]; this scenario is backed both by theoretical investigations and by experimental evidence. According to this conjecture, only two bifurcations are precursive to chaos—one standard Hopf bifurcation and one subsequent bifurcation to a two-frequency torus. Then chaos is more likely than bifurcation to a 3-torus. When the third frequency is about to appear, simultaneously a strange attractor arises, since a three-frequency flow is destroyed by certain small perturbations. Bifurcation into a 3-torus is not generic [160]. The Ruelle-Takens-Newhouse route to chaos has been observed in Bénard-type experiments (see Section 2.7) and in the Taylor experiment [339]. The scenario is stable with respect to small external noise [69, 84].

9.3.2 Period Doubling Route

As noted in Section 7.4.2, there are models exhibiting a sequence of period doubling bifurcations. In case the sequence is infinite, the periods quickly approach infinity. The "speed" of how rapidly chaos is reached (in terms of λ) is given by the scaling law in Eq. (7.14). For instance, in the logistic map in Eq. (9.2), the entire cascade of period doublings takes place for $3 \le \lambda < 3.5699\ldots$; the latter value is the accumulation point of the period doubling sequence [90, 121]. Noise affects the sequence of period doublings in that higher-order subharmonics are truncated. The period doubling route to chaos has been supported by many experiments, including chemical reactions [165, 168, 325], driven nonlinear oscillations [342], and Rayleigh-Bénard cells [103]. We note in passing that inside the chaotic regime sequences of strange attractors were discovered which resemble a sequence of period doublings [68, 87]. Frequency divisions other than doubling were reported in [214]. Chaos may be interrupted by "windows" of regular behavior.

9.3.3 Intermittency

Pomeau and Manneville [258] proposed three mechanisms for the onset of chaos, which are related to intermittency. The term *intermittency* refers to oscillations that are periodic for certain time intervals, interrupted by bursts of aperiodic oscillations of finite duration. After the burst stops, a new periodic phase starts, and so on. Since there are three ways in which a stable periodic oscillation can lose its stability (Section 7.4), three kinds of intermittency are distinguished. The number of aperiodic bursts varies with λ; for λ entering the chaotic range, the duration of chaotic intervals increases. Intermittency has been observed, for instance, in Rayleigh-Bénard cells [34].

The above three routes to chaos are not the only scenarios. Recent investigations have revealed new routes to chaos [115, 169]. All the different hypotheses have in common that the onset of chaos is initiated by bifurcations. The different scenarios do not contradict each other; they may evolve concurrently in different regions of the phase space [84]. For the circle map, chaos is initiated by overlapping phase-locked intervals; here the ω values without resonance form a Cantor set.

9.4 CHARACTERIZATION OF STRANGE ATTRACTORS

Frequently one has calculated or measured some oscillation $y(t)$ and wants to decide whether the oscillation is periodic or aperiodic. A seemingly irregular oscillation might be periodic, with a period so large that it can hardly be assessed. For example, the Duffing equation in Eq. (9.4) possesses for $\omega = 0.38$ a subharmonic orbit of order 11, which has the period $T = 181.88$ and looks irregular. By merely inspecting a graphical output of a time history, it is often difficult to decide whether an oscillation is chaotic. Hence we require instruments that facilitate a decision. Among the tools that are helpful in this context are fractal dimensions, Liapunov exponents, and power spectra.

Before discussing these instruments, let us comment on the method of inspecting a Poincaré map. In Section 9.2, a two-dimensional Poincaré map was evaluated easily because two variables were immediately available. Frequently, in measurements, only one scalar variable is observable. In this case, a useful two-dimensional illustration of attractors can be constructed as follows: Suppose we are given a sequence of scalar measurements

$$y(t_0),\ y(t_0 + \Delta t),\ y(t_0 + 2\Delta t),\ y(t_0 + 3\Delta t),\ \ldots, \tag{9.5}$$

which is obtained by sampling a continuous output. The time lag Δt can be chosen freely but must be kept constant. A two-dimensional representation of an underlying attractor is obtained by plotting the pairs of consecutive numbers

$$(y(t_0 + j\Delta t),\ y(t_0 + (j + 1)\Delta t)), \qquad j = 0, 1, 2, \ldots. \tag{9.6}$$

This is a simplification of a general result obtained by Takens [340], which states that an m-vector $z(t)$, formed by m consecutive values

$$y(t),\ y(t + \Delta t),\ y(t + 2\Delta t),\ \ldots,$$

embeds the underlying flow. The dimension m is related to the dimension of the flow, which is usually unknown when the time series are derived from measurements. The set of pairs in Eq. (9.6) results from a projection of the embedding onto a plane.

9.4.1 Fractal Dimension

Throughout this book, we have met only integer dimensions. We know that a point is zero-dimensional, a curve is one-dimensional, and a surface is two-dimensional. Regular attractors of ODEs can be distinguished accordingly:

stationary point: dimension zero

limit cycle: dimension one

two-frequency torus ($n = 3$): dimension two.

Hence, following the route to chaos that is based on primary and secondary Hopf bifurcation, we encounter an increasing of the attractor's dimension,

$$0 \rightarrow 1 \rightarrow 2 \rightarrow ?$$

The question arises, What dimension has a resulting strange attractor? Since the volume of the strange attractor is zero, its dimension must be smaller than the dimension n of the state space. Consider, for instance, the Duffing equation in Eq. (9.4) (Figure 9.6) or the Lorenz equation in Eq. (2.18), which both have a three-dimensional state space. Here a zero volume of the strange attractor means that its dimension must be smaller than three. Hence, one may expect a noninteger dimension.

Fractal dimensions require refined concepts of dimension [222, 251]. A widely used definition of dimension is named after Hausdorff: Suppose that $N(\epsilon)$ is the number of small n-dimensional cubes of side length ϵ that are required to cover a region of the state space. Then the *fractal dimension* or *Hausdorff dimension* of this region is defined by

$$D = \lim_{\epsilon \rightarrow 0} \frac{\ln(N(\epsilon))}{-\ln \epsilon}. \tag{9.7}$$

To illustrate this concept of dimension, we calculate two simple examples.

Example 9.2. Unit Square.
To cover the unit square, we take small squares with side length ϵ. Then exactly

$$N(\epsilon) = \frac{1}{\epsilon^2}$$

of the small squares are required to cover the unit square. The dimension is now readily calculated,

$$D = \lim_{\epsilon \rightarrow 0} \frac{-2 \ln \epsilon}{-\ln \epsilon} = 2.$$

Here the Hausdorff dimension specializes to the standard integer dimension.

Example 9.3. Cantor Set.
In Figure 9.8, $N(\epsilon) = 4$ short lines are required to cover the structure, each of length $\epsilon = \frac{1}{9}$. In the j-th stage of the set that converges to the Cantor set,

$$N(\epsilon) = 2^j \text{ with } \epsilon = 3^{-j}$$

of these short lines are needed. Hence,

$$D = \lim_{j\to\infty} [\ln 2^j / \ln 3^j] = \frac{\ln 2}{\ln 3} = 0.6309 \ldots .$$

Fractal dimensions are characteristic for strange attractors. Some examples are listed in Table 9.1 (fractal dimensions, from [347]). The first column of Table 9.1 displays the integer dimension of the hosting state space; this number n is to be related to the fractal dimension D of the strange attractor. For instance, the Hénon attractor (Exercise 9.3) has a line structure, with lines densely folded in a Cantor set structure. The richness of the line structure leads to a dimension $D > 1$ ($D = 1$ for an "ordinary" line). On the other hand, the lines are sparse in the plane, thus $D < 2$; a nonzero volume (here nonzero area) would mean $D = 2$. Similarly, the Duffing strange attractor with its leaved structure appears "thicker" than an ordinary surface, but still has volume zero; hence, $2 < D < 3$. In higher-dimensional spaces, strange attractors are possible with higher dimensionality [251]. Table 9.1 shows for the Duffing equation in Eq. (9.4) how the complexity of a strange attractor can be measured by its fractal dimension. The related Poincaré sections are shown in Figure 9.9. Further figures that illustrate the growing com-

TABLE 9.1

n	Example	D
1	Cantor set	0.6309
2	Hénon map, Eq. (9.3)	1.265
3	Duffing eq., Eq. (9.4)	
	$\omega = 0.04$	2.167
	$\omega = 0.2$ (Figure 9.7)	2.381
	$\omega = 0.32$	2.526

plexity of the Duffing attractor for increasing ω can be found in [318]. Bear in mind that the actual attractor has a leaved structure; imagine that Figure 9.9 illustrates cuts through growing lettuce plants. An exciting application of fractal dimensions was discovered recently: Time series obtained from an electroencephalogram have revealed that brain activities are characterized by strange attractors [24]; jumps in the dimensionality are observed when an epileptic seizure occurs.

We close this introduction to fractal dimensions with the remark that, apart from the Hausdorff dimension Eq. (9.7), there are other concepts of fractal dimensions. An infinite sequence of dimensions can be defined, all of them being less or equal to the Hausdorff dimension [144]. For instance, the *information dimension* is defined by

$$\lim_{\epsilon\to 0} \left[\sum_{i=0}^{N(\epsilon)} p_i \ln(p_i)/\ln \epsilon \right] .$$

The notion of ϵ and $N(\epsilon)$ is the same as for the Hausdorff dimension in Eq. (9.7); p_i is the probability that a point of the attractor lies within the i-th cell.

9.4.2 Liapunov Exponents

As pointed out in Section 9.1, a strange attractor can be seen as the result of an infinite number of stretchings in one direction and contractions in another direction, combined with foldings. Let us look at how to measure a stretching or contraction in case of a one-dimensional map

$$y(i + 1) = f(y(i)).$$

The factor $\exp(L)$ measures an exponential stretching of a distance δ to a distance δ',

$$\delta' = \delta e^L.$$

Because in general the stretching varies with y, the exponent L depends on where the distance δ is measured on the y-axis,

$$| f(y + \delta) - f(y)| = | \delta | e^L.$$

Iterating the map N times means

$$| f^N(y + \delta) - f^N(y)| = | \delta | e^{NL},$$

which can be written

$$L = \frac{1}{N} \ln |(f^N(y + \delta) - f^N(y))/\delta |.$$

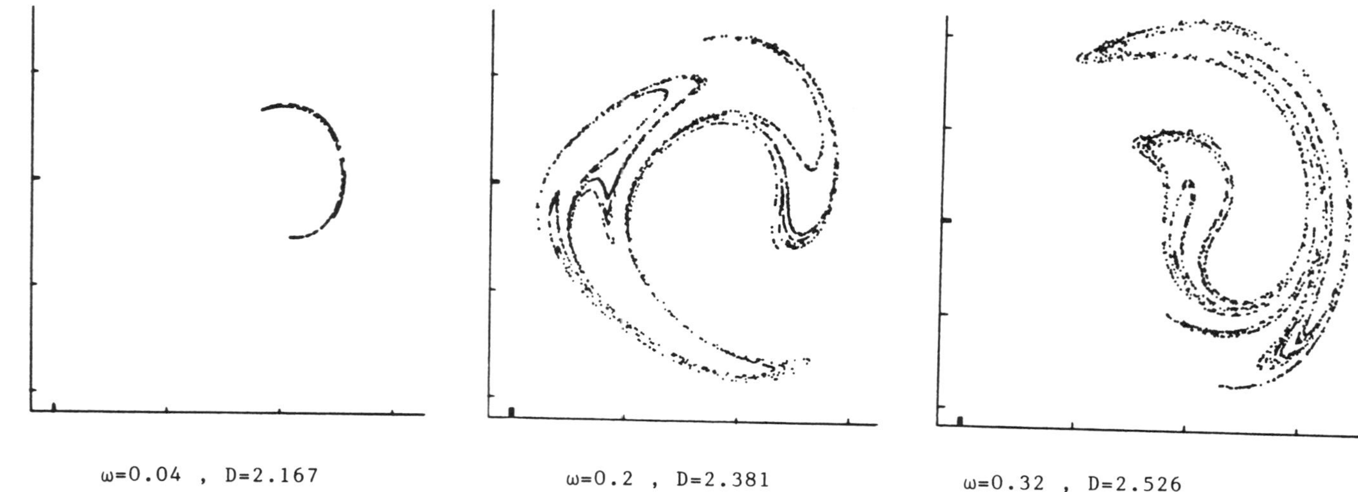
ω=0.04 , D=2.167
ω=0.2 , D=2.381
ω=0.32 , D=2.526
(axes are scaled by 0.5 ; zeros are accentuated)

Figure 9.9

This attempt to define a stretching measure L still suffers from being dependent on N and δ. Hence, we consider the stretching of an infinitesimal small distance $\delta \to 0$ after an infinite number of iterations $N \to \infty$. This leads to the definition

$$L(y) = \lim_{N\to\infty} \lim_{\delta\to 0} \frac{1}{N} \ln |(f^N(y+\delta) - f^N(y))/\delta| \tag{9.8}$$

which, by definition of the derivative via the limit of difference quotients, is equivalent to

$$L(y) = \lim_{N\to\infty} \frac{1}{N} \ln \left| \frac{df^N(y)}{dy} \right| . \tag{9.8$'$}$$

This number is called the *Liapunov exponent* of a map, it serves as a measure for exponential divergence ($L > 0$) or contraction ($L < 0$).

For systems of ODEs (n scalar equations), the derivation of a Liapunov exponent is similar [250]. Imagine a small test volume around the starting point of a flow line—for instance, take a ball with the radius ρ. The flow deforms the ball toward an ellipsoid-like object; for dissipative systems the volume eventually shrinks to zero. We want to know how the principle axes $\rho_i(t)$ $(i = 1, \ldots, n)$ evolve with time. Accordingly, a Liapunov exponent is defined by

$$L_i = \lim_{t\to\infty} \lim_{\rho\to 0} \frac{1}{t} \ln | \rho_i(t)/\rho | \tag{9.9}$$

(see the analogy to Eq. (9.8), $t \leftrightarrow N$, $\rho \leftrightarrow \delta$). The sum of all Liapunov exponents is the average volume contraction rate. Hence, for dissipative systems, the sum of all Liapunov exponents must be negative,

$$L_1 + L_2 + \cdots + L_n < 0.$$

Chaos is characterized by at least one positive Liapunov exponent, which reflects a stretching into one or more directions while the overall volume is shrinking. That is, after ordering the Liapunov exponents

$$L_1 > L_2 > \cdots > L_n,$$

a criterion for chaos is given by $L_1 > 0$.

Kaplan and Yorke [170] conjectured that a fractal dimension D is related to the Liapunov exponents L_i by the equation

$$D = j - \left(\sum_{i=1}^{j} L_i \right) \Big/ L_{j+1}. \tag{9.10}$$

In Eq. (9.10), j is the maximum index such that

$$L_1 + L_2 + \cdots + L_j \geq 0;$$

the Liapunov exponents are again assumed to be ordered. For chaotic flows such a j exists, because $L_1 > 0$. Since there is no exponential shrinking or growth along the trajectory, one of the Liapunov exponents vanishes. Hence, for a periodic attractor, $L_1 = 0$. In the special case of a strange attractor that arises for an ODE system with $n = 3$, L_3 must be negative, since $L_1 > 0$, $L_2 = 0$, and since the average volume contraction rate is negative. Consequently, $j = 2$, and Eq. (9.10) specializes to

$$D = 2 - L_1/L_3.$$

The conjecture in Eq. (9.10) is not yet proved, but it appears to be trustworthy. The relation offers a way to calculate a fractal dimension of a strange attractor. In ODE problems, it has been shown to be more convenient to calculate the Liapunov exponents than to count small boxes as required by Eq. (9.7). Accordingly, the fractal dimensions reported in Figure 9.9 (from [347]) were calculated via Liapunov exponents, exploiting the conjecture in Eq. (9.10).

It remains to discuss how to calculate the Liapunov exponents in Eq. (9.9). For the analytical background, see Section 7.2. Assuming that the initial test volume is infinitesimally small, its deformation by the flow is described by the linearization

$$\dot{\mathbf{h}} = \mathbf{f_y}\mathbf{h}.$$

Here the Jacobian is evaluated along the particular flow line investigated. For any initial deviation $\mathbf{h}(0)$, the evolution of the solution $\mathbf{h}(t)$ can be expressed by means of the fundamental solution matrix $\Phi(t)$ from Eq. (7.9),

$$\mathbf{h}(t) = \Phi(t)\mathbf{h}(0).$$

Hence, for an appropriately chosen $\mathbf{h}(0)$, the expression

$$\frac{|\,\Phi(t)\mathbf{h}(0)\,|}{|\,\mathbf{h}(0)\,|} \tag{9.11}$$

is a measure for

$$\lim_{\rho\to 0} |\,\rho_i(t)/\rho\,|$$

from Eq. (9.9). The coefficient Eq. (9.11) has been called the coefficient of expansion in the direction of $\mathbf{h}(0)$ [250]. In Eq. (9.11), $|\quad|$ must be understood as a vector norm. A numerical computation of Eq. (9.11) by integrating Eq. (7.9) may suffer from overflow because the linearization has at least one exponentially diverging solution. Some care must be taken in choosing $\mathbf{h}(0)$ and how to renormalize the fundamental matrix during the integration [321]. Multidimensional Liapunov exponents can be defined and calculated; their existence is established in [250]. Hints on the numerical realization and an application to the Lorenz equation are given in [321].

For the study of divergence of nearby trajectories, the entropy-like quantities

$$q_N = \frac{1}{N\Delta t} \sum_{i=1}^{N} \ln(|\mathbf{d}_i|/|\mathbf{d}|) \tag{9.12}$$

can be calculated. Here Δt is a sampling interval and $|\mathbf{d}|$ and $|\mathbf{d}_i|$ are norms of distances that are defined as follows (cf. Figure 9.10): Denote by

$$\mathbf{x}_i = \mathbf{y}(i\Delta t)$$

points calculated along the particular trajectory that is being investigated. These points are evaluated at integral multiples of the sampling rate Δt (chosen rather arbitrarily, but fixed). At these regularly occurring time instances, the behavior of nearby trajectories is measured. To this end, select a small perturbation $\mathbf{d}_0$ with norm $|\mathbf{d}_0| = |\mathbf{d}|$ (with, for instance, the Euclidian vector norm $|\mathbf{d}|$; see Appendix 2), and calculate

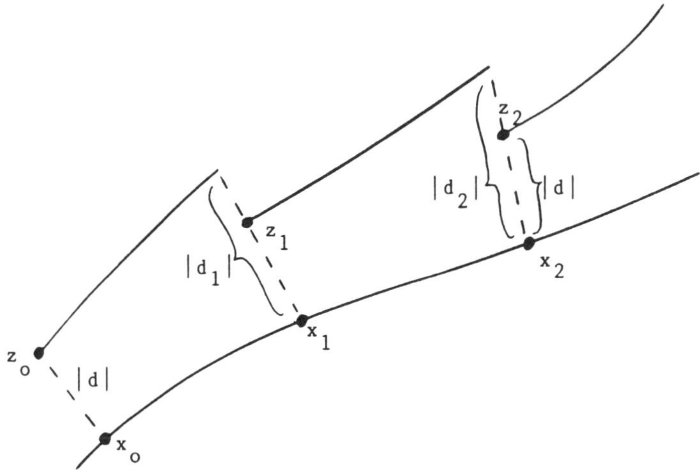

Figure 9.10

a part of the trajectory that emanates at

$$\mathbf{z}_0 = \mathbf{x}_0 + \mathbf{d}_0.$$

After the time Δt is elapsed, the difference

$$\mathbf{d}_1 = \varphi(\Delta t;\mathbf{z}_0) - \mathbf{x}_1$$

gives a first hint on the divergence behavior. The process of calculating distances $\mathbf{d}_i$ is repeated again and again, always starting with a perturbation obtained by normalizing the current distance to $|\mathbf{d}|$. In summary, one selects Δt and and $\mathbf{d}_0$ with small $|\mathbf{d}_0| = |\mathbf{d}|$, and calculates the sequence

$$\begin{aligned}
\mathbf{z}_i &= \mathbf{x}_i + \mathbf{d}_i \,|\mathbf{d}|/|\mathbf{d}_i| \\
\mathbf{x}_{i+1} &= \varphi(\Delta t;\mathbf{x}_i) \\
\mathbf{d}_{i+1} &= \varphi(\Delta t;\mathbf{z}_i) - \mathbf{x}_{i+1}
\end{aligned} \tag{9.13}$$

for $i = 0, 1, 2, \ldots$. This establishes the calculation of the numbers q_N from Eq. (9.12). As observed in [33, 56], the limit of q_N exists and can be identified with the maximum Liapunov exponent. The reported experiments indicate that the limit is independent of Δt and $\mathbf{d}_0$ as long as $|\mathbf{d}_0| = |\mathbf{d}|$ is small enough.

9.4.3 Power Spectra

Consider a time series

$$y(0), y(1), y(2), \ldots$$

of a scalar variable. Upon integrating systems of ODEs, a time series can be obtained by sampling one component $y_k(t)$ at sampling intervals Δt,

$$y_k(0), y_k(\Delta t), y_k(2\Delta t), \ldots.$$

The power spectrum of such a time series is usually calculated by the *fast Fourier transformation*. Corresponding routines are available on most computers. The result of applying a fast Fourier transformation consists of a sequence of frequencies with associated amplitudes (energies). This set of pairs of numbers is best presented graphically, displaying amplitude versus frequency. Such a diagram (see Figures 9.11 through 9.14) is referred to as a *power spectrum*. Different periodic, quasiperiodic, and aperiodic states can be distinguished by their power spectra. This is illustrated qualitatively in what follows by means of four characteristic figures.

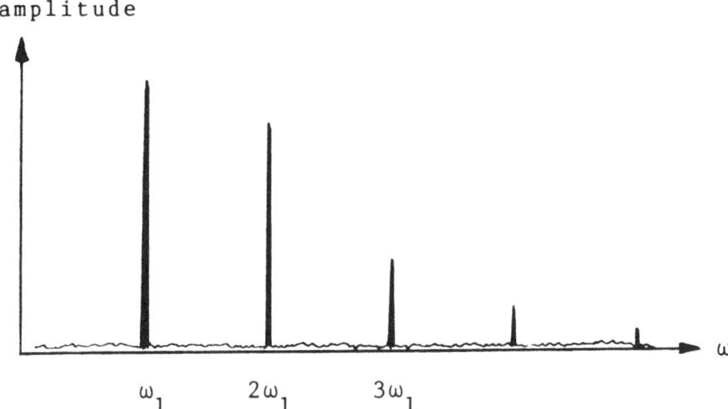

Figure 9.11

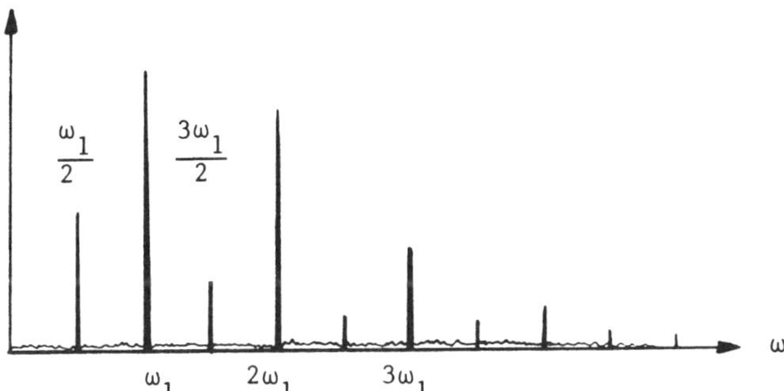

Figure 9.12

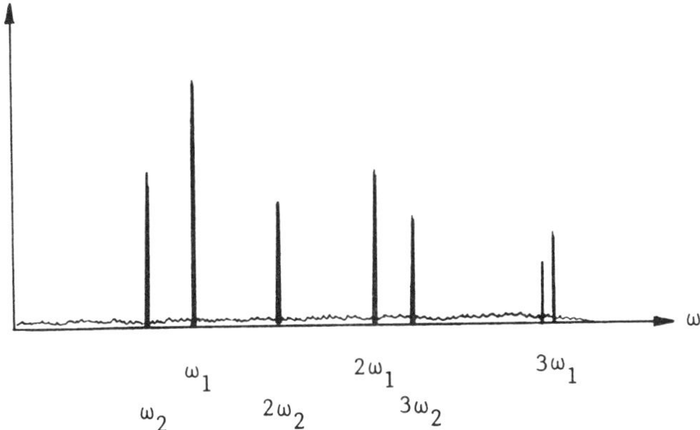

Figure 9.13

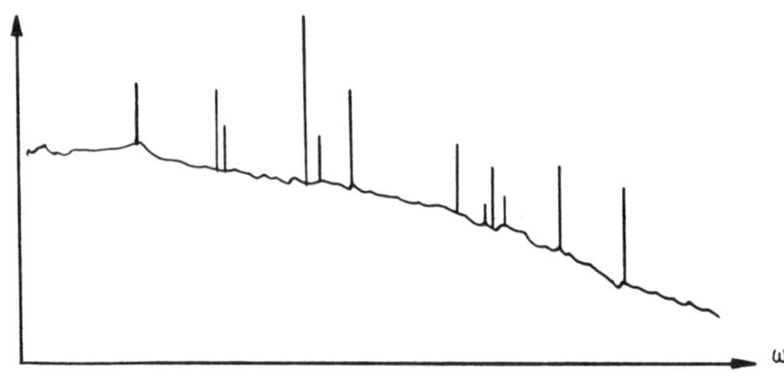

Figure 9.14

First, we illustrate a power spectrum of a periodic time series (Figure 9.11). The fundamental frequency ω_1 and the harmonics $2\omega_1$, $3\omega_1$, . . . are visible as sharp peaks. Apart from these lines, there may be some small-amplitude noise, as indicated in Figure 9.11; this background can be due to external noise. When period doubling takes place, subharmonics occur with the frequency $\omega_1/2$ and integer multiples (compare Figure 9.12). Similarly, a second period doubling contributes peaks with "distance" $\omega_1/4$.

A bifurcation from periodic motion to quasiperiodic motion on a 2-torus adds a fundamental frequency ω_2. This new frequency is generally not commensurable with ω_1. With two fundamental frequencies, a power spectrum may resemble Figure 9.13. Both periodic motions (Figures 9.11 and 9.12) and the quasiperiodic motion (Figure 9.13) are characterized by power spectra with small-amplitude noise and more or less discrete peaks.

This is in contrast to the situation of chaos, which is characterized by a broad-band noise (Figure 9.14). There may be peaks sitting on top of the broad-band noise, which are more or less accentuated. The construction of power spectra has shown to be the standard means for characterizing chaos. Almost all the references quoted in the context of chaos illustrate their results with power spectra. Thereby, the onset of broad-band noise serves as confirmation of chaotic phenomena.

Appendixes

APPENDIX 1. SOME ELEMENTARY FACTS FROM ODEs

Here we report on a few elementary facts from ordinary differential equations. For further background, refer to textbooks on ODEs, such as [15, 19, 41, 63, 124, 125, 140, 157].

A scalar ODE of the order m is given by an equation of the type

$$F(t,y(t),\dot{y}(t), \ldots, y^{(m)}(t)) = 0.$$

The dot and the superscript represent differentiation with respect to the independent variable t. Solutions are implicitly defined by this equation. We are interested mainly in explicit differential equations of the first order—namely,

$$\dot{\mathbf{y}}(t) = \mathbf{f}(t,\mathbf{y}(t)).$$

This differential equation may be a scalar equation or a vector equation. In the latter case, this equation stands for a system consisting of n scalar components,

$$\begin{pmatrix} \dot{y}_1 \\ \dot{y}_2 \\ \vdots \\ \dot{y}_n \end{pmatrix} = \begin{pmatrix} f_1(t,y_1, \ldots, y_n) \\ f_2(t,y_1, \ldots, y_n) \\ \vdots \\ f_n(t,y_1, \ldots, y_n) \end{pmatrix}.$$

An equation (system) of the type

$$\dot{\mathbf{y}} = \mathbf{f}(\mathbf{y}),$$

in which the independent variable t does not occur explicitly, is called *autonomous*. A nonautonomous equation can be transformed into an extended autonomous system by means of

$$y_{n+1} = t.$$

This leads to a system with $n + 1$ components,

$$\begin{pmatrix} \dot{y}_1 \\ \vdots \\ \dot{y}_n \\ \dot{y}_{n+1} \end{pmatrix} = \begin{pmatrix} f_1(y_{n+1},y_1, \ldots, y_n) \\ \vdots \\ f_n(y_{n+1},y_1, \ldots, y_n) \\ 1 \end{pmatrix}.$$

Linear ODE System with Constant Coefficients

We here review the general solution of a system of n homogeneous linear differential equations

$$\dot{\mathbf{y}} = \mathbf{A}\mathbf{y}.$$

The n^2 matrix $\mathbf{A}$ is constant and consists of n rows and n columns—that is, $\mathbf{A}$ has n^2 entries. A general solution is obtained by linear superposition of n linearly independent solutions $\mathbf{y}^k(t)$,

$$\mathbf{y}(t) = \sum_{k=1}^{n} c_k \mathbf{y}^k(t)$$

(the superscript is no exponent). The constants c_k are determined by initial conditions or boundary conditions. The solution structure is most simple when $\mathbf{A}$ has n linear independent *eigenvectors* $\mathbf{w}^k$ ($k = 1, \ldots, n$) associated with the *eigenvalues* μ_k. Then the *basis functions* $\mathbf{y}^k(t)$ take the form

$$\mathbf{y}^k(t) = \exp(\mu_k t)\mathbf{w}^k.$$

Eigenvalues μ_k and eigenvectors $\mathbf{w}^k \neq 0$ are defined as solutions to the system

$$(\mathbf{A} - \mu\mathbf{I})\mathbf{w} = \mathbf{0}.$$

The matrix $\mathbf{I}$ denotes the n^2 identity matrix. All its entries are zero, except for the diagonal elements, which are unity. The basis functions $\mathbf{y}^k(t)$ can be taken as real even when μ_k is complex: Let one pair of eigenvalues be

$$\mu_k = \alpha_k + i\beta_k, \quad \mu_{k+1} = \alpha_k - i\beta_k$$

(i: imaginary unit) with eigenvectors

$$\text{Re}(\mathbf{w}^k) \pm i\,\text{Im}(\mathbf{w}^k)$$

(Re: real part, Im: imaginary part). Then two linearly independent basis functions are

$$\mathbf{y}^k(t) = \exp(\alpha_k t)\{\text{Re}(\mathbf{w}^k)\cos\beta_k t - \text{Im}(\mathbf{w}^k)\sin\beta_k t\}$$

$$\mathbf{y}^{k+1}(t) = \exp(\alpha_k t)\{\text{Re}(\mathbf{w}^k)\sin\beta_k t + \text{Im}(\mathbf{w}^k)\cos\beta_k t\}.$$

Bernoulli Differential Equation

In Chapter 1 of this book we mentioned the Bernoulli differential equation. The Bernoulli differential equation is of the form

$$\dot{y} = a(t)y + b(t)y^m.$$

This equation is nonlinear for $m \neq 1$, $m \neq 0$. The transformation

$$y = u^{1/(1-m)}$$

leads to the differential equation

$$\dot{u} = (1 - m)au + (1 - m)b.$$

This version is linear in u; calculation of a closed-form solution is elementary.

APPENDIX 2. IMPLICIT FUNCTION THEOREM

For ease of notation, we use the symbol R^n for the set of all vectors with n real components. The superscript n is the dimension of the "space" R^n. Writing $\mathbf{y} \in R^n$ means that $\mathbf{y}$ is a vector with n components. The length (*Euclidian vector norm*) of $\mathbf{y}$ is defined by

$$|\mathbf{y}| = (y_1^2 + y_2^2 + \cdots + y_n^2)^{1/2}.$$

Interpreting the components as coordinates, we see that R^1 represents a line, R^2 a plane, and R^3 our everyday life space. The equation

$$\mathbf{0} = \mathbf{f}(\mathbf{y},\lambda)$$

with $\mathbf{y} \in R^n$, $\lambda \in R$, $\mathbf{f}(\mathbf{y},\lambda) \in R^n$ consists of n components; $\mathbf{0}$ stands for the zero vector. The function $\mathbf{f}$ is often defined only on a subset of the $(n + 1)$-dimensional space. We exclude equations with an empty solution set. In general, the equation $\mathbf{0} = \mathbf{f}(\mathbf{y},\lambda)$ defines implicitly one or more curves in the $(n + 1)$-dimensional $(\mathbf{y},\lambda)$-space. A simple example is furnished by the scalar equation

$$0 = -1 + y^2 + \lambda^2,$$

which defines the unit circle in the (y,λ)-plane. The question is whether an equation $0 = f(y,\lambda)$ defines implicitly a function

$$y = F(\lambda).$$

The above equation defining the unit circle has two implicitly defined functions. In this simple example, we encounter the rare situation that the functions are known explicitly,

$$y = F(\lambda) = +(1 - \lambda^2)^{1/2} \text{ and } y = F(\lambda) = -(1 - \lambda^2)^{1/2}.$$

Both functions are defined only for $-1 \leq \lambda \leq 1$.

The general assertion is as follows:

Implicit Function Theorem. Assume that

(1) $f(y^*,\lambda^*) = \mathbf{0}$
(2) f *is continuously differentiable on its domain*
(3) $f_y(y^*,\lambda^*)$ *is nonsingular.*

Then there is an interval $\lambda_1 < \lambda^* < \lambda_2$ *about* λ^*, *in which a function* $\mathbf{y} = F(\lambda)$ *is defined by* $\mathbf{0} = f(y,\lambda)$ *with the following properties holding for all* λ *with* $\lambda_1 < \lambda < \lambda_2$:

(a) $f(F(\lambda),\lambda) = \mathbf{0}$
(b) $F(\lambda)$ *is unique with* $\mathbf{y}^* = F(\lambda^*)$
(c) $F(\lambda)$ *is continuously differentiable*
(d) $f_y(y,\lambda)dy/d\lambda + f_\lambda(y,\lambda) = 0$.

REMARKS

(1) If in assumption (3) a higher degree of smoothness is assumed, the same smoothness is guaranteed in statement (c).
(2) An implicit function theorem was proved in [146]; other proofs can be found in any analysis textbook.
(3) Analog theorems apply for more general spaces than R^n (function spaces, Banach spaces).

For smooth functions $\mathbf{f}$, the implicitly defined function $F(\lambda)$ can be extended until $\mathbf{f_y}$ gets singular. That is, for λ_1 and λ_2 one has the property

$$\mathbf{f_y}(F(\lambda_1),\lambda_1) \text{ singular}, \quad \mathbf{f_y}(F(\lambda_2),\lambda_2) \text{ singular;}$$

here is $\lambda_1 < \lambda < \lambda_2$ the maximum interval such that the hypotheses of the theorem are valid. In the simple example of the unit circle, one has $\lambda_1 = -1$ and $\lambda_2 = +1$.

APPENDIX 3. SOME BASIC FACTS FROM LINEAR ALGEBRA

Assume that $\mathbf{z}^1, \mathbf{z}^2, \ldots, \mathbf{z}^m$ are vectors with n components. A useful vector is the unit vector $\mathbf{e}_k$; all its components are zero except for the k-th component, which is unity. A *linear combination* of m vectors $\mathbf{z}^i$ is a sum

$$\sum_{i=1}^{m} c_i \mathbf{z}^i,$$

the c_i are arbitrary real numbers. A vector $\mathbf{z} \in \mathrm{R}^n$ is called *linear dependent* of $\mathbf{z}^1$, $\mathbf{z}^2, \ldots, \mathbf{z}^m$ if $\mathbf{z}$ can be represented as a linear combination of these vectors $\mathbf{z}^i$; otherwise, $\mathbf{z}$ is called linear independent of $\mathbf{z}^1, \ldots, \mathbf{z}^m$. The *rank* of a matrix is the maximum number of linear independent columns (equivalently, rows). A square matrix $\mathbf{A}$ ($m = n$) is called *nonsingular* if $\text{rank}(\mathbf{A}) = n$. The following statements are equivalent:

$\mathbf{A}$ is singular
$\Leftrightarrow$ $\text{rank}(\mathbf{A}) < n$
$\Leftrightarrow$ $\mathbf{A}$ has 0 as eigenvalue
$\Leftrightarrow$ there exists a nonzero vector $\mathbf{z}$ with $\mathbf{Az} = \mathbf{0}$
$\Leftrightarrow$ $\det(\mathbf{A}) = 0$.

The *range* of a matrix $\mathbf{A}$ with column vectors $\mathbf{z}^1, \ldots, \mathbf{z}^n$ is the set of all linear combinations,

$$\text{range}(\mathbf{A}) = \{\mathbf{Ac}, \text{ for all vectors } \mathbf{c} \in \mathrm{R}^n\}$$
$$= \left\{ \sum_{i=1}^{n} c_i \mathbf{z}^i, \text{ for arbitrary } c_i \in \mathrm{R} \right\}.$$

The system of linear equations represented by the equation $\mathbf{Az} = \mathbf{b}$ for $\mathbf{b} \in \mathrm{R}^n$ is solvable by definition if and only if

$$\mathbf{b} \in \text{range}(\mathbf{A}).$$

This is equivalent to

$$\text{rank}(\mathbf{A}) = \text{rank}(\mathbf{A} \mid \mathbf{b}),$$

$(\mathbf{A} \mid \mathbf{b})$ is the rectangular matrix that consists of the n columns of $\mathbf{A}$ and the column vector $\mathbf{b}$. The *null space* or *kernel* of a matrix $\mathbf{A}$ is the solution set for $\mathbf{b} = 0$; the null space has the dimension

$$n - \text{rank}(\mathbf{A}).$$

This implies that the null space of a "full-rank" matrix has dimension 0 and consists only of the zero vector. In the case of $\text{rank}(\mathbf{A}) = n - 1$, there is one nonzero vector $\mathbf{h} \in \mathrm{R}^n$ that *spans* the null space,

$$\text{null}(\mathbf{A}) = \{\gamma\mathbf{h} \text{ for all } \gamma \in \mathrm{R}\}.$$

The null space is the linear combination of (n-rank($\mathbf{A}$)) vectors.

APPENDIX 4. RUNGE-KUTTA-FEHLBERG METHODS

Among the many successful codes designed to integrate initial-value problems of ODEs, Runge-Kutta type methods appear to be most popular [294, 206]. Combined with error formulas of Fehlberg [88, 89], these methods have proved to be robust and widely applicable. Runge-Kutta type methods have been adopted to integrate stiff differential equations as well (cf., for instance, [270]). We focus on the nonstiff situation and present the formulas on which a Runge-Kutta-Fehlberg method of order four is based.

Suppose that the initial-value problem

$$\mathbf{y}' = \mathbf{f}(t,\mathbf{y}),\ \mathbf{y}(t_0) = \mathbf{y}_0$$

is to be integrated. A typical integration step approximates $\mathbf{y}$ at $t = t_0 + \Delta$; Δ is the step length. The formulas are

$$\mathbf{y} = \mathbf{y}_0 + \Delta \sum_{k=0}^{4} c_k \mathbf{f}^{(k)}$$

$$\bar{\mathbf{y}} = \mathbf{y}_0 + \Delta \sum_{k=0}^{5} \bar{c}_k \mathbf{f}^{(k)}$$

TABLE A4.1

k	α_k	β_{k0}	β_{k1}	β_{k2}	β_{k3}	β_{k4}	c_k	$\bar{c}_k$
0	0	0					$\frac{25}{216}$	$\frac{16}{135}$
1	$\frac{1}{4}$	$\frac{1}{4}$					0	0
2	$\frac{3}{8}$	$\frac{3}{32}$	$\frac{9}{32}$				$\frac{1408}{2565}$	$\frac{6656}{12825}$
3	$\frac{12}{13}$	$\frac{1932}{2197}$	$-\frac{7200}{2197}$	$\frac{7296}{2197}$			$\frac{2197}{4104}$	$\frac{28561}{56430}$
4	1	$\frac{439}{216}$	-8	$\frac{3680}{513}$	$-\frac{845}{4104}$		$-\frac{1}{5}$	$-\frac{9}{50}$
5	$\frac{1}{2}$	$-\frac{8}{27}$	2	$-\frac{3544}{2565}$	$\frac{1859}{4104}$	$-\frac{11}{40}$		$\frac{2}{55}$

with

$$\mathbf{f}^{(0)} = \mathbf{f}(t_0, \mathbf{y}_0)$$

$$\mathbf{f}^{(k)} = \mathbf{f}(t_0 + \alpha_k \Delta, \mathbf{y}_0 + \Delta \sum_{j=0}^{k-1} \beta_{kj} \mathbf{f}^{(j)}).$$

Both $\mathbf{y}$ and $\bar{\mathbf{y}}$ approximate the exact solution, $\bar{\mathbf{y}}$ is the approximation of higher order. The difference $\mathbf{y} - \bar{\mathbf{y}}$ serves as an estimate of the error in $\mathbf{y}$. The coefficients from [89] are given in Table A4.1.

APPENDIX 5. TRANSFORMATION INTO STANDARD FORM

A significant part of this book is devoted to a system of n nonlinear scalar equations, written in vector form as

$$\mathbf{f}(\mathbf{y}, \lambda) = \mathbf{0}.$$

In what follows, we write the *standard form* of such a finite-dimensional system as

$$\mathbf{F}(\mathbf{Y}, \lambda) = \mathbf{0},$$

and show how various problems can be transformed into this standard form. A review of transformation techniques is given in [21]; further techniques are explained in this book. Hence, in this appendix, we confine ourselves to giving a list of transformations.

Problem with $m > 1$ Parameters

Consider the problem $\mathbf{f}(\mathbf{y},\boldsymbol{\Lambda}) = \mathbf{0}$ with a parameter vector $\boldsymbol{\Lambda}$ consisting of $m > 1$ parameters. In most applications, one can formulate $m - 1$ relations between the m parameters. These relations can be written as $\mathbf{g}(\boldsymbol{\Lambda}) = \mathbf{0}$. In the simplest case, $\mathbf{g}$ fixes $\Lambda_1, \ldots, \Lambda_{m-1}$ to take constant values. We introduce a new vector by

$$(\mathbf{Y},\lambda)^{\mathrm{tr}} = (\mathbf{y},\boldsymbol{\Lambda})^{\mathrm{tr}},$$

that is, $\lambda = \boldsymbol{\Lambda}_m$ and $\mathbf{Y}$ is a vector with $n + m - 1$ components. Then the original problem can be written in standard form:

$$\mathbf{F}(\mathbf{Y},\lambda) = \begin{pmatrix} \mathbf{f}(\mathbf{y},\boldsymbol{\Lambda}) \\ \mathbf{g}(\boldsymbol{\Lambda}) \end{pmatrix} = \mathbf{0}.$$

Solving a "Difficult" Problem by Homotopy

Assume that the problem $\mathbf{f}(\mathbf{y}) = \mathbf{0}$ is difficult to solve because of a lack of an initial guess sufficiently close to the solution. Assume further that the problem $\mathbf{g}(\mathbf{y}) = \mathbf{0}$ is easy to solve and that $\mathbf{f}$ and $\mathbf{g}$ fit together (same dimension). Then one can solve, for example, the family of problems

$$\mathbf{F}(\mathbf{y},\lambda) = \lambda\mathbf{f}(\mathbf{y}) + (1 - \lambda)\mathbf{g}(\mathbf{y}),\ 0 \le \lambda \le 1,$$

starting at $\lambda_a = 0$ with the simple problem and ending (hopefully) at $\lambda_b = 1$ with the more difficult problem. A simple but successful choice of $\mathbf{g}$ is

$$\mathbf{g} = \mathbf{y} - \mathbf{z}$$

for some constant vector $\mathbf{z}$ [361]. This is one version of a *homotopy method*. For other types of *embedding* see, for instance, [8, 179, 359].

Boundary-Value Problems

Consider the boundary-value problem

$$\mathbf{y}' = \mathbf{f}(t,\mathbf{y},\lambda), \quad \mathbf{r}(\mathbf{y}(a),\mathbf{y}(b)) = \mathbf{0}.$$

As pointed out in Section 1.4, the boundary-value problem can be transformed into standard form by means of shooting methods:

$$\mathbf{F}(\mathbf{Y},\lambda) = \mathbf{r}(\mathbf{Y},\boldsymbol{\varphi}(b;\mathbf{Y})) = \mathbf{0}.$$

Here $\boldsymbol{\varphi}(t;\mathbf{Y})$ denotes the solution of the initial-value problem

$$\mathbf{y}' = \mathbf{f}(t,\mathbf{y},\lambda), \quad \mathbf{y}(a) = \mathbf{Y}.$$

Other possibilities for approximating a boundary-value problem by an equation in standard form are furnished by finite differences or finite elements. This includes the "reformulation" of PDEs as a (large) system of type $\mathbf{F}(\mathbf{Y},\lambda) = \mathbf{0}$.

Periodic Solutions of ODEs

Consider the system of autonomous ordinary differential equations with n components

$$\mathbf{y}' = \mathbf{f}(\mathbf{y},\lambda).$$

The task of calculating periodic solutions can be transformed to a standard boundary-value problem and thus to $\mathbf{F}(\mathbf{Y},\lambda) = 0$. Details are in Section 7.1.

Following Critical Boundaries

As pointed out in Section 2.9, there are critical boundaries that play a role in multiparameter problems. The calculation of critical boundaries can be reformulated into equations in standard form. This is discussed in Section 5.7 for two-parameter problems; one possibility for $\mathbf{F}$ is given by Eq. (5.60).

APPENDIX 6. NUMERICAL SOFTWARE AND PACKAGES

General Numerical Methods

There are so many computer programs of numerical software that it is impossible to give an exhaustive review, so we shall mention briefly only some of the standard algorithms and packages. As far as numerical linear algebra is concerned, the main reference is the handbook [367], which includes ALGOL programs. FORTRAN versions of these outstanding procedures are included in LINPACK (solving of linear equations, [82]) and EISPACK [326]; the latter package is devoted especially to eigenvalue problems. The software collection MINPACK offers programs for solving systems of nonlinear equations. ELLPACK contains programs for solving elliptic PDEs [275]. Apart from packages, single programs have been developed. For instance, several shooting codes exist. Much of the material presented in this book has been calculated by means of BOUNDS (from [49], based on [47]), used along with integrators from [48, 88]. General software collections are offered by IMSL and NAG. Any computing center has at least some of the mentioned packages implemented, together with assistance on how to get access to other programs.

Algorithms Devoted to Nonlinear Phenomena

Many aspects of a bifurcation and stability analysis can be calculated by means of the package BIFPACK [316]. The diagrams and figures in this book have been obtained using programs of this package. BIFPACK has been designed as an interface between standard routines of numerical analysis and the particular problem of the user. That is, many of the transformations into standard forms described in this book are carried out automatically; BIFPACK acts as a driving routine. This open design makes it possible to run BIFPACK along with the latest versions of standard software or with versions the user is familiar with. Another package designed to handle general nonlinear phenomena is AUTO [80].

Apart from these packages with general applicability, there are other programs for specific tasks. In particular, in the context of continuation we mention the codes from [195, 274, 315]. The codes announced in [11, 12, 155] are specifically designed for periodic solutions. A package for the calculation of Hopf bifurcation points has been established by [135, 136]. Programs for the calculation of test functions τ and initial guess $\overline{\mathbf{h}}$ (see Sections 5.3 and 6.2) have been published in [310]. The calculation of bifurcation points of integral equations is treated in [54]. Programs for a direct calculation of Hopf bifurcation points in parabolic PDEs are offered in [283].

References

1. Abbott, J. P. 1977. Numerical continuation methods for nonlinear equations and bifurcation problems. Ph.D. diss., Australian National University.
2. Abbott, J. P. 1978. An efficient algorithm for the determination of certain bifurcation points. *J. Comput. Appl. Math.* 4:19–27.
3. Abraham, R., and J. E. Marsden. 1967. *Foundations of Mechanics.* Elmsford, N.Y.: Benjamin.
4. Abraham, R. H., and C. D. Shaw. 1984. *Dynamics: The Geometry of Behavior,* 4 parts. Santa Cruz, Calif.: Aerial.
5. Albrecht, J., L. Collatz, and K. Kirchgässner (eds.). 1979. *Constructive Methods for Nonlinear Boundary Value Problems and Nonlinear Oscillations.* Proceedings of a Conference in Oberwolfach 1978, Basel: Birkhäuser. *ISNM* 48.
6. Al-Humadi, A., and N. D. Kazarinoff. 1985. Hopf bifurcation in the Watt steam engine. *J. Inst. Math. Appl.* 21:133–136.
7. Allgower, E. L., and K. Böhmer. 1986. Resolving singular nonlinear equations. Preprint, Department of Mathematics, Colorado State University.
8. Allgower, E., and K. Georg. 1980. Simplicial and continuation methods for approximating fixed points and solutions to systems of equations. *SIAM Review* 22:28–85.
9. Allgower, E., K. Glashoff, and H.-O. Peitgen (eds.). 1981. *Numerical Solution of Nonlinear Equations: Proceedings, Bremen 1980.* Lecture Notes in Math, 878. Berlin: Springer.
10. Allgower, E. L., and P. H. Schmidt. 1985. An algorithm for piecewise-linear approximation of an implicitly defined manifold. *SIAM J. Numer. Anal.* 22:322–346.
11. Aluko, M., and H.-C. Chang. 1984. PEFLOQ: An algorithm for the bifurcation analysis of periodic solutions of autonomous systems. *Computers and Chemical Engineering* 8:355–365.
12. Aluko, M. 1984. *ABODE: A Computer Package for the Bifurcation Analyses of Autonomous Ordinary Differential Equations.*
13. Anselone, P. M., and R. H. Moore. 1966. An extension of the Newton-Kantorovic method for solving nonlinear equations with an application to elasticity. *J. Math. Anal. Appl.* 13:476–501.
14. Antonelli, P. L., K. D. Fuller, and N. D. Kazarinoff. 1987. A study of large

amplitude periodic solutions in a model for starfish predation of coral. *IMA J. Math. Applied in Medicine and Biology*, to appear.

15. Arnol'd, V. I. 1973. *Ordinary Differential Equations.* Cambridge: MIT Press.
16. Arnol'd, V. I. 1983. *Geometrical Methods in the Theory of Ordinary Differential Equations.* New York: Springer.
17. Arnol'd, V. I. 1984. *Catastrophe Theory.* Berlin: Springer.
18. Aronson, D. G., R. P. McGehee, I. G. Kevrekidis, and R. Aris. 1986. Entrainment regions for periodically forced oscillators. *Physical Review* A33:2190–2192.
19. Aronson, D. G., and H. F. Weinberger. 1975. Nonlinear diffusion in population genetics, combustion, and nerve propagation. *Lecture Notes in Math,* 446. Berlin: Springer.
20. Arrowsmith D. K., and C. M. Place. 1982. *Ordinary Differential Equations.* London: Chapman & Hall.
21. Ascher, U., and R. D. Russell. 1981. Reformulation of boundary value problems into "standard" form. *SIAM Review* 23:238–254.
22. Ascher, U. M., and R. D. Russell (eds.). 1985. *Numerical Boundary Value Problems, Proceedings, Vancouver 1984. Progress in Scientific Computing,* vol. 5. Boston: Birkhäuser.
23. Aulbach, B. 1984. Trouble with linearizations. In H. Neunzert (ed.), *Mathematics in Industry.* Stuttgart: Teubner.
24. Babloyantz, A., and A. Destexhe. 1986. Low-dimensional chaos in an instance of epilepsy. *Proc. Natl. Acad. Sci.* USA 83:3513.
25. Balakotaiah, V., and D. Luss. 1982. Structure of the steady-state solutions of lumped-parameter chemically reacting systems. *Chem. Eng. Sci.* 37:1611–1623.
26. Bazley, N. W., and G. C. Wake. 1978. The disappearance of criticality in the theory of thermal ignition. *ZAMP* 29:971–976.
27. Bazley, N. W., and G. C. Wake. 1981. Criticality in a model for thermal ignition in three or more dimensions. *J. Appl. Math. Phys. (ZAMP)* 32:594–601.
28. Becker, K.-H., and R. Seydel. 1981. A Duffing equation with more than 20 branch points. In [9].
29. Beeler, G. W., and H. Reuter. 1977. Reconstruction of the action potential of ventricular myocardial fibres. *J. Physiol.* 268:177–210.
30. Bell, G. I. 1973. Predator-prey equations simulating an immune response. *Math. Biosciences* 16:291–314.
31. Bénard, H. 1901. Les tourbillons cellulaires dans une nappe liquide transportant de la chaleur par convection en regime permanent. *Ann. Chim. Phys.* 7 (ser. 23):62.
32. Benjamin, T. B. 1978. Bifurcation phenomena in steady flows of a viscous fluid. *I. Theory. Proc. R. Soc. Lond.* A359:1–26.
33. Benettin, G., L. Galgani, and J.-M. Strelcyn. 1976. Kolmogorov entropy and numerical experiments. *Physical Review* A14:2338–2345.
34. Bergé, P., M. Dubois, P. Manneville, and Y. Pomeau. 1980. Intermittency in Rayleigh-Benard convection. *J. Phys. Lett.* 41:341.

35. Beyn, W.-J. 1980. On discretizations of bifurcation problems. *ISNM* 54:46–73.
36. Beyn, W.-J. 1984. Defining equations for singular solutions and numerical applications. *ISNM* 70:42–56. In [203].
37. Beyn, W.-J. 1985. Zur numerischen Berechnung mehrfacher Verzweigungspunkte. *ZAMM* 65:T370–T371.
38. Beyn, W.-J., and E. Doedel. 1981. Stability and multiplicity of solutions to discretizations of nonlinear ordinary differential equations. *SIAM J. Sci. Stat. Comput.* 2:107–120.
39. Bohl, E. 1979. On the bifurcation diagram of discrete analogues of ordinary bifurcation problems. *Math. Meth. Appl. Sci.* 1:566–571.
40. Bouc, R., M. Defilippi, and G. Iooss. 1978. On a problem of forced nonlinear oscillations: Numerical example of bifurcation into an invariant torus. *Nonlinear Analysis* 2:211–224.
41. Brady, J. F., and A. Acrivos. 1981. Steady flow in a channel or tube with an accelerating surface velocity: An exact solution to the Navier-Stokes equations with reverse flow. *J. Fluid Mech.* 112:127–150.
42. Braun, M. 1978. *Differential Equations and Their Applications.* New York: Springer.
43. Brent, R. P. 1973. Some efficient algorithms for solving systems of nonlinear equations. *SIAM J. Numer. Anal.* 10:327–344.
44. Brezzi, F., J. Rappaz, and P.-A. Raviart. 1981. Finite dimensional approximation of nonlinear problems, part 3: Simple bifurcation points. *Numer. Math.* 38:1–30.
45. Brown, K. M., and W. B. Gearhardt. 1971. Deflation techniques for the calculation of further solutions of a nonlinear system. *Numer. Math.* 16:334–342.
46. Broyden, C. G. 1965. A class of methods for solving nonlinear simultaneous equations. *Math. Comput.* 19:577–593.
47. Bulirsch, R. 1971. Die Mehrzielmethode zur numerischen Lösung von nichtlinearen Randwertproblemen und Aufgaben der optimalen Steuerung. Report der Carl-Cranz-Gesellschaft.
48. Bulirsch, R., and J. Stoer. 1966. Numerical treatment of ordinary differential equations by extrapolation methods. *Numer. Math.* 8:1–13.
49. Bulirsch, R., J. Stoer, and P. Deuflhard. 1977. BOUNDSOL, a FORTRAN code for a numerical solution of boundary-value problems of ODEs. Technische Universität, München.
50. Busch, W. 1962. *Max und Moritz* (1865). Facsimile. Hannover: W. Busch Gesellschaft.
51. Busse, F. H. 1977. Mathematical problems of dynamo theory. In [264].
52. Busse, F. H. 1981. Transition to turbulence in Rayleigh-Bénard convection. In [338].
53. Busse, F. H. 1982. Transition to turbulence in thermal convection. In [372].
54. Caluwaerts, R. 1985. An automatic procedure for the calculation of bifurcation points of integral equations. *J. Comput. Appl. Math.* 12:207–215.
55. Carr, J. 1981. *Applications of Center Manifold Theory.* New York: Springer.

56. Casartelli, M., E. Diana, L. Galgani, and A. Scotti. 1976. Numerical computations on a stochastic parameter related to the Kolmogorov entropy. *Phys. Rev.* A13:1921–1930.
57. Chan, T. F. 1985. An approximate Newton method for coupled nonlinear systems. *SIAM J. Numer. Anal.* 22:904–913.
58. Chan, T. F. C., and H. B. Keller. 1982. Arclength continuation and multigrid techniques for nonlinear elliptic eigenvalue problems. *SIAM J. Sci. Stat. Comput.* 3:173–194.
59. Chang, Ch.J., and R. A. Brown. 1984. Natural convection in steady solidification: Finite element analysis of a two phase Rayleigh-Bénard problem. *J. Comp. Phys.* 53:1–27.
60. Chay, T. R., and Y. S. Lee. 1985. Phase resetting and bifurcation in the ventricular myocardium. Manuscript. Department of Biological Sciences, University of Pittsburgh.
61. Chay, T. R., and J. Rinzel. 1985. Bursting, beating, and chaos in an excitable membrane model. *Biophys. J.* 47:357–366.
62. Chow, S.-N., and J. K. Hale. 1982. *Methods of Bifurcation Theory.* New York: Springer.
63. Clever, R. M., and F. H. Busse. 1974. Transition to time-dependent convection. *J. Fluid Mech.* 65:625–645.
64. Coddington, E. A., and N. Levinson. 1955. *Theory of Ordinary Differential Equations.* New York: McGraw-Hill.
65. Crandall, M. G., and P. H. Rabinowitz. 1971. Bifurcation from simple eigenvalues. *J. Functional Analysis* 8:321–340.
66. Crandall, M. G., and P. H. Rabinowitz. 1977–78. The Hopf bifurcation theorem in infinite dimensions. *Arch. Rat. Mech. Anal.* 67:53–72.
67. Crawford, J. D., and S. Omohundro. 1984. On the global structure of period doubling flows. *Physica* 13D:161–180.
68. Crutchfield, J., D. Farmer, N. Packard, R. Shaw, G. Jones, and R. J. Donnelly. 1980. Power spectral analysis of a dynamical system. *Phys. Lett.* 76A:1–4.
69. Crutchfield, J. P., and B. A. Huberman. 1980. Fluctuations and the onset of chaos. *Phys. Lett.* 77A:407–410.
70. Curry, J. H. 1978. A generalized Lorenz system. *Comm. Math. Phys.* 60:193–204.
71. Cvitanovic, P. (ed.). 1984. *Universality in Chaos: A Reprint Selection.* Bristol: A. Hilger.
72. Dahlquist, G., and A. Björck. 1974. *Numerical Methods.* Englewood Cliffs, N.J.: Prentice-Hall.
73. Davidenko, D. F. 1953. On a new method of numerical solution of systems of nonlinear equations. *Dokl. Akad. Nauk. SSSR* 88:601–602. *Mathematical Reviews* 14:906.
74. Decker, D. W., and H. B. Keller. 1981. Path following near bifurcation. *Comm. Pure Appl. Math.* 34:149–175.
75. Decker, D. W., H. B. Keller, and C. T. Kelley. 1983. Convergence rates for Newton's method at singular points. *SIAM J. Numer. Anal.* 20:296–314.

76. Deist, F. H., and L. Sefor. 1967. Solution of systems of non-linear equations by parameter variation. *Computer J.* 10:78–82.
77. Den Heijer, C., and W. C. Rheinboldt. 1981. On steplength algorithms for a class of continuation methods. *SIAM J. Numer. Anal.* 18:925–948.
78. Dennis, J. E., and J. J. Moré. 1977. Quasi-Newton methods, motivation, and theory. *SIAM Review* 19:46–89.
79. Deuflhard, P. 1979. A stepsize control for continuation methods and its special application to multiple shooting techniques. *Numer. Math.* 33:115–146.
80. Doedel, E. 1981. AUTO: A program for the automatic bifurcation analysis of autonomous systems. *Congressus Numerantium* 30:265–284.
81. Doedel, E. J., and R. F. Heinemann. 1983. Numerical computation of periodic solution branches and oscillatory dynamics of the stirred tank reactor with A→B→C reactions. *Chem. Eng. Sc.* 38:1493–1499.
82. Dongarra, J. J., J. R. Bunch, C. B. Moler, and G. W. Stewart. 1979. *LINPACK User's Guide.* Philadelphia: SIAM Publications.
83. Duffing, G. 1918. *Erzwungene Schwingungen bei veränderlicher Eigenfrequenz.* Braunschweig: Vieweg.
84. Eckmann, J. P. 1981. Roads to turbulence in dissipative dynamical systems. *Reviews of Modern Physics* 53:643–654.
85. Epstein, I. R. 1983. Oscillations and chaos in chemical systems. *Physica* 7D:47–56.
86. Euler, L. 1952 (1744). *De Curvis Elasticis, Methodus Inveniendi Lineas Curvas Maximi Minimive Proprietate Gaudentes. Additamentum I* (1744). In *Opera Omnia I,* vol. 24, pp. 231–297. Zürich.
87. Farmer, J. D. 1981. Spectral broadening of period-doubling bifurcation sequences. *Phys. Rev. Lett.* 47:179–182.
88. Fehlberg, E. 1969. Klassische Runge-Kutta Formeln fünfter und siebenter Ordnung mit Schrittweitenkontrolle. *Computing* 4:93–106.
89. Fehlberg, E. 1970. Klassische Runge-Kutta-Formeln vierter und niedrigerer Ordnung mit Schrittweiten-Kontrolle und ihre Anwendung auf Wärmeleitungsprobleme. *Computing* 6:61–71.
90. Feigenbaum, M. J. 1980. Universal behavior in nonlinear systems. *Los Alamos Science* 1:4–27.
91. Field, R. J., and R. M. Noyes. 1974. Oscillations in chemical systems, IV: Limit cycle behavior in a model of a chemical reaction. *J. Chem. Physics* 60:1877–1884.
92. Fife, P. C. 1978. Asymptotic states for equations of reaction and diffusion. *Bull. AMS* 84:693–726.
93. Fife, P. C. 1979. *Mathematical Aspects of Reacting and Diffusing Systems.* Lecture Notes in Biomathematics 28. Berlin: Springer.
94. Fisher, R. A. 1937. The wave of advance of advantageous genes. *Ann. Eugenics* 7:355–369.
95. FitzHugh, R. 1961. Impulses and physiological states in theoretical models of nerve membrane. *Biophys. J.* 1:445–466.
96. Flockerzi, D. 1985. Resonance and bifurcation of higher-dimensional tori. *Proceedings of the AMS* 94:147–157.

97. Flockerzi, D. 1985. On the $T^k \to T^{k+1}$ bifurcation problem. In S. N. Pnevmatikos (ed.), *Singularities and Dynamical Systems.* New York: Elsevier.
98. Flockerzi, D. 1987. Codimension two bifurcations near an invariant n-torus. In [205].
99. Franceschini, V., and C. Tebaldi. 1979. Sequences of infinite bifurcations and turbulence in a five-mode truncation of the Navier-Stokes equations. *J. Phys.* 21:707–726.
100. Georg, K. 1980 (1981). Numerical integration of the Davidenko equation. In [9].
101. Georg, K. 1981. On tracing an implicitly defined curve by quasi-Newton steps and calculating bifurcation by local perturbations. *SIAM J. Sci. Stat. Comput.* 2:35–50.
102. Ghil, M., and J. Tavantzis. 1983. Global Hopf bifurcation in a simple climate model. *SIAM J. Appl. Math.* 43:1019–1041.
103. Giglio, M., S. Musazzi, and U. Perini. 1981. Transition to chaotic behavior via reproducible sequences of period-doubling bifurcations. *Phys. Rev. Lett.* 47:243.
104. Gill, W. N., and R. Seydel. 1985. Calculation of Marangoni instabilities. State University of New York at Buffalo.
105. Gilmore, R. 1981. *Catastrophe Theory for Scientists and Engineers.* New York: John Wiley.
106. Glansdorff, P., and I. Prigogine. 1971. *Thermodynamic Theory of Structure, Stability, and Fluctuations.* London: Wiley-Interscience.
107. Glass, L. 1975. Combinatorial and topological methods in nonlinear chemical kinetics. *J. Chem. Phys.* 63:1325–1335.
108. Goldberg, J. E., and R. H. Richards. 1963. Analysis of nonlinear structures. *J. Struct. Div., Proc. ASCE* 89:333.
109. Golub, G. H., and C. F. van Loan. 1983. *Matrix Computations.* Baltimore: Johns Hopkins University Press.
110. Golubitsky, M., B. L. Keyfitz, and D. G. Schaeffer. 1981. A singularity theory analysis of a thermal-chainbranching model for the explosion peninsula. *Comm. Pure Appl. Math.* 34:433–463.
111. Golubitsky, M., and W. F. Langford. 1981. Classification and unfoldings of degenerate Hopf bifurcations. *J. Diff. Eq.* 41:375–415.
112. Golubitsky, M., and D. Schaeffer. 1979. A theory for imperfect bifurcation via singularity theory. *Comm. Pure Appl. Math.* 32:21–98.
113. Golubitsky, M., and D. G. Schaeffer. 1985. *Singularities and Groups in Bifurcation Theory,* vol. 1. New York: Springer.
114. Graff, M., R. Scheidl, H. Troger, and E. Weinmüller. 1985. An investigation of the complete post-buckling behavior of axisymmetric spherical shells. *J. Appl. Math. Phys. (ZAMP)* 36:803–821.
115. Grebogi, C., E. Ott, and J. A. Yorke. 1983. Crises, sudden changes in chaotic attractors, and transient chaos. *Physica* 7D:181–200.
116. Greenberg, J. M., B. D. Hassard, and S. P. Hastings. 1978. Pattern formation and periodic structures in systems modeled by reaction-diffusion equations. *Bull. Amer. Math. Soc.* 84:1296.

117. Griewank, A. 1985. On solving nonlinear equations with simple singularities or nearly singular solutions. *SIAM Review* 27:537–563.
118. Griewank, A., and G. W. Reddien. 1983. The calculation of Hopf bifurcation points by a direct method. *IMA J. Numer. Anal.* 3:295–303.
119. Griewank, A., and G. W. Reddien. 1984. Characterization and computation of generalized turning points. *SIAM J. Numer. Anal.* 21:176–185.
120. Griewank, A., and G. W. Reddien. 1984. Computation of generalized turning points and two-point boundary value problems. *ISNM* 70:385–400. In [203].
121. Grossmann, S., and S. Thomae. 1977. Invariant distributions and stationary correlation functions of one-dimensional discrete processes. *Z. Naturforschung* 32A:1353–1363.
122. Guckenheimer, J. 1978. The catastrophe controversy. *Mathem. Intelligencer* 1:15–20.
123. Guckenheimer, J., and Ph. Holmes. 1983. *Nonlinear Oscillations, Dynamical Systems, and Bifurcation of Vector Fields.* New York: Springer.
124. Guevara, M. R., L. Glass, and A. Shrier. 1981. Phase locking, period-doubling bifurcations, and irregular dynamics in periodically stimulated cardiac cells. *Science* 214:1350–1353.
125. Hadeler, K. P., U. an der Heiden, and K. Schumacher. 1976. Generation of nervous impulse and periodic oscillations. *Biol. Cybernetics* 23:211–218.
126. Hadeler, K. P., P. de Mottoni, and K. Schumacher. 1980. Dynamic models for animal orientation. *J. Mathem. Biol.* 10:307–332.
127. Hadeler, K. P., and F. Rothe. 1975. Travelling fronts in nonlinear diffusion equations. *J. Math. Biol.* 2:251–263.
128. Hahn, W. 1967. *Stability of Motion.* Berlin: Springer.
129. Haken, H. 1977. *Synergetics: An Introduction.* Berlin: Springer.
130. Haken, H. 1983. *Advanced Synergetics.* New York: Springer.
131. Hale, J. K. 1969. *Ordinary Differential Equations.* New York: Wiley-Interscience.
132. Hartmann, P. 1964. *Ordinary Differential Equations.* New York: Wiley.
133. Haselgrove, C. B. 1961. The solution of nonlinear equations and of differential equations with two-point boundary conditions. *Comput. J.* 4:255–259.
134. Hassard, B. D. 1978. Bifurcation of periodic solutions of the Hodgkin-Huxley model for the squid giant axon. *J. Theor. Biol.* 71:401–420.
135. Hassard, B. D. 1980. BIFOR2: Analysis of Hopf bifurcation in an ordinary differential system. Computer program, State University of New York at Buffalo.
136. Hassard, B. D., N. D. Kazarinoff, and Y. H. Wan. 1981. *Theory and Applications of Hopf Bifurcation.* Cambridge: Cambridge University Press.
137. Hastings, S. P. 1975. Some mathematical problems from neurobiology. *Amer. Math. Monthly* 82:881–895.
138. Hastings, S. P. 1982. Single and multiple pulse waves for the FitzHugh-Nagumo equations. *SIAM J. Appl. Math.* 42:247–260.
139. Hastings, S. P., and J. D. Murray. 1975. The existence of oscillatory solutions in the Field-Noyes model for the Belousov-Zhabotinskii reaction. *SIAM J. Appl. Math.* 28:678–688.

140. Heinemann, R. F., K. A. Overholser, and G. W. Reddien. 1979. Multiplicity and stability of premixed laminar flames: An application of bifurcation theory. *Chem. Eng. Sci.* 34:833–840.
141. Heinemann, R. F., and A. B. Poore. 1981. Multiplicity, stability, and oscillatory dynamics of the tubular reactor. *Chem. Eng. Sci.* 36:1411–1419.
142. Helleman, R. H. G. 1980. Self-generated chaotic behavior in nonlinear mechanics. In E. G. D. Cohen (ed.), *Fundamental Problems in Statistical Mechanics,* vol. 5. Amsterdam: North Holland.
143. Hénon, M. 1976. A two-dimensional mapping with a strange attractor. *Commun. Math. Phys.* 50:69.
144. Hentschel, H. G. E., and I. Procaccia. 1983. The infinite number of dimensions of probabilistic fractals and strange attractors. *Physica* 8D:435.
145. Hess, B., and M. Markus. 1985. The diversity of biochemical time patterns. *Berichte Bunsen-Gesellschaft Physik. Chem.* 89:642–651.
146. Hildebrandt, T. H., and L. M. Graves. 1927. Implicit functions and their differentials in general analysis. *A.M.S. Transactions* 29:127–153.
147. Hirsch, M. W., and S. Smale. 1974. *Differential Equations, Dynamical Systems, and Linear Algebra.* New York: Academic Press.
148. Hlavacek, V., M. Kubicek, and J. Jelinek. 1970. Modeling of chemical reactors, XVIII: Stability and oscillatory behaviour of the CSTR. *CES* 25:1441–1461.
149. Hlavacek, V., M. Marek, and M. Kubicek. 1968. Modelling of chemical reactors, X: Multiple solutions of enthalpy and mass balances for a catalytic reaction within a porous catalyst particle. *Chem. Eng. Sci.* 23:1083–1097.
150. Holder, E. J., and D. Schaeffer. 1984. Boundary conditions and mode jumping in the von Kármán equations. *SIAM J. Math. Anal.* 15:446–458.
151. Hodgkin, A. L., and A. F. Huxley. 1952. A quantitative description of membrane current and its application to conduction and excitation in nerve. *J. Physiol.* 117:500–544.
152. Holmes, P. J. 1977. Bifurcations to divergence and flutter in flow-induced oscillations: A finite dimensional analysis. *J. Sound Vibration* 53:471–503.
153. Holmes, Ph. 1979. A nonlinear oscillator with a strange attractor. *Phil. Trans. of the Royal Soc. London* A292:419–448.
154. Holodniok, M., and M. Kubicek. 1984 (1981). New algorithms for evaluation of complex bifurcation points in ordinary differential equations: A comparative numerical study. Preprint 1981. *Appl. Math. Comput.* 15:261–274.
155. Holodniok, M., and M. Kubicek. 1984. DERPER: An algorithm for the continuation of periodic solutions in ordinary differential equations. *J. Comput. Phys.* 55:254–267.
156. Honerkamp, J., G. Mutschler, and R. Seitz. 1985. Coupling of a slow and a fast oscillator can generate bursting. *Bull. Mathem. Biology* 47:1–21.
157. Hopf, E. 1942. Abzweigung einer periodischen Lösung von einer stationären Lösung eines Differentialsystems. *Bericht der Math.-Phys. Klasse der Sächsischen Akademie der Wissenschaften zu Leipzig* 94.
158. Huang, N. C., and W. Nachbar. 1968. Dynamic snap-through of imperfect viscoelastic shallow arches. *J. Appl. Mech.* 35:289–296.

159. Iooss, G., and D. D. Joseph. 1981. *Elementary Stability and Bifurcation Theory.* New York: Springer.
160. Iooss, G., and W. F. Langford. 1980. Conjectures on the routes to turbulence via bifurcations. In *Nonlinear Dynamics,* ed. H. G. Helleman. New York: New York Academy of Sciences.
161. Janssen, R., V. Hlavacek, and P. van Rompay. 1983. Bifurcation pattern in reaction-diffusion dissipative systems. *Z. Naturforschung* 38A:487–492.
162. Jensen, K. F., and W. H. Ray. 1982. The bifurcation behavior of tubular reactors. *Chem. Eng. Sci.* 37:199–222.
163. Jepson, A. D. 1981. Numerical Hopf bifurcation. Thesis, California Institute of Technology.
164. Jordan, D., and P. Smith. 1977. *Nonlinear Ordinary Differential Equations.* Oxford: Oxford University Press.
165. Jorgensen, D. V., and A. Rutherford. 1983. On the dynamics of a stirred tank with consecutive reactions. *Chem. Eng. Sci.* 38:45–53.
166. Joseph, D. D. 1976. *Stability of Fluid Motions I and II.* Berlin: Springer.
167. Joseph, D. D., and D. H. Sattinger. 1972. Bifurcating time periodic solutions and their stability. *Arch. Rat. Mech. Anal.* 45:79–109.
168. Kahlert, C., O. E. Rössler, and A. Varma. 1981. Chaos in a CSTR with two consecutive first-order reactions, one exo, one endothermic. In K. H. Ebert et al. (eds.): *Modelling of Chemical Reaction Systems.* Berlin: Springer.
169. Kaplan, J. L., and J. A. Yorke. 1979. Preturbulence: A regime observed in a fluid flow model of Lorenz. *Comm. Math. Phys.* 67:93–108.
170. Kaplan, J., and J. Yorke. 1979. Chaotic behavior of multidimensional difference equations. In [253].
171. Käser, G. 1985. *Direkte Verfahren zur Bestimmung von Umkehrpunkten nichtlinearer parameterabhängiger Gleichungssysteme.* Würzburg: Diplomarbeit.
172. Kazarinoff, N. D. 1985. Pattern formation and morphogenetic fields. In P. L. Antonelli (ed.), *Mathematical Essays on Growth and the Emergence of Form.* Alberta: University of Alberta Press.
173. Kazarinoff, N. D., and P. van der Driessche. 1978. A model predator-prey system with functional response. *Math. Biosciences* 39:125–134.
174. Kazarinoff, N. D., and R. Seydel. 1986. Bifurcations and period doubling in E. N. Lorenz's symmetric fourth order system. *Physical Review* A34:3387–3392.
175. Kearfott, R. B. 1983. Some general bifurcation techniques. *SIAM J. Sci. Stat. Comput.* 4:52–68.
176. Keener, J. P. 1979. Secondary bifurcation and multiple eigenvalues. *SIAM J. Appl. Math.* 37:330–349.
177. Keener, J. P. 1980. Waves in excitable media. *SIAM J. Appl. Math.* 39:528–548.
178. Keller, H. B. 1977. Numerical solution of bifurcation and nonlinear eigenvalue problems. In [264].
179. Keller, H. B. 1978. Global homotopies and Newton methods. In *Recent Advances in Numerical Analysis,* ed. C. deBoor et al. New York: Academic Press.

180. Keller, H. B. 1981. Geometrically isolated nonisolated solutions and their approximation. *SIAM J. Num. Anal.* 18:822–838.
181. Keller, H. B. 1981. Continuation methods in computational fluid dynamics. In T. Cebeci (ed.), *Numerical and Physical Aspects of Aerodynamic Flows.* Berlin: Springer.
182. Keller, J. B., and S. Antmann. 1969. *Bifurcation Theory and Nonlinear Eigenvalue Problems.* New York: Benjamin.
183. Kelley, A. 1967. The stable, center-stable, center, center-unstable, and unstable manifolds. *J. Diff. Eqs.* 3:546–570.
184. Kelley, C. T., and R. Suresh. 1983. A new acceleration method for Newton's method at singular points. *SIAM J. Numer. Anal.* 20:1001–1009.
185. Kevrekidis, I. G., R. Aris, L. D. Schmidt, and S. Pelikan. 1985. Numerical computation of invariant circles of maps. *Physica* 16D:243–251.
186. Kirchgässner, K. 1975. Bifurcation in nonlinear hydrodynamic stability. *SIAM Review* 17:652–683.
187. Kirchgässner, K., and P. Sorger. 1969. Branching analysis for the Taylor problem. *Quart. J. Mech. Appl. Math.* 22:183–190.
188. Kirchgraber, U., and E. Stiefel. 1978. *Methoden der analytischen Störungsrechnung und ihre Anwendungen.* Stuttgart: Teubner.
189. Klopfenstein, R. W. 1961. Zeros of nonlinear functions. *J. ACM* 8:366–373.
190. Kobayashi, N. 1984. Computer simulation of the steady flow in a cylindrical floating zone under low gravity. *J. Crystal Growth* 66:63–72.
191. Kopell, N., and L. N. Howard. 1973. Plane wave solutions to reaction-diffusion equations. *Studies Appl. Math.* 52:291–328.
192. Krasnosel'skii, M. A. 1964. *Topological Methods in the Theory of Nonlinear Integral Equations.* New York: Pergamon Press.
193. Krug, H.-J., and L. Kuhnert. 1985. Ein oszillierendes Modellsystem mit autokatalytischem Teilschritt. *Z. Phys. Chemie* 266:65–73.
194. Kubicek, M. 1975. Evaluation of branching points for nonlinear boundary-value problems based on GPM technique. *Appl. Math. Comput.* 1:341–352.
195. Kubicek, M. 1976. Algorithm 502: Dependence of solution on nonlinear systems on a parameter. *ACM Trans. of Math. Software* 2:98–107.
196. Kubicek, M., and V. Hlavacek. 1971. Solution of nonlinear boundary value problems—III. A novel method: Differentiation with respect to an actual parameter. *Chem. Eng. Sci.* 26:705.
197. Kubicek, M., H. Hofmann, and V. Hlavacek. 1979. Modelling of chemical reactors, XXXII: Non-isothermal nonadiabatic tubular reactor. One-dimensional model—detailed analysis. *Chem. Eng. Sci.* 34:593.
198. Kubicek, M., and M. Holodniok. 1984. Evaluation of Hopf bifurcation points in parabolic equations describing heat and mass transfer in chemical reactors. *Chem. Eng. Sci.* 39:593–599.
199. Kubicek, M., and M. Holodniok. 1984. Numerical determination of bifurcation points in steady-state and periodic solutions: Numerical algorithms and examples. *ISNM* 70:247–270. In [203].
200. Kubicek, M., and A. Klic. 1983. Direction of branches bifurcating at a bi-

furcation point: Determination of starting points for a continuation algorithm. *Appl. Math. Comput.* 13:125–142.

201. Kubicek, M., and M. Marek. 1979. Evaluation of limit and bifurcation points for algebraic equations and nonlinear boundary-value problems. *Appl. Math. Comput.* 5:253–264.
202. Kubicek, M., V. Ryzler, and M. Marek. 1978. Spatial structures in a reaction-diffusion system—detailed analysis of the "Brusselator." *Biophys. Chem.* 8:235–246.
203. Küpper, T., H. D. Mittelmann, and H. Weber (eds.). 1984. *Numerical Methods for Bifurcation Problems. Proceedings of a Conference in Dortmund, 1983. ISNM* 70. Basel: Birkhäuser.
204. Küpper, T., and B. Kuszta. 1984. Feedback stimulated bifurcation. *ISNM* 70:271–284. In [203].
205. Küpper, T., R. Seydel, and H. Troger (eds.). 1987. *Bifurcation: Analysis, Algorithms, Applications. Proceedings of a Conference in Dortmund, 1986.* Basel: Birkhäuser. *ISNM 79.*
206. Kutta, W. 1901. Beitrag zur näherungsweisen Integration totaler Differentialgleichungen. *Z. Math. Phys.* 46:435–453.
207. Landau, L., and E. Lifshitz. 1959. *Fluid Mechanics.* Oxford: Pergamon.
208. Langford, W. F. 1977. Numerical solution of bifurcation problems for ordinary differential equations. *Numer. Math.* 28:171–190.
209. Langford, W. F. 1979. Periodic and steady-state mode interactions lead to tori. *SIAM J. Appl. Math.* 37:22–48.
210. Langford, W. F. 1981. A review of interactions of Hopf and steady-state bifurcations. In G. I. Barenblatt et al. (eds.), *Nonlinear Dynamics and Turbulence.* Boston: Pitman.
211. Langford, W. F. 1984. Numerical studies of torus bifurcations. *ISNM* 70:285–295. In [203].
212. Levin, S. A., and L. A. Segel. 1985. Pattern generation in space and aspect. *SIAM Review* 27:45–67.
213. Liapunov, A. M. 1966. *Stability of Motion.* New York: Academic Press.
214. Libchaber, A., and J. Maurer. 1982. A Rayleigh Benard experiment: Helium in a small box. *Proceedings of the NATO Advanced Studies Institute.*
215. Lichtenberg, A. J., and M. A. Liebermann. 1982. *Regular and Stochastic Motion.* Heidelberg and New York: Springer.
216. List, S. E. 1978. Generic bifurcation with application to the von Kármán equations. *J. Diff. Eqs.* 30:89–118.
217. Lorenz, E. N. 1963. Deterministic nonperiodic flow. *J. Atmospheric Sci.* 20:130–141.
218. Lorenz, E. N. 1984. The local structure of a chaotic attractor in four dimensions. *Physica* 13D:90–104.
219. Lory, P. 1980. Enlarging the domain of convergence for multiple shooting by the homotopy method. *Numer. Math.* 35:231–240.
220. Lyberatos, G., B. Kuszta, and J. E. Bailey. 1984. Discrimination and identification of dynamic catalytic reaction models via introduction of feedback. *Chem. Eng. Sci.* 39:739–750.

221. Lyberatos, G., B. Kuszta, and J. E. Bailey. 1984. Steady-state multiplicity and bifurcation analysis via the Newton polyhedron approach. *Chem. Eng. Sci.* 39:947–960.
222. Mandelbrot, B. B. 1983. *The Fractal Geometry of Nature.* New York: Freeman.
223. Marathe, A. G., and V. K. Jain. 1977. Parametric differentiation technique applied to a combustion problem. *AIAA J.* 15:732.
224. Margolis, S. B., and B. J. Matkowsky. 1983. Nonlinear stability and bifurcation in the transition from laminar to turbulent flame propagation. *Combustion Science and Technology* 34:45–77.
225. Margolis, S. B., and B. J. Matkowsky. 1985. Flame propagation in channels: Secondary bifurcation to quasi-periodic pulsations. *SIAM J. Appl. Math.* 45:93–129.
226. Margolis, S. B., and G. I. Sivashinsky. 1984. Flame propagation in vertical channels: Bifurcation to bimodal cellular flames. *SIAM J. Appl. Math.* 44:344–368.
227. Markus, M., D. Kuschmitz, and B. Hess. 1985. Properties of strange attractors in yeast glycolysis. *Biophys. Chem.* 22:95–105.
228. Marsden, J. E., and M. McCracken. 1976. *The Hopf Bifurcation and Its Applications.* New York: Springer.
229. Matkowsky, B. J., and G. I. Sivashinsky. 1978. Propagation of a pulsating reaction front in solid fuel combustion. *SIAM J. Appl. Math.* 35:465–478.
230. Matthies, H. G., and Chr. Nath. 1985. Dynamic stability of periodic solutions of large scale nonlinear systems. *Computer Meth. Appl. Mech. Engng* 48:191–202.
231. May, R. M. 1976. Simple mathematical models with very complicated dynamics. *Nature* 261:459–467.
232. McKean, H. P. 1970. Nagumo's equation. *Advances in Math.* 4:209–223.
233. Melhem, R. G., and W. C. Rheinboldt. 1982. A comparison of methods for determining turning points of nonlinear equations. *Computing* 29:201–226.
234. Menzel, R. 1984. Numerical determination of multiple bifurcation points. *ISNM* 70:310–318. In [203].
235. Menzel, R., and H. Schwetlick. 1978. Zur Lösung parameterabhängiger nichtlinearer Gleichungen mit singulären Jacobi-Matrizen. *Numer. Math.* 30:65–79.
236. Meyer-Spasche, R. 1975. Numerische Behandlung von elliptischen Randwertproblemen mit mehreren Lösungen und von MHD Gleichgewichtsproblemen. Report 6/141 of the Institut für Plasmaphysik, Garching/München.
237. Meyer-Spasche, R., and H. B. Keller. 1980. Numerical study of Taylor-vortex flows between rotating cylinders. *J. Comput. Physics* 35:100–109.
238. Minorsky, N. 1962. *Nonlinear Oscillations.* New York: Van Nostrand.
239. Mittelmann, H. D., and H. Weber (eds.). 1980. Bifurcation problems and their numerical solution. Proceedings of a conference in Dortmund, 1980. Basel: Birkhäuser, *ISNM* 54.
240. Moon, F. C. 1980. Experiments on chaotic motions of a forced nonlinear oscillator: Strange attractors. *J. Appl. Mech.* 47:638–644.

241. Moore, G. 1980. The numerical treatment of non-trivial bifurcation points. *Numer. Funct. Anal. Optim.* 2:441–472.
242. Moore, G., and A. Spence. 1980. The calculation of turning points of nonlinear equations. *SIAM J. Numer. Anal.* 17:567–576.
243. Moser, J. 1966. On the theory of quasiperiodic motions. *SIAM Review* 8:145–172.
244. Nath, G. 1973. Solution of nonlinear problems in Magneto fluid dynamics and non-Newtonian fluid mechanics through parametric differentiation. *AAIA J.* 11:1429.
245. Nayfeh, A. H., and D. T. Mook. 1979. *Nonlinear Oscillations.* New York: John Wiley.
246. Newhouse, S., D. Ruelle, and F. Takens. 1978. Occurrence of strange axiom-A attractors near quasiperiodic flow on T^n, $n \leq 3$. *Comm. Math. Phys.* 64:35.
247. Odeh, F. 1967. Existence and bifurcation theorems for the Ginzburg-Landau equations. *J. Math. Phys.* 8:2351–2356.
248. Olmstead, W. E. 1977. Extent of the left branching solution to certain bifurcation problems. *SIAM J. Math. Anal.* 8:392–401.
249. Ortega, J. M., and W. C. Rheinboldt. 1970. *Iterative Solution of Nonlinear Equations in Several Variables.* New York: Academic Press.
250. Oseledec, V. I. 1968. A multiplicative ergodic theorem. Ljapunov characteristic numbers for dynamical systems. *Trans. Moscow Math. Soc.* 19:197–231.
251. Ott, E. 1981. Strange attractors and chaotic motion of dynamical systems. *Rev. Modern Phys.* 53:655–671.
252. Pearson, J. R. A. 1958. On convection cells induced by surface tension. *J. Fluid Mech.* 4:489–500.
253. Peitgen, H.-O., and M. Prüfer. 1979. The Leray-Schauder continuation method is a constructive element in the numerical study of nonlinear eigenvalue and bifurcation problems. In H.-O. Peitgen and H. O. Walther (eds.), *Functional Differential Equations and Approximation of Fixed Points.* Lecture Notes in Math. 730. Berlin: Springer.
254. Peitgen, H.-O., D. Saupe, and K. Schmitt. 1981. Nonlinear elliptic boundary value problems versus their finite difference approximations: Numerically irrelevant solutions. *J. Reine Angew. Math.* 322:74–117.
255. Peterson, J., K. A. Overholser, and R. F. Heinemann. 1981. Hopf bifurcation in a radiating laminar flame. *Chem. Eng. Sci.* 36:628–631.
256. Petrillo, G. A., and L. Glass. 1984. A theory for phase locking of respiration in cats to a mechanical ventilator. *Am. J. Physiol.* 246:R311–R320.
257. Poincaré, H. 1885. Sur l'équilibre d'une masse fluide animée d'un mouvement de rotation. *Acta Mathematica* 7:259–380.
258. Pomeau, Y., and P. Manneville. 1980. Intermittent transition to turbulence in dissipative dynamical systems. *Commun. Math. Phys.* 74:189–197.
259. Pönisch, G. 1985. Computing simple bifurcation points using a minimally extended system of nonlinear equations. *Computing* 35:277–294.
260. Pönisch, G., and H. Schwetlick. 1977 (1982). Ein lokal überlinear konvergentes Verfahren zur Bestimmung von Rückkehrpunkten implizit definierter

Raumkurven. Report 07-30-77. TU Dresden, 1977. *Numer. Math.* 38:455–465.

261. Pönisch, G., and H. Schwetlick. 1981. Computing turning points of curves implicitly defined by nonlinear equations depending on a parameter. *Computing* 26:107–121.
262. Poston, T., and I. Stewart. 1978. *Catastrophe Theory and Its Applications.* London: Pitman.
263. Preisser, F., D. Schwabe, and A. Scharmann. 1983. Steady and oscillatory thermocapillary convection in liquid columns with free cylindrical surface. *J. Fluid Mech.* 126:545–567.
264. Rabinowitz, P. H. (ed.). 1977. *Applications of Bifurcation Theory.* Academic Press: New York.
265. Rauch, J., and J. Smoller. 1978. Qualitative theory of the FitzHugh-Nagumo equations. *Advances in Math.* 27:12–44.
266. Ray, W. H. 1977. Bifurcation phenomena in chemically reacting systems. In [264].
267. Rayleigh, Lord. 1916. On convection currents in a horizontal layer of fluid, when the higher temperature is on the under side. *Philos. Mag.* 32:529–546.
268. Reddien, G. W. 1978. On Newton's method for singular problems. *SIAM J. Numer. Anal.* 15:993–996.
269. Renardy, M. 1979. Hopf-Bifurkation bei Lasern. *Math. Meth. Appl. Sci.* 1:194–213.
270. Rentrop, P. 1985. Partitioned Runge-Kutta methods with stiffness detection and stepsize control. *Numer. Math.* 47:545–564.
271. Rheinboldt, W. C. 1978. Numerical methods for a class of finite dimensional bifurcation problems. *SIAM J. Numer. Anal.* 15:1–11.
272. Rheinboldt, W. C. 1980. Solution fields of nonlinear equations and continuation methods. *SIAM J. Numer. Anal.* 17:221–237.
273. Rheinboldt, W. C. 1982. Computation of critical boundaries on equilibrium manifolds. *SIAM J. Numer. Anal.* 19:653–669.
274. Rheinboldt, W. C., and J. v. Burkardt. 1983. A locally parametrized continuation process. *ACM Trans. of Math. Software* 9:215–235.
275. Rice, J., and R. F. Boisvert. 1985. *Solving Elliptic Problems Using ELLPACK.* New York: Springer.
276. Riks, E. 1972. The application of Newton's method to the problem of elastic stability. *Trans. ASME, J. Appl. Mech.* 1060–1065 (December 1972).
277. Rinzel, J., and R. FitzHugh. 1984. Private communication.
278. Rinzel, J., and J. P. Keener. 1983. Hopf bifurcation to repetitive activity in nerve. *SIAM J. Appl. Math.* 43:907–922.
279. Rinzel, J., and R. N. Miller. 1980. Numerical calculation of stable and unstable periodic solutions to the Hodgkin-Huxley equations. *Mathem. Biosciences* 49:27–59.
280. Rinzel, J., and D. Terman. 1982. Propagation phenomena in a bistable reaction-diffusion system. *SIAM J. Appl. Math.* 42:1111–1137.
281. Rinzel, J., and W. C. Troy. 1982. Bursting phenomena in a simplified Oregonator flow system model. *J. Chem. Phys.* 76:1775–1789.

282. Rössler, O. E. 1976. Different types of chaos in two simple differential equations. *Z. Naturforschung* 31A:1664–1670.
283. Roose, D. 1984. FORTRAN-programs for the computation of Hopf bifurcation points of systems of parabolic diffusion-reaction equations. Department of Computer Science, University of Leuven, Leuven, Belgium.
284. Roose, D. 1985. An algorithm for the computation of Hopf bifurcation points in comparison with other methods. *J. Comput. Appl. Math.* 12 & 13:517–529.
285. Roose, D., and V. Hlavacek. 1983. Numerical computation of Hopf bifurcation points for parabolic diffusion-reaction differential equations. *SIAM J. Appl. Math.* 43:1075–1085.
286. Roose, D., and V. Hlavacek. 1985. A direct method for the computation of Hopf bifurcation points. *SIAM J. Appl. Math.* 45:879–894.
287. Roose, D., and R. Piessens. 1985. Numerical computation of nonsimple turning points and cusps. *Numer. Math.* 46:189–211.
288. Roose, D., R. Piessens, V. Hlavacek, and P. van Rompay. 1984. Direct evaluation of critical conditions for thermal explosion and catalytic reaction. *Combustion and Flame* 55:323–329.
289. Rosenblat, S., S. H. Davis, and G. M. Homsy. 1982. Nonlinear Marangoni convection in bounded layers, Part 1: Circular cylindric containers. *J. Fluid Mech.* 120:91–122.
290. Rubbert, P. E., and M. T. Landahl. 1967. Solution of nonlinear flow problems through parametric differentiation. *Phys. Fluids* 10:831.
291. Ruelle, D. 1980. Strange attractors. *Mathem. Intelligencer* 2:126–137.
292. Ruelle, D., and F. Takens. 1971. On the nature of turbulence. *Comm. Math. Phys.* 20:167–192.
293. Ruhe, A. 1973. Algorithms for the nonlinear eigenvalue problem. *SIAM J. Numer. Anal.* 10:674–689.
294. Runge, C. 1895. Über die numerische Auflösung von Differential Gleichungen. *Math. Ann.* 46:167–178.
295. Sattinger, D. H. 1979. *Group Theoretic Methods in Bifurcation Theory.* Berlin: Springer.
296. Sattinger, D. H. 1980. Bifurcation and symmetry breaking in applied mathematics. *Bull. Amer. Math. Soc.* 3:779–819.
297. Schaffer, W. M., and M. Kot. 1985. Do strange attractors govern ecological systems? *BioSciences* 35:342–350.
298. Scheidl, R., H. Troger, and K. Zeman. 1983. Coupled flutter and divergence bifurcation of a double pendulum. *Int. J. Non-Linear Mech.* 19:163–176.
299. Scheurle, J. 1977. Ein selektives Projektions-Iterationsverfahren und Anwendungen auf Verzweigungsprobleme. *Numer. Math.* 29:11–35.
300. Scheurle, J., and J. Marsden. 1986. Bifurcation to quasi-periodic tori in the interaction of steady state and Hopf bifurcation. *SIAM J. Math. Anal.* 15:1055–1074.
301. Schneider, F. W. 1985. Periodic perturbations of chemical oscillators: Experiments. *Ann. Rev. Phys. Chem.* 36:347–378.
302. Schuster, H. G. 1984. *Deterministic Chaos.* Weinheim: Physik-Verlag.

303. Schwetlick, H. 1984. Algorithms for finite-dimensional turning point problems from viewpoint to relationships with constrained optimization problems. *ISNM* 70:459–479. In [203].
304. Scriven, L. E., and C. V. Sternling. 1960. The Marangoni effects. *Nature* 187:186–188.
305. Seydel, R. 1975. Bounds for the lowest critical value of the nonlinear operator $-u''+u^3$. *ZAMP* 26:714–720.
306. Seydel, R. 1977. Numerische Berechnung von Verzweigungen bei gewöhnlichen Differentialgleichungen. Ph.D. thesis. Report TUM 7736, Mathem. Institut, Technical Univ. Munich.
307. Seydel, R. 1979. Numerical computation of branch points in ordinary differential equations. *Numer. Math.* 32:51–68.
308. Seydel, R. 1979. Numerical computation of primary bifurcation points in ordinary differential equations. *ISNM* 48:161–169. In [5].
309. Seydel, R. 1979. Numerical computation of branch points in nonlinear equations. *Numer. Math.* 33:339–352.
310. Seydel, R. 1980. Programme zur numerischen Behandlung von Verzweigungsproblemen bei nichtlinearen Gleichungen und Differentialgleichungen. In [239].
311. Seydel, R. 1981. Numerical computation of periodic orbits that bifurcate from stationary solutions of ordinary differential equations. *Appl. Math. Comput.* 9:257–271.
312. Seydel, R. 1981. *Neue Methoden zur numerischen Berechnung abzweigender Lösungen bei Randwertproblemen gewöhnlicher Differentialgleichungen.* Munich: Habilitationsschrift.
313. Seydel, R. 1983. Branch switching in bifurcation problems of ordinary differential equations. *Numer. Math.* 41:93–116.
314. Seydel, R. 1983. Efficient branch switching in systems of nonlinear equations. *ISNM* 69:181–191.
315. Seydel, R. 1984. A continuation algorithm with step control. *ISNM* 70:480–494. In [203].
316. Seydel, R. 1983, 1985. *BIFPACK: A program package for calculating bifurcations.* State University of New York at Buffalo. (Second version 1985.)
317. Seydel, R. 1985. Calculating the loss of stability by transient methods, with application to parabolic partial differential equations. In [22].
318. Seydel, R. 1985. Attractors of a Duffing equation: Dependence on the exciting frequency. *Physica* 17D:308–312.
319. Sharp, G. H., and R. W. Joyner. 1980. Simulated propagation of cardiac action potentials. *Biophys. J.* 31:403–423.
320. Shaw, R. 1981. Strange attractors, chaotic behavior, and information flow. *Zeitschr. Naturforschung* 36A:80–112.
321. Shimada, I., and T. Nagashima. 1979. A numerical approach to ergodic problem of dissipative dynamical systems. *Progress of Theoretical Phys.* 61:1605–1616.
322. Simpson, R. B. 1975. A method for the numerical determination of bifur-

cation states of nonlinear systems of equations. *SIAM J. Numer. Anal.* 12:439–451.

323. Sivaneri, N. T., and W. L. Harris. 1980. Numerical solutions of transonic flow by parametric differentiation and integral equations technique. *AIAA J.* 18:1534.
324. Smale, S. 1967. Differentiable dynamical systems. *Bull. Amer. Math. Soc.* 73:747–817.
325. Smith, C. B., B. Kuszta, G. Lyberatos, and J. E. Bailey. 1983. Period doubling and complex dynamics in an isothermal chemical reaction system. *Chem. Eng. Sci.* 38:425–430.
326. Smith, B. T., J. M. Boyle, J. J. Dongarra, B. S. Garbow, Y. Ikebe, V. C. Klema, and C. B. Moler. 1976. *Matrix Eigensystem Routines—EISPACK Guide.* Lecture Notes in Comp. Sci. 6. Berlin: Springer.
327. Smith, M. K., and S. H. Davis. 1983. Instabilities of dynamic thermocapillary liquid layers, Part 1: Convective instabilities. *J. Fluid Mech.* 132:119–144.
328. Smooke, M. D., J. A. Miller, and R. J. Kee. 1985. Solution of premixed and counterflow diffusion flame problems by adaptive boundary value methods. In [22].
329. Sparrow, C. 1982. *The Lorenz Equations: Bifurcations, Chaos, and Strange Attractors.* Berlin: Springer.
330. Spence, A., and B. Werner. 1982. Non-simple turning points and cusps. *IMA J. Numer. Anal.* 2:413–427.
331. Spence, A., and A. D. Jepson. 1984. The numerical calculation of cusps, bifurcation points, and isola formation points in two parameter problems. *ISNM* 70:502–514. In [203].
332. Spirig, F. 1983. Sequence of bifurcations in a three-dimensional system near a critical point. *ZAMP* 34:259–276.
333. Stakgold, I. 1971. Branching of solutions of nonlinear equations. *SIAM Review* 13:289–332.
334. Stoer, J., and R. Bulirsch. 1980. *Introduction to Numerical Analysis.* New York: Springer.
335. Stoker, J. J. 1950. *Nonlinear Vibrations.* New York: Interscience.
336. Stuart, J. T. 1977. Bifurcation theory in non-linear hydrodynamic stability. In [264].
337. Suchy, H., H. Troger, and R. Weiss. 1985. A numerical study of mode jumping of rectangular plates. *ZAMM* 65:71–78.
338. Swinney, H. L., and J. P. Gollub (eds.). 1981. *Hydrodynamic Instabilities and the Transition to Turbulence.* Berlin: Springer.
339. Swinney, H. L., and J. P. Gollub. 1978. The transition to turbulence. *Phys. Today* 31(8):41–49.
340. Takens, F. 1981. Detecting strange attractors in turbulence. In D. A. Rand et al. (eds.), *Dynamical Systems and Turbulence.* Lecture Notes in Math. 898. New York: Springer.
341. Taylor, G. I. 1923. Stability of a viscous liquid contained between two rotating cylinders. *Philos. Trans. Roy. Soc. London* A223:289–343.

342. Testa, J., J. Pérez, and C. Jeffries. 1982. Evidence for chaotic universal behavior of a driven nonlinear oscillator. *Phys. Rev. Lett.* 48:714.
343. Thom, R. 1972. *Stabilité Structurelle et Morphogénèse.* New York: Benjamin.
344. Thompson, J. M. T., and G. W. Hunt (eds.). 1983. Collapse: The buckling of structures in theory and practice. (IUTAM Symp., 1982.) Cambridge: Cambridge University Press.
345. Troger, H. 1982. Über chaotisches Verhalten einfacher mechanischer Systeme. *ZAMM* 62:T18–T27.
346. Troger, H. 1984. Application of bifurcation theory to the solution of nonlinear stability problems in mechanical engineering. *ISNM* 70:525–546. In [203].
347. Troger, H., V. Kazani, and A. Stribersky. 1985. Zur Dimension Seltsamer Attraktoren. *ZAMM* 65:T109–T111.
348. Troger, H., and K. Zeman. 1984. A nonlinear analysis of the generic types of loss of stability of the steady state motion of a tractor-semitrailer. *Vehicle System Dynamics* 13:161–172.
349. Ueda, Y. 1980. Steady motions exhibited by Duffing's equation: A picture book of regular and chaotic motions. Report, Institut of Plasma Physics, Nagoya University, IPPJ-434.
350. Ueda, Y., and N. Akamatsu. 1981. Chaotically transitional phenomena in the forced negative-resistance oscillator. IEEE Trans. *Circuits and Systems* CAS-28:217–223.
351. Uppal, A., W. H. Ray, and A. B. Poore. 1974. On the dynamic behavior of continuous stirred tank reactors. *Chem. Eng. Sci.* 29:967–985.
352. Vaganov, D. A., N. G. Samoilenko, and V. G. Abramov. 1978. Periodic regimes of continuous stirred tank reactors. *Chem. Eng. Sci.* 33:1133–1140.
353. Vainberg, M. M. 1964. *Variational Methods for the Study of Nonlinear Operators.* San Francisco: Holden-Day.
354. Vainberg, M. M., and V. A. Trenogin. 1974. *Theory of Branching of Solutions of Nonlinear Equations.* Leyden: Noordhoff.
355. Van der Pol, B. 1927. Forced oscillations in a circuit with nonlinear resistance (receptance with reactive triode). *London, Edinburgh, and Dublin Phil. Mag.* 3:65–80.
356. Velte, W. 1964. Stabilitätsverhalten und Verzweigung stationärer Lösungen der Navier-Stokesschen Gleichungen. *Arch. Rat. Mech. Anal.* 16:97–125.
357. Velte, W. 1966. Stabilität und Verzweigung stationärer Lösungen der Navier-Stokesschen Gleichungen beim Taylor Problem. *Arch. Rat. Mech. Anal.* 22:1–14.
358. Volterra, V. 1931. *Leçons sur la Théorie Mathématique de la Lutte pour la Vie.* Paris: Gauthier-Villars.
359. Wacker, H. (ed.). 1978. *Continuation Methods.* New York: Academic Press.
360. Wan, Y.-H. 1978. Bifurcation into invariant tori at points of resonance. *Arch. Rat. Mech. Anal.* 68:343–357.
361. Watson, L. T. 1979. An algorithm that is globally convergent with probability one for a class of nonlinear two-point boundary value problems. *SIAM J. Numer. Anal.* 16:394–401.

362. Weinitschke, H. J. 1970. Die Stabilität elliptischer Zylinderschalen bei reiner Biegung. *ZAMM* 50:411–422.

363. Weiss, R. 1975. Bifurcation in difference approximations to two-point boundary value problems. *Math. Comput.* 29:746–760.

364. Weber, H. 1979. Numerische Behandlung von Verzweigungsproblemen bei gewöhnlichen Differentialgleichungen. *Numer. Math.* 32:17–29.

365. Werner, B. 1984. Regular systems for bifurcation points with underlying symmetries. *ISNM* 70:562–574. In [203].

366. Werner, B., and A. Spence. 1984. The computation of symmetry-breaking bifurcation points. *SIAM J. Numer. Anal.* 21:388–399.

367. Wilkinson, J. H., and C. Reinsch. 1971. *Linear Algebra: Handbook for Automatic Computation.* Springer: Berlin.

368. Winfree, A. T. 1975. Unclockwise behaviour of biological clocks. *Nature* 253:315–319.

369. Yamamoto, N. 1983. Newton's method for singular problems and its application to boundary value problems. *J. Mathem. Tokushima Univ.* 17:27–88.

370. Yorke, J. A., E. D. Yorke, and J. Mallet-Paret. 1985. Lorenz-like chaos in a partial differential equation for a heated fluid loop. Manuscript.

371. Zeeman, E. C. 1977. *Catastrophe Theory: Selected Papers.* Reading, Mass.: Addison-Wesley.

372. Zierep, J., and H. Oertel (eds.). 1982. *Convective Transport and Instability Phenomena.* Karlsruhe: G. Braun.

Index

I

J

K

L

M

Q

R

S